Frequency Measurement and Control

Microwave Technology Series

The *Microwave Technology Series* publishes authoritative works for professional engineers, researchers and advanced students across the entire range of microwave devices, sub-systems, systems and applications. The series aims to meet the reader's needs for relevant information useful in practical applications. Engineers involved in microwave devices and circuits, antennas, broadcasting communications, radar, infra-red and avionics will find the series an invaluable source of design and reference information.

Series editors:
Michel-Henri Carpentier
Professor in 'Grandes Écoles', France
Fellow of the IEEE, and President of the French SEE

Bradford L. Smith
International Patents Consultant and Engineer
with the Alcatel group in Paris, France
and a Senior Member of the IEEE and French SEE

Titles available

1. The Microwave Engineering Handbook Volume 1
    Microwave components
    Edited by Bradford L. Smith and Michel-Henri Carpentier

2. The Microwave Engineering Handbook Volume 2
    Microwave circuits, antennas and propagation
    Edited by Bradford L. Smith and Michel-Henri Carpentier

3. The Microwave Engineering Handbook Volume 3
    Microwave systems and applications
    Edited by Bradford L. Smith and Michel-Henri Carpentier

4. Solid-state Microwave Generation
    J. Anastassiades, D. Kaminsky, E. Perea and A. Poezevara

5. Infrared Thermography
    G. Gaussorgues

6. Phase Locked Loops
    J. B. Encinas

7. Frequency Measurement and Control
    Chronos Group

8. Microwave Integrated Circuits
    Edited by I. Kneppo

Frequency Measurement and Control

Chronos Group,
French National Observatory, and
National Centre of Scientific Research,
France

CHAPMAN & HALL
London · Glasgow · New York · Tokyo · Melbourne · Madras

。6125803

PHYSICS

Published by Chapman & Hall, 2–6 Boundary Row, London SE1 8HN

Chapman & Hall, 2–6 Boundary Row, London SE1 8HN, UK

Blackie Academic & Professional, Wester Cleddens Road, Bishopbriggs, Glasgow G64 2NZ, UK

Chapman & Hall Inc., One Penn Plaza, 41st Floor, New York NY10119, USA

Chapman & Hall, Japan, Thomson Publishing Japan, Hirakawacho Nemoto Building, 6F, 1-7-11 Hirakawa-cho, Chiyoda-ku, Tokyo 102, Japan

Chapman & Hall Australia, Thomas Nelson Australia, 102 Dodds Street, South Melbourne, Victoria 3205, Australia

Chapman & Hall India, R. Seshadri, 32 Second Main Road, CIT East, Madras 600 035, India

First English language edition 1994

© 1994 Chapman & Hall

Original French language edition – La Mesure de la Fréquence des Oscillateurs – © 1991, Masson et CNET-ENST

Typeset by S. Chomet

Printed in England by Clays Ltd, St Ives plc

ISBN 0 412 48270 3

∞ Printed on acid-free text paper, manufactured in accordance with ANSI/NISO Z39.48-1992 (Permanence of Paper).

Contents

Members of the Chronos Group
Chronos or Kronos?
Foreword

Chronos or Kronos?

Chronos ($\chi\rho o\nu o\sigma$) is the time that flows inexorably – a model of irreversibility. The main purpose of the measurement of frequency is to mark this flow. It is therefore appropriate to offer this book under the collective pseudonym of Chronos. Moreover, the evolution of a book, between conception and publication, is also an irreversible process that is definite but slow.

An oscillator, on the other hand, is governed by the laws of physics, with time as a reversible parameter. It comes and goes endlessly, always returning to its initial state. It allows the establishment of order...

Kronos ($\kappa\rho o\nu o\sigma$) is the son of Gaea and Uranus (whom he will castrate). He is the father of Zeus (whom he will attempt to kill and who will disown him). This god of Greek mythology seems unconnected with the flow of time. And yet, in the theogony of Hesiod, Kronos is the master of the order of things. He regulates their successive arrangement and disposition in time. He establishes order, just like an oscillator can do.....

Nonetheless, the behaviour of this god in his family environment seems highly antipathetic and we prefer to ignore him by adopting Chronos as a pseudonym.

This homonymy of Chronos and Kronos is disturbing; some Greek authors confuse the two names. It seems that Chronos created chaos, and that Kronos, by establishing order, took the universe out of this chaos.

The word *chaos* now has a very precise meaning in science; it appears in the context of certain oscillators. Perhaps those who study mythology should bear this in mind.

This book is the result of fruitful collaboration between a number of individuals who refer to themselves collectively as *Chronos*. They wish to record their gratitude to Christiane Hisleur for her excellent work on the word processor, Robert Pretot for producing the illustrations and Gérard Tinelli for successfully liaising with the Publishers.

Foreword

Periodical phenomena or, more precisely, *quasiperiodical* phenomena, occupy a central position in physics. For a long time, their most important parameter has been their period. However, nowadays, we are much more interested in their frequency, and the many reasons for this are discussed in this book.

Throughout history, evaluations of time have been based on periodical phenomena such as the apparent motion of the Sun. Indeed, the oldest unit of time is the *day*. The apparent motion of the Moon and of the celestial sphere, including changes in the appearance of the former, provided longer units, namely, *week, month and year*. All these periodical phenomena – the natural clocks – were obviously well suited to the observation and prediction of the evolution of nature with its seasonal rhythm.

The gnomon and the clepsydra gave reasonably precise subdivisions of the day that could be used in timing human activities, so long as they were mostly agricultural. The invention of the pendulum and of balance-wheel clocks marked the dawn of industrial civilisation, which soon demanded measurements of time with ever increasing precision over shorter and shorter periods.

We know how to determine – by comparison – the regular or irregular variations in the running of certain clocks. By comparing the solar clock with the lunar and celestial clocks, our ancestors readily identified seasonal variations in the former. More recently, and again by comparison with the lunar clock, unforeseen variations in the solar clock have been detected. But it is really through the development of man-made clocks that the measurement of time intervals has become a crucial element of advanced scientific and technical work.

On the other hand, it has become clear during the last few decades that *frequency* plays a more fundamental role than duration, at least in the physical phenomena on the atomic scale. We shall return to this later.

All clocks – from the solar clock to atomic clocks – are essentially oscillators. They are useful to the extent that their frequency is stable, i.e., independent of time. From the very outset, we are thus confronted by a paradox: a phenomenon that evolves with time is expected to exhibit time independent behaviour. Moreover, whilst the duration of our observations is necessarily finite, these oscillators are expected to reproduce indefinitely the same cycle of successive states. Finally, one of them is expected to define the time by itself.

How do we choose a clock as a standard of time? As before, the procedure involves comparison of available clocks among themselves, so that we can judge whether some of them remain in agreement whilst others display 'changes of mood' that are incompatible with the status of a standard. In the past, such comparisons relied on astronomical observations performed over many years. In the case of balance-wheel clocks, which have a period of the order of one second and a frequency of the order of one hertz, a few days or months of observation reveal unavoidable drifts in their operation. In the case of quartz oscillators or atomic clocks, whose frequencies lie between, say, 1 MHz and 10 GHz, significant information on their relative stability can be obtained in a few minutes. Many months or decades were necessary to obtain the same information on the relative stability of balance-wheel clocks (assuming that such efforts could be justified, which was not the case).

All this would be sufficient to convince us of the importance of oscillator frequency comparisons. However, there are many other reasons for the continuing interest in frequency measurements. We shall mention a few more without going into technical details.

Frequency is the only physical quantity that can be transmitted from one point to another with practically no change, so long as the two points remain fixed relative to one another. The propagation of sound in air was exploited by man (and by some animals) well before the concept of frequency became familiar to us. The telephone, and subsequently radio, have extended the reach of the human voice to the entire globe. Voice and message recognition is based on the analysis – by ear and brain – of a very complex group of superimposed oscillations of different frequency. The fact that a voice or a message are perfectly recognisable at distances of thousands of kilometres illustrates the efficacy of the transmission of frequencies.

The information that we receive from the Sun, planets and stars is coded according to the characteristic frequencies of the atoms and molecules present in their vicinity. Of course, in the case of these celestial bodies and artificial satellites, the relative motion of the emitter relative to a terestrial observer translates into a frequency shift that is well known in acoustics under the name of the Doppler effect. We know how to account for this very accurately, so that some of the most refined modern measurements rely on the transmission of signals by artificial satellites.

From another point of view, the behaviour of an oscillator cannot in general be characterised by a single frequency. Even in simple cases, the elementary oscillation is never purely sinusoidal, i.e., it always contains harmonics whose frequencies are multiples of the fundamental frequency (these harmonics are responsible for the tonal quality of musical instruments).

An oscillator may also be disturbed by different factors that temporarily modify its frequency. It may support parasitic oscillations. The mere fact that it does not oscillate indefinitely and that observations have a limited duration creates a group of neighbouring frequencies. These effects combine to produce a frequency spectrum which we know how to analyse. The prism that splits white light into the colours of the rainbow was probably the first frequency analyser. The very language of frequency measurements bears witness to it: we speak about *spectrum, spectral purity* and *white noise*.

Finally, and this is probably the main reason for the importance of frequency measurements, the success of Planck's quantum theory shows that frequency plays a fundamental role in physics.

The stationary states of atoms and molecules, and the transfers of energy between them and their environment, are discrete and are subject to quantum rules in which we always find the same fundamental relationship: quantum energy equals frequency times Planck's constant. This has been one of the most fruitful concepts in physics since the turn of the century. It has been responsible for advances in atomic and molecular spectroscopy at optical and radio frequencies, and for spectacular techniques such as the medical applications of nuclear magnetic resonance. It has also been fundamental to modern theories of the properties of solids, molecules, atoms, nuclei and so on. For example, we encounter Planck's law in the Josephson effect, which is the basis for the best reference standard for the volt and involves the exposure of a suitable junction between two superconducting metals to microwaves with high spectral purity and precisely known stable frequency.

Advances in atomic physics, especially in atomic spectrometry, have provided us with atomic oscillators that function as frequency standards in practical clocks in permanent operation. By exploiting the techniques and resources of spectral analysis in the comparison of clocks, we have been able to choose caesium-133 as the basic oscillator that acts as the standard of frequency and hence of time. Since 1967, it has been used in the definition of the unit of time – *the second*. This frequency standard allows previously unattainable precision in studies of new astronomical objects such as pulsars, i.e., sources of radio-wave pulses with extraordinarily stable frequency (at least in some cases and if we ignore the disturbing long-term drift).

The numerous methods and techniques used to produce, stabilise, add, subtract, multiply and divide frequencies are the indispensable tools of frequency measurement and spectral analysis. They have been extended from the lowest frequencies to visible radiation with frequencies of the order of 5×10^{14} Hz. Nevertheless, it is not as yet possible to compare directly any two frequencies over this entire range: frequency comparisons are possible between a few terahertz and a few 10^{12} Hz. Thereafter, wavelength comparisons take over until x-rays and γ-rays are reached. It has nevertheless been possible to compare, point by point, a few frequencies in the visible range with radio frequencies, in particular with the frequency of the caesium-133 standard. These measurements of optical frequencies, and measurements of the wavelengths of the same radiations, provided the basis in around 1975 for the best determinations of the speed of light. The process was then inverted and, in 1983, the *metre* ceased to be defined in terms of wavelength. Instead, it was defined in terms of the distance travelled by light in a certain fraction of a second. This means that the wavelength of radiation can be deduced from measurement of its frequency, and hence the standard of length can be based on a measurement of frequency.

If we consider that frequency measurements run from approximately 10^{-9} Hz (a period of a few tens of years) to approximately 10^{12} Hz or even 10^{15} Hz, the range covered is more than respectable. It is striking that spectral purity, noise and stability characteristics persist up to the highest frequencies, and even beyond the frequencies that are directly measurable, whatever the nature of the phenomenon considered (astronomical, mechanical, acoustic, atomic or electromagnetic).

The idea of gathering under the common title of 'frequency measurement' the contributions of eminent specialists in fields as diverse as those covered in the present book seems to me to be a guarantee of well-deserved success. Technicians, engineers and researchers are thus provided with a review of information that is normally scattered across the entire literature of the subject.

We must thank the authors for agreeing to to make the effort necessary to harmonise their contributions and to produce a review that is both up to date and accessible to many potential readers.

P. Giacomo
Honorary Director
International Bureau of Weights and Measures

Members of the Chronos Group

Claude Audouin
Director of Research
CNRS Laboratoire de l'Horloge Atomique (Orsay)

Michel Yves Bernard
Professor
Conservatoire National des Arts et Métiers
and
President
Comité de Direction du Laboratoire de Physique et Métrologie des Oscillateurs

Raymond Besson
Professor
l'Ecole Nationale Supérieure de Mécanique et des Microtechniques
and
Director
Laboratoire de Chronométrie, Electronique et Piézoélectricité (Besançon)

Jean-Jacques Gagnepain
Director of Research
CNRS Laboratoire de Physique et Métrologie des Oscillateurs (Besançon)

Jacques Groslambert
Research Engineer
CNRS Laboratoire de Physique et Métrologie des Oscillateurs

Michel Granveaud
Director
Laboratoire Primaire du Temps et des Fréquences (Observatoire de Paris)

Jean-Claude Neau
Lecturer
Conservatoire National des Arts et Métiers

Marcel Olivier
Professor
l'Université de Besançon
Laboratoire de Physique et Métrologie des Oscillateurs

Jacques Rutman
Technical Director
Féderation Nationale des Industries Electriques et Electroniques

1

Measurement of frequency

1.1 PERIOD COUNTING: TIME-DOMAIN ANALYSIS

1.1.1 Basic principles

The basic principles of frequency measurement may be summarised as follows. A periodic signal is observed and the number of consecutive transits across a given point in the direction of propagation of the signal is counted. A chronometer is switched on at the beginning of the counting process and stopped after n pulses have crossed the chosen point. This gives the time lapse τ in seconds. The frequency of the periodic signal (in hertz, i.e., cycles per second) is then given by

$$f = \frac{n}{\tau} \tag{1.1}$$

and the period (in seconds) is

$$T = \frac{\tau}{n} = \frac{1}{f} \tag{1.2}$$

This elementary experiment is included in all courses of basic physics and is probably the way Galileo concluded that the small oscillations of a pendulum have a constant period.

The same method is used by physicians to measure the heart-beat frequency and by astronomers to determine the frequency of a pulsar.

Frequency measurements are essential for the characterisation of the countless oscillatory phenomena encountered in nature and in the appliances (mechanical, electronic and so on) produced by modern technology.

Whilst the measurement of frequency is an operation that seems simple, at least in principle, the interpretation of the results requires some sophistication. Indeed, in equation (1.1), n can be different in different measurements and τ is subject to experimental uncertainty. We therefore begin by listing the difficulties hiding behind the above apparent simplicity of the measurement of f.

1.1.2 Statistics of measurement

A single measurement has little significance. The operation has to be repeated many times in order to obtain a table of values $f_1, f_2, f_3, ...f_N$ that has to be processed by statistical methods. The most elementary step is to calculate the mean of N measurements

$$\langle f \rangle = \frac{1}{N} \sum_{k=1}^{N} f_k \tag{1.3}$$

and the mean square deviation (the *variance*)

$$\sigma_f^2 = \frac{1}{N} \sum_{k=1}^{N} (f_k - \langle f \rangle)^2 = \langle f_k^2 \rangle - \langle f \rangle^2 \tag{1.4}$$

The variance of the relative frequency change can be introduced in a similar way and is denoted by σ_y^2, where

$$y = \frac{f - \langle f \rangle}{\langle f \rangle} \tag{1.5}$$

It is clear that σ_f^2 and σ_y^2 depend on the number N and the duration τ of each measurement, which is the reason why they are better written as $\sigma_f^2(N, \tau)$ and $\sigma_y^2(N, \tau)$. The square root of the variance, e.g., $\sigma_y(N, \tau)$, is the standard deviation and is a measure of the relative frequency stability (often written, somewhat simplistically, as $\delta f / f$).

This type of statistical analysis is illustrated in Fig. 1.1 which shows $\sigma_y(\tau)$ as a function of τ. This indicates the important effect of the duration τ of measurement on the final result. The rather complex shape of this curve will be interpreted later; for the moment, we note that each element of this curve is the sum of contributions of one or more sources of frequency fluctuation, which shows that the statistical calculation is not only a means of comparing oscillators, but also a way of analysing their behaviour.

It is tempting to say that N must be as large as possible. Indeed, only an infinite N will give a perfectly deterministic σ; conversely, σ is uncertain to some degree when N is finite, and the uncertainty increases with decreasing N. This suggests the idea of the true variance ($N \to \infty$), which the mathematicians know how to take advantage of in statistical analysis (Chapter 3). Unfortunately, we know nothing, in the first instance, about the existence of such a limit for σ as $N \to \infty$, or about the choice of N that allows the determination of this limit with given precision. Another difficulty is that the measurement actually takes time, i.e., a time $N\tau$ is needed to obtain a value for the variance. What is interesting is the way in which σ varies with τ, i.e., several values of σ are needed. Indeed, the curve of Fig. 1.1 requires many measurements. Finally, it is clear that such measurements are only possible for τ that is neither too small (a few tens of oscillator periods) nor too large. Mathematicians, who know how to relate $\sigma(N, \tau)$ to other statistical quantities, often need to integrate over all values of τ. Of course, one can attempt to extrapolate, but the shape of the curve does not render this operation obvious.

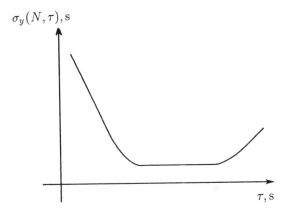

Fig. 1.1 Typical graph of $\sigma_y(N, \tau)$ as a function of τ for an oscillator.

1.1.3 The measurement of τ

In practice, the period-counting device is a logic circuit. The system starts, runs and stops once it has counted n oscillations, and we have to know with precision the time interval τ between the beginning and the end of the process. The only way of doing this is to count, during the same time interval, the number n_0 of oscillations of a reference oscillator that is effectively a clock. In fact, any measurement is a comparison between an unknown quantity and a reference quantity. In our case, n periods of the frequency being measured are compared with n_0 periods of the reference frequency.

If the logic system performing the counting is perfect, fluctuations in the ratio n/n_0 depend only on fluctuations in each of the two oscillators. Meaningful results can be obtained in only two cases:

- one of the oscillators is much more stable than the other, and is called a clock (or a reference or standard oscillator); fluctuations in n/n_0 provide information on fluctuations of the tested oscillator of period T

- the two oscillators have the same structure and have been constructed with the same care, i.e., they are supposed identical, so that their fluctuations can be regarded as statistically comparable and the variance of either one of them can be obtained by dividing the total variance of the measurement by 2 and the standard deviation by $\sqrt{2}$.

The metrology of frequency will therefore involve two areas of activity. One of them is concerned with the creation of standard clocks which have very low standard deviation σ (10^{-13} or 10^{-14} or a even few units of 10^{-15}) over a long period of time (several hours). The development of such clocks relies on good knowledge of the physics of oscillators, which will be considered briefly in Sections 1.4 and 1.5, and in more detail in Chapters 5 and 6.

The other area is concerned with the comparison of oscillators among themselves, or with a supposedly much better clock, whenever this is possible. Here everything rests on the quality of the method of measurement and the statistical treatment of the results. These questions will be considered in Chapters 2, 3 and 4.

In the current state of metrology, the primary standard clock employs an oscillator that relies on quantum transitions between two atomic energy levels of an isotope of caesium. The frequency of this reference transition is

$$f_0 = 9\,192\,631\,770\,\text{Hz}.$$

The principle and properties of the caesium clock are described in Chapter 6.

1.2 FREQUENCY-DOMAIN ANALYSIS

1.2.1 The sinusoidal signal

A periodic phenomenon can be regarded as a signal $v(t)$ such that

$$v\,(t + mT) \equiv v\,(t) \tag{1.6}$$

where T is the period and m an integer.

If we know $v(t)$ over a period T, this is sufficient for a complete description of the signal by a Fourier series:

$$v\,(t) = \sum_{n=-\infty}^{+\infty} V_n \exp\left(jn\frac{2\pi t}{T}\right) \tag{1.7}$$

where

$$V_n = \frac{1}{T} \int_{t_0}^{t_0+T} v\,(t) \exp\left(-jn\frac{2\pi t}{T}\right) dt \tag{1.8}$$

The set of complex numbers V_n defines the spectrum of the signal. This formula dates back to the middle of the eighteenth century, but until the 1950s numerical evaluations were almost impossible, except for very simple signals. Today, numerical techniques, known as fast Fourier transforms, can be used in rapid evaluations of the coefficients V_n of different signals. Numerical calculations of this kind require a sampling of the signal. If ϑ is the sampling period, Shanon's theorem shows that the highest frequency that can be explored in the spectrum is of $F_{\max} = 1/2\vartheta$, so that the coefficients V_n can be found up to

$$n_{\max} = \frac{T}{2\vartheta} \tag{1.9}$$

The Fourier series is the basis of mathematical methods of signal analysis known collectively as *harmonic analysis* or *frequency analysis*.

An essential tool in frequency analysis is the band-pass filter, a device that stops all the components of the spectrum except those whose frequencies lie within a chosen band.

A band-pass filter that passes only the frequency $1/T_0$, provides a sinusoidal signal of the form

$$v = V_0 \sin(\omega_0 t + \Phi_0) \tag{1.10}$$

where

$$\omega_0 = \frac{2\pi}{T_0} = 2\pi f_0 \tag{1.11}$$

This is simplest periodic signal; V_0 is its amplitude, Φ_0 the phase at the origin and ω_0 the pulsatance in radians per second.

1.2.2 Comparison of sinusoidal signals

The measurement of frequency involves all the techniques of comparison between two sinusoidal signals, one arriving from a clock and the other from the oscillator under investigation. The procedures employed rely heavily on the theory and application of electronic circuits, which will be covered in detail in Chapters 2 and 4 and be recalled briefly below.

The frequencies f_1 and f_2 of two sinusoidal oscillators can be compared by mixing them (e.g., by adding the signals in a suitable circuit) and then using a nonlinear device (diode) to extract the composite frequency

$$f_{mn} = mf_1 + nf_2 \tag{1.12}$$

where n, m are integers.

This is the classical principle of the *heterodyne*. The particular composite frequency and the amplitude of the corresponding signal depend on the nonlinearity. A diode with a square-law characteristic produces only $f_1 + f_2$ and $f_1 - f_2$. Filtering can then be used to extract one of these frequencies, which may be easier to measure than f_1 or f_2. For example, signals produced by two lasers of similar frequency gives rise to a hypersignal whose frequency we know how to measure. The difficulty lies in the exact determination of the integer coefficients m and n and, especially, in the design of the circuitry that must be able to handle simultaneous laser signals at 500 THz and hypersignals at, say, 10 GHz. These circuits are described in Chapter 4 and can be used to cover the entire spectrum between 0 and 600 THz, by linking all the frequencies by rigorously known physical laws.

The above method is readily implemented if a large number of fixed points is known on the frequency scale. These reference frequencies are provided by high-grade oscillators or are obtained by synthesis from f_0, using the formula

$$f = \frac{m}{n} f_0 \qquad (1.13)$$

where m and n are integers, obtained with the help of electrical circuits that carry out mathematical operations on frequencies. They include

- a nonlinear circuit that converts the standard frequency f_0 to the harmonics $m f_0$ (multiplication by an integer); a filter then extracts the desired harmonic

- a period-counting circuit, driven by a signal of this type, that records every n-th impulse and generates a signal of frequency $m f_0/n$ (division by an integer)

- a heterodyne circuit connected to a multiplier (by a factor K) that generates the sum frequency

$$f = m \frac{f_0}{n} + K f_0 \qquad (1.14)$$

In Chapter 2 we shall discuss the devices used to perform frequency synthesis, i.e., calibrate the frequency scale against a standard clock of frequency f_0.

Let us suppose, finally, that the two oscillators do not have exactly the same frequency and the phase difference

$$\Delta\Phi = (2\pi f_1 t + \Phi_1) - (2\pi f_2 t + \Phi_2) \qquad (1.15)$$

varies slowly. The signal $\Delta\dot{\Phi}$ is equal to $2\pi(f_1 - f_2)$, i.e., a constant, and is used to control the frequency of one of the oscillators (e.g., by acting on a variable capacitance, which affects the frequency). When $\Delta\dot{\Phi}$ is zero, we have $f_1 = f_2$ and the two oscillators are locked together. One of them can be the frequency under investigation and the other a standard clock (or a source of frequency $m f_0/n$, synthesised from a clock). This phase locking feedback loop system is another basic tool of the metrology of frequency (Chapter 2).

1.2.3 The quasi-sinusoidal signal and the phase spectrum

The signal produced by a real oscillator is more complicated than the simple sinusoid would suggest. Random phenomena occur and modify the amplitude and the phase from one oscillation to another, and the signal falls to zero a little sooner or a little later than predicted by the true sinusoid. When these variations are small, we can represent them by the quasi-sinusoidal signal

$$v(t) = V_0 [1 + \varepsilon(t)] \sin [\omega_0 t + \Phi(t)] \qquad (1.16)$$

where $\varepsilon(t)$ and $\Phi(t)$ are random functions of time, but can include a constant component (usually a drift). The function $\varepsilon(t)$ represents the amplitude fluctuations and $\Phi(t)$ the phase fluctuations; $\varepsilon(t)$ and $\Phi(t)$ can be correlated.

The next step is to introduce a working assumption that will be verified by comparing calculations with experiment. We assume the existence of ω_0, the mean beat frequency obtained as described above, so that $\Phi(t)$ is a random signal with zero mean. To extract $\Phi(t)$, the signal (1.16) is mixed with that of a variable frequency oscillator (in the manner just described) and the two oscillators are locked together with a certain time constant (Fig. 1.2).

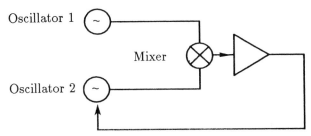

Fig. 1.2 The extraction of phase noise.

Once the loop is closed, the two mean frequencies are equal and the phase difference between the two signals can be generated. If the phase of the output of the standard oscillator is constant (or varies very slightly), this approach provides information about the phase fluctuations of the oscillator under investigation. It is therefore possible, by frequency translation, to take $\Phi(t)$ out of the assigned term, i.e., to demodulate. It is also possible to envisage the measurement of the amplitude noise.

The signal $\Phi(t)$ is not periodic and its Fourier representation must be completed before we can proceed further. The line spectrum is no longer suitable, but it can be used to divide the frequency axis of a filter into narrow bands of width dF. The power spectrum is then described by the quantity dP/dF, the power spectral density, which is a function of the frequency F where, by definition, power is equal to the square of amplitude. Figure 1.3 illustrates the principle of the spectrum analyser – another tool essential to the metrology of frequency. This simple diagram helps us to appreciate the difficulties that arise:

8 *Measurement of frequency*

- the integrator must function for a time Θ that is long enough to take us as close as possible to the ideal theoretical condition $\Theta \to \infty$; this presupposes the existance of the limit, and if this is correct, the signal $\Phi(t)$ will be stationary in the second order
- the filter window must be narrow $(dF \ll F)$ and it must be tunable; this is achieved by heterodyning against an auxiliary generator whose frequency fluctuation must be negligible
- F must be scanned across the entire frequency band in which the signal $\Phi(t)$ appears; in practice, it is necessary to go up to a few dozen kilohertz (sometimes up to a few megahertz).

The spectral analysis of $\Phi(t)$ leads to the LF phase spectrum $S_\Phi(F)$, normalized to the total power P of the signal put out by the oscillator.

It is also possible to perform a spectral analysis of the derivative $\dot\Phi(t)$ of the signal, which characterises the instantaneous frequency fluctuation, defined as the derivative of the total phase

$$\Psi = \omega_0 t + \Phi(t) \tag{1.17}$$

i.e.,

$$f(t) = \frac{1}{2\pi}\frac{d\Psi}{dt} = f_0 + \frac{\dot\Phi(t)}{2\pi} \tag{1.18}$$

where $f_0 = \omega_0/2\pi$.

It is useful to introduce the dimensionless quantity

$$y = \frac{f - f_0}{f_0} = \frac{\dot\Phi(t)}{2\pi f_0} \tag{1.19}$$

which represents the relative frequency fluctuation. The corresponding spectrum will be denoted by $S_y(f)$. The spectral densities of the signals $\Phi(t)$ and $y(t)$ are related by

$$S_y(F) = \frac{F^2}{f_0^2}S_\Phi(F) \tag{1.20}$$

Finally, the results of time-domain analysis can be linked to those of frequency-domain analysis. In fact, if n zero crossings are counted in the same direction for a time τ, the same phase is found to the nearest $2\pi n$, i.e.,

$$2\pi f_0(t+\tau) + \Phi(t+\tau) - 2\pi f_0 t - \Phi(t) = 2\pi n \tag{1.21}$$

and

$$f(t,\tau) = \frac{n}{\tau} = f_0 + \frac{1}{2\pi n}[\Phi(t+\tau) - \Phi(t)] \tag{1.22}$$

The statistical analysis of $f(t,\tau)$, which yields the variance $\sigma_f^2(N,\tau)$, is therefore related to the statistical analysis of Φ, which yields the spectral

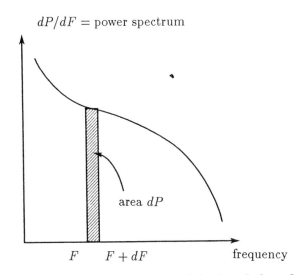

Fig. 1.3 Measurement of the power spectral density of phase fluctuations of a system that is stationary to the second order.

density $S_\Phi(F)$. The results of time-domain analysis can be compared with those of the frequency-domain analysis in order to obtain the spectrum from the measured variance, and *vice versa*. This will be described in Chapter 3.

1.3 RADIOFREQUENCY SPECTRUM OF A SIGNAL

1.3.1 Linewidth and coherence time

We have just noted the difficulties encountered in the frequency-domain analysis of a real, quasi-sinusoidal signal delivered by an oscillator. In practice, if we assume that the signal is stationary in the second order, it is possible to determine the mean frequency and the spectra of the phase and frequency fluctuations. The next problem to consider is the direct analysis of the signal $v(t)$ generated by the oscillator.

The Wiener - Khintchine theorem states that the power spectrum and the correlation function $\Gamma(\tau)$ are related by the Fourier transform

$$\Gamma(\tau) = \int_{-\infty}^{+\infty} \frac{dP}{dF} \exp\left(j2\pi F\tau\right) dF \qquad (1.23)$$

$$\frac{dP}{dF} = \int_{-\infty}^{+\infty} \Gamma(\tau) \exp\left(-j2\pi F\tau\right) d\tau = S(F) \qquad (1.24)$$

All this presupposes that the correlation function

$$\Gamma(\tau) = \lim_{\vartheta \to \infty} \frac{1}{\vartheta} \int_{t}^{t+\vartheta} v(t)v(t-\tau)\, dt \qquad (1.25)$$

exists i.e., that the operation of going to the limit can be carried out. This is so if the signal $v(t)$ is stationary in the second order, and this constitutes an essential assumption.

The formulas are readily applied to a strictly sinusoidal signal, in which case the spectrum is found to consist of two Dirac functions (Fig. 1.4). We thus have

$$S(F) = \frac{dP}{dF} = \frac{1}{2} P_0 \left[\delta\left(F - f_0\right) + \delta\left(F + f_0\right)\right] \qquad (1.26)$$

and

$$\Gamma(\tau) = P_0 \cos 2\pi f_0 t \qquad (1.27)$$

The fluctuation-free signal maintains its memory indefinitely. A very precise result is then always obtained and is determined in advance by the initial conditions; there is always a correlation between the current signal and its very old values.

In the case of a real, quasi-sinusoidal signal, the fluctuations $\varepsilon(t)$ and $\Phi(t)$ give rise to a loss of memory. In the long term, the phase is no longer related to the initial phase which is erased under the influence of the fluctuations. This phenomenon translates into the declining correlation function indicated in Fig. 1.4.

A possible model of this type of correlation is

$$\Gamma(\tau) = P_0 \exp\left(-\pi\Delta F |\tau|\right) \cos 2\pi f_0 \tau \qquad (1.28)$$

in which the loss of memory is shown by the exponential decrease in the amplitude of $\Gamma(\tau)$.

The power spectrum is then described by (1.24) and the bell-shaped curve of Fig. 1.4:

$$S(F) = \frac{P_0}{\pi\Delta F} \left\{ \frac{1}{1 + 4\left[(F - f_0)/\Delta F\right]^2} + \frac{1}{1 + 4\left[(F + f_0)/\Delta F\right]^2} \right\} \qquad (1.29)$$

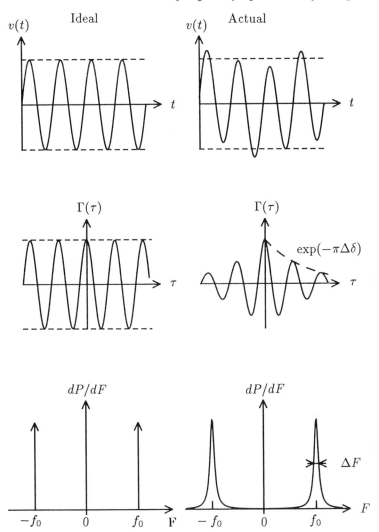

Fig. 1.4 Amplitude and phase fluctuations that lead to a loss of signal memory. The correlation diminishes and the spectrum broadens.

where P_0 is the mean power and ΔF is the linewidth at $-3\,$dB. The effect of the fluctuations is thus described by a loss of memory or by line broadening ΔF at frequency f_0.

The correlation function decreases and the quantity $1/\Delta F$ gives the order of magnitude of the delay τ necessary for a reduction in $\Gamma(\tau)$ from 1 to 0.1. This is the coherence time during which the signal remembers its initial phase. The memory is lost when τ is long in comparison with $1/\Delta F$.

1.3.2 The radiofrequency spectrum

The radiofrequency (RF) spectrum, is the power spectral density of the signal $v(t)$ from an oscillator. This spectrum is evidently governed by $\Phi(t)$ or $y(t)$ and the associated spectral densities $S_\Phi(F)$ and $S_y(F)$, respectively. The latter are sometimes called the LF spectra in contrast to the RF spectrum of $v(t)$.

Of course, the RF spectrum is also governed by the amplitude fluctuations $\varepsilon(t)$, which can be measured by bridge or square-law detection techniques. It is generally considered that they are negligible in the case of very high stability oscillators. This approximation is sometimes too extreme and it is in any case unsatisfactory for oscillators of average quality, e.g., those operating at microwave frequencies.

Direct measurement of the RF spectrum is not a simple matter. It involves the scanning of a narrow-band filter in the neighbourhood of a high frequency f_0, and this is very difficult. Neither is it possible to measure the correlation function with a sufficient delay to obtain significant measurements.

The classical formulas for the phase and amplitude modulation of a sinusoidal signal of frequency f_0 suggest the way towards the determination of the RF spectrum, but much progress remains to be made.

The spectrum of the amplitude-modulated signal

$$v(t) = V_0 (1 + \varepsilon_0 \sin 2\pi F_1 t) \sin 2\pi f_0 t \tag{1.30}$$

consists of a line of amplitude V_0 at the carrier frequency f_0, flanked by two side lines, with amplitudes $V_0 \varepsilon_0 / 2$, at frequencies $f_0 \pm F_1$.

The spectrum of the frequency-modulated signal

$$v(t) = V_0 \sin (2\pi f_0 t + \Phi_0 \sin 2\pi F_2 t) \tag{1.31}$$

consists of a line with amplitude V_0 corresponding to the carrier and a series of lines at frequencies $f_0 \pm m F_2$ with relative amplitudes $V_0 J_m(\Phi_0)$ where J_m denotes a Bessel function of order m.

Modulation of the amplitude and phase by a sinusoidal signal broadens the frequency band occupied by the spectrum around the carrier. If Φ_0 is small, $J_1(\Phi_0) \approx 1/2\Phi_0$ and the other Bessel functions are practically negligible. The spectra are reduced to only two side lines.

Figure 1.5 shows the one-sided spectra in which everything is collected on the positive frequency side. In order to retain the description of the above section, it is preferable to use the two-sided LF spectrum, covering all frequencies, as indicated in Figure 1.4.

The two-sided phase modulation spectrum is thus given by

$$S_\Phi(F) = \frac{1}{4} \Phi_0^2 [\delta(F - F_2) + \delta(F + F_2)] \tag{1.32}$$

where S_Φ is normalised to the total power of the signal, which amounts to taking $V_0 = 1$ [equation (1.26)].

Similarly, the two-sided amplitude modulation spectrum is

$$S_\varepsilon(F) = \frac{1}{4}\varepsilon_0^2 \left[\delta(F - F_1) + \delta(F + F_1)\right] \qquad (1.33)$$

Collecting these results together, and remembering that the amplitude noise and the phase noise are not correlated, we find that the one-sided spectrum of the signal becomes

$$S_{RF}(F) = P_0 \left|\delta(F - f_0) + S_\varepsilon(F - f_0) + S_\Phi(F - f_0)\right| \qquad (1.34)$$

which is normalised, as usual, to the power of the quasi-sinusoidal signal.

When the phase and amplitude modulated spectra are continuous, because the modulation is random, we can consider developing a comparable theory, illustrated in Fig. 1.5. This leads to a single formula, identical to (1.34), if we suppose that the fluctuations are small. We know how to measure $S_\phi(F)$, and if we neglect the amplitude fluctuations, we find that, to a first approximation,

$$S_{RF} \approx P_0 \left[\delta(F - f_0) + S_\Phi(F - f_0)\right] \qquad (1.35)$$

where P_0 is the total power.

The RF spectrum of a quasi-sinusoidal signal is the superposition of a Dirac function, at frequency f_0, broadened at its base by the two-sided phase spectrum and shifted from zero frequency to f_0. We could perform an identical operation for a known amplitude spectrum $S_\varepsilon(F)$. We shall take care from now on to distinguish between the one-sided spectrum (convenient for measurements) and the two-sided spectrum (introduced in the theory): the ordinates differ by a factor of two, and the spectrum is completed by using the symmetry with respect to the zero frequency.

1.4 FREQUENCY STANDARDS: THE RESONATOR

1.4.1 The ideal loss-free resonator

We shall discuss frequency stability and drift by analysing the operation of an oscillator designed to generate a periodic signal. This will allow us to specify the conditions for the creation of a standard clock.

We begin by considering a mechanical clock, which will introduce a useful formalism to be used later on. The clock contains a resonator in the form of a pendulum, which is the essential component that fixes the frequency. The resonator is coupled to a driving system (a weight and escapement

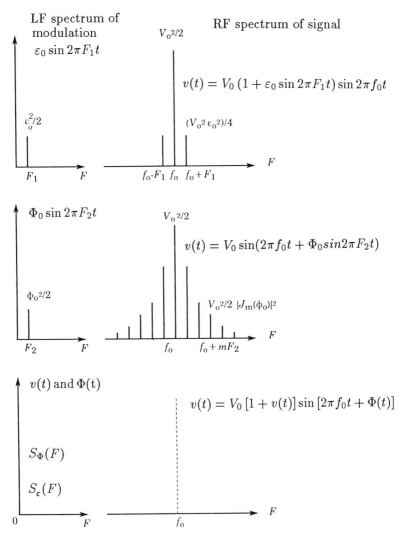

Fig. 1.5 RF spectra of a sinusoidal signal modulated in amplitude and in phase.

mechanism) which compensates for losses, but also has an effect on the determination of frequency, as we shall see later (Section 1.5).

The simplest mechanical pendulum (Fig. 1.6) consists of an object of mass m, suspended from an arm of length l. The pendulum is initially at rest ($\vartheta = 0$) and is given an energy W by displacing it from its equilibrium position; when released, it oscillates. The oscillation involves the continuous sharing of the total energy W between kinetic and potential energies.

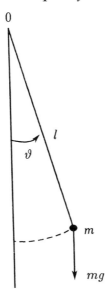

Fig. 1.6 The pendulum: a mechanical resonator.

When the angle of deflection ϑ is small, the potential energy is

$$W_p = mgl\left(1 - \cos\vartheta\right) \approx mgl\vartheta^2/2 \tag{1.36}$$

The kinetic energy W_k is related to the angular velocity $\dot\vartheta = d\vartheta/dt$ by

$$W_k = \frac{1}{2}ml^2\dot\vartheta^2 \tag{1.37}$$

and since total energy is conserved, we have

$$W = W_k + W_p \tag{1.38}$$

so that

$$l\dot\vartheta^2 + g\vartheta^2 = g\vartheta_0^2$$

and

$$\ddot\vartheta = -(g/l)\vartheta \tag{1.39}$$

which is the equation of simple harmonic motion, the solution of which is

$$\vartheta = \vartheta_0 \sin\left(\sqrt{g/l}\,t + \Phi_0\right) \tag{1.40}$$

where ϑ_0 is the maximum deflection.

For small deflections, the mechanical pendulum is therefore an oscillator that can serve as a standard clock. Its constant frequency is given by

$$f_0 = \frac{1}{2\pi}\sqrt{g/l} \tag{1.41}$$

The solution given by (1.40) creates a time scale because the variable t is related to the variable ϑ at all times. We therefore define time in terms of the law of mechanics that relates velocity to kinetic energy. Measurement of angle ϑ gives the time t, i.e., the mechanical time.

In order to advance towards a frequency standard, we have to introduce the technology which ensures that g and l remain constant, so that there is no frequency drift. For l to remain constant, temperature variations have to minimised; g remains constant if the clock stays in the same place. These constraints can be removed by replacing the gravitational potential energy by the elastic potential energy of a loaded spring (this is the transition from the pendulum clock to the balance wheel clock). Nevertheless, no technology will ensure absolute constancy, if only because any working mechanism is subject to wear.

Several identical resonators are therefore necessary and have to be constantly compared with one another. A system that has drifted too far must be reset to the mean of the others, or replaced by a new device. Progress towards the standard oscillator implies a search for a resonator that is easily reproducible. A natural system, readily available in nature, is preferred to an artificial system. And it must operate as continuously as possible.

In practice, we take advantage of the properties of two types of resonator in order to construct standard oscillators. Some are classical macroscopic objects whose frequency depends on their dimensions, whereas others are sets of atoms that behave as natural resonators. We shall see in Chapter 6 how the latter can be exploited to construct atomic clocks. The artificial resonators, which will be studied in Chapter 5, fall into two categories:

- Vibrating solids, in which the total energy oscillates between elastic energy (due to deformation) and kinetic energy (associated with motion). A mechanical signal is difficult to use, so that piezoelectric strip transducers are used to produce an electric signal. The quartz resonator, which can be used up to the gigahertz range, is an example.

- Electromagnetic cavities in which waves are trapped in the space between reflecting walls. The total energy oscillates between electrical and magnetic forms. These resonators are used in the microwave range (from gigahertz up to hundreds of gigahertz) and at optical frequencies (a few hundred of terahertz), in which case they are simply classical Fabry-Perot cavities, first used in spectrometers at the beginning of the century.

1.4.2 The real damped resonator

A real, totally isolated resonator is worthless because it cannot transmit information. The useful resonator is always coupled to the outside world with which it exchanges energy i.e., W is not constant. This coupling causes losses that damp the oscillations. In the case of the pendulum clock, the losses are due to air resistance (the system can be placed in vacuum in order to reduce these losses) and to the transmission of oscillations to the pendulum support (this can be reduced by suitable design).

Furthermore, the resonator must communicate information, for example, by advancing a pointer by one division for each oscillation. This leads to further losses. The losses are unavoidable: there is no such thing as the perfect, lossless resonator.

It is convenient to characterise losses by the energy ΔW given up to the environment during one period T_0. We therefore introduce the quantity Q (Q for quality)

$$Q = \frac{2\pi W}{\Delta W} \tag{1.42}$$

In order to be of interest, a resonator must have a high Q (of the order of a thousand for mechanical resonators, a million for quartz resonators and a billion for atomic resonators). The Q is generally independent of the oscillation amplitude ϑ_0 which decreases slowly. By considering ϑ_0 as a continuous variable, we have

$$W = \frac{1}{2}mgl\vartheta_0^2 \tag{1.43}$$

and

$$\Delta W = -\frac{dW}{dt}\frac{2\pi}{\omega_0} = \frac{2\pi W}{Q} \tag{1.44}$$

so that

$$\frac{\dot{\vartheta}_0}{\vartheta_0} = -\frac{\omega_0}{2Q} \tag{1.45}$$

The solution of this is an exponentially decreasing amplitude:

$$\vartheta_0 = \vartheta_i \exp\left(-\frac{\omega_0}{2Q}t\right) \tag{1.46}$$

where ϑ_i is the initial displacement. A real damped resonator produces a signal of the form

$$\vartheta(t) = \vartheta_i \exp\left(-\frac{\omega_0}{2Q}t\right)\sin\left(\omega_0 t + \Phi_0\right) \tag{1.47}$$

The question is: does this signal provide a frequency suitable, for example, for the measurement of time? The answer is yes, if we count the oscillations as the resonator crosses the point $\vartheta = 0$. The amplitude variation does not play a significant role. We can therefore introduce a frequency standard, but not if we look for the spectral energy distribution by calculating the spectrum. The signal produced by a damped resonator that has been set running by an impulse is given by

$$\vartheta(t) = \int_{-\infty}^{+\infty} C(F)\exp\left(j2\pi Ft\right)dF \tag{1.48}$$

where

$$C(F) = \int_{-\infty}^{+\infty} \vartheta(t) \exp(-j2\pi Ft)dt \qquad (1.49)$$

Parseval's theorem now allows the total energy of the impulse (proportional to the square of the displacement) to be distributed along the frequency axis. We thus have

$$\int_{-\infty}^{+\infty} [\vartheta(t)]^2 dt = \int_{-\infty}^{+\infty} |C(F)|^2 dF \qquad (1.50)$$

where $|C(F)|^2 dF$ is the energy contained within the frequency interval $F, F + dF$. In the case of the above impulse, we have

$$|C(F)|^2 = \frac{\vartheta_i^2 f_0^2}{\left[(f_0/2Q)^2 + (F - f_0)^2\right]\left[(f_0/2Q)^2 + (F + f_0)^2\right]} \qquad (1.51)$$

This spectral energy density is illustrated in Fig. 1.7. It is spread over all the frequencies, but there is a well-defined resonance (if Q is high) at the frequency f_0. The linewidths are of the order of

$$\Delta F = \frac{f_0}{Q} \qquad (1.52)$$

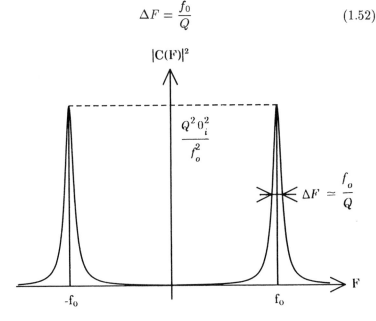

Fig. 1.7 Spectral energy density of a damped resonator.

Damping broadens the spectrum. The concept of a well-defined frequency (corresponding to a Dirac function in the case of the lossless resonator) has gone because the signal is damped. A damped resonator still allows frequency to be measured by counting oscillations as they cross the point $\vartheta = 0$, but this process has practical limitations because the amplitude tends to zero.

The periods can be counted as long as the amplitude remains higher than a certain fraction of the inital amplitude; this means that measurement is possible for coherence times of the order of $10Q/\omega_0$ (if we assume that measurement is possible for ϑ greater than $0.01\vartheta_i$). The number of oscillations that can be counted is therefore of the order of Q.

A further problem in this case is that the frequency of the resonator depends on the oscillation amplitude. This phenomenon is associated with the nonlinear characteristics of the device and is encountered to some degree in all resonating physical systems (Chapter 5).

If energy losses can be compensated by the injection of energy into each oscillation, Q can be made to tend to infinity and, hence, the linewidths to zero. If fluctuations, drift and wear were put aside, we would seem to be well on the way towards a standard oscillator. However, other complications now come into play (Section 1.5).

1.4.3 Simple model of a resonator

We now consider a simple electrical circuit that will be used frequently later. First, we have to introduce components capable of storing energy in two different forms, since the oscillation results from the exchange of energy between these two forms. An inductor of self inductance L and a capacitor of capacitance C can be used in this representation. It is then sufficient to add a parallel resistance R to obtain a simple but effective model of a resonator.

The signal is the applied voltage v (Fig. 1.8) for which we have the equation

$$C\frac{dv}{dt} + Gv + \frac{1}{L}\int v\,dt = i(t) \tag{1.53}$$

where $i(t)$ can be the input current representing exchanges between the resonator and the outside world [$i(t)$ can be a random signal representing noise, or a controlled signal used, for example, to synchronise the resonator with an external signal].

Harmonic analysis of the circuit yields the transfer function

$$Z(j\omega) = \frac{v}{i} = \frac{Z_0}{1 + jQ(\omega/\omega_0 - \omega_0/\omega)} \tag{1.54}$$

where $Z_0 = R$, $\omega_0^2 LC = 1$ and $Q = R\sqrt{C/L}$.

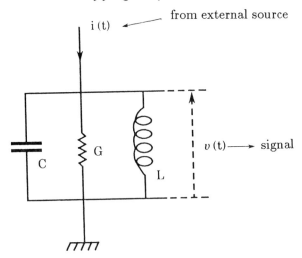

Fig. 1.8 A simple but effective model of a resonator.

Figure 1.9 shows $|Z|^2$ and arg Z as functions of pulsatance. The pass band at $-3\,\mathrm{dB}$ is found to be $\Delta\omega = \omega_0/Q$. In the neighbourhood of resonance, we have

$$Z(j\omega) \approx \frac{Z_0}{1 + j2Q(\omega - \omega_0)/\omega_0} \qquad (1.55)$$

Any circuit for which harmonic analysis results in curves of this type is a resonator that can be described by the above simple model. The measurement of $\omega_0, Q,$ and Z_0 yields the values of the three parameters of the model, namely, L, C, R.

1.4.4 Control of an oscillator by the response of a resonator

When a very good resonator is produced, it is usually kept as isolated as possible. Still, it must be used. For example, it can be integrated in an active circuit designed for loss compensation, so that we can pass from the resonator to the oscillator. This is an effective method, but it reduces the isolation of the resonator.

We can confine ourselves to the interrogation of the resonator with a very small sinusoidal signal generated by an auxiliary oscillator of variable and controllable frequency f. The response of the resonator to the test signal depends on the separation between the resonance frequency f_0 and the signal frequency f. The second curve of Fig. 1.9 illustrates the frequency-phase conversion: the phase passes through zero, and changes sign, at $f = f_0$. All that is necessary is to supply, from a supplementary circuit, the voltage controlling the oscillator, so that the resonator response is always centred on the point of zero phase. The frequency f is then said to be

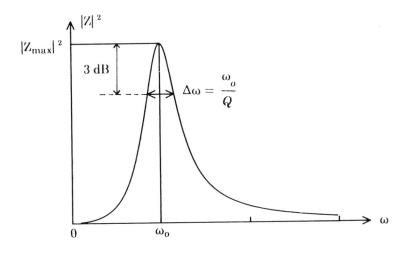

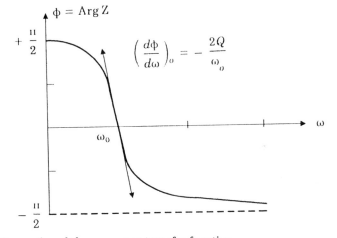

Fig. 1.9 Properties of the resonator transfer function.

locked to f_0. The response curves of natural resonators such atoms are also found to have singular points to which the frequency of a controlling oscillator can be locked. The standard resonator can then be used in the optimum regime to protect it from external perturbations.

1.5 FROM THE RESONATOR TO THE OSCILLATOR

1.5.1 Maintaining the oscillations of a pendulum

To produce an oscillator, we have to couple the damped resonator to a source of energy that compensates losses in each oscillation.

The recoil escapement of a mechanical clock is an example of an oscillation-maintaining system. A weight that goes down, or a spring that is being released, drives a cog wheel (Fig. 1.10). The anchor is rigidly attached to the pendulum and one of its pallets blocks a cog on the wheel; the attendant friction increases losses. However, as the equilibrium position is approached, the cog escapes and the other pallet blocks the next cog. During the escape, the tip of the cog driven by the weight slides on the sloping face of the anchor, which is shaped in such a way that the cog pushes the anchor in the direction of its movement by transmitting to it the energy supplied by the weight as its descends between successive cog escapes. This mechanical system works at frequencies of the order of a few cycles per second. It wears out quickly even if the pallets anchor are lined with ruby. It is subject to short-term fluctuations because the escape process is complex and depends on the state of the surface and the exact shape of the cog tip. This has nevertheless been the only driving system available for two centuries. It has ensured that the measurement of time and of other associated quantities (such as those used in geodesy and in navigation) attained much higher precision by the end of the nineteenth century.

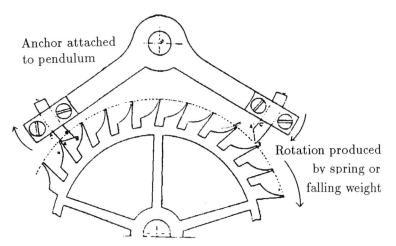

Anchor attached to pendulum

Rotation produced by spring or falling weight

Fig. 1.10 Recoil escapement was for two centuries the only system providing a quantity of energy W for each oscillation, thus compensating for the resonator loss.

1.5.2 Maintaining electrical oscillations

The advent of the valve (1905) and then of the transistor (1948) led to the development of electronic systems that were much more convenient than mechanical devices. The resonator can be a LCR circuit (such as the model of Fig. 1.8) and was the basis for the first radiofrequency oscillators (1919). It can also be a vibrating elastic blade made from a piezoelectric material (e.g., quartz), or a metal blade with a magnet at each end. The magnets are allowed to oscillate in a coil and induce an electromotive force (a tuning fork is a possible example).

Figure 1.11 illustrates the principle used to maintain the oscillations of a resonator. The signal $v(t)$ controls a field effect transistor whose output current is fed into the resonator. This closed-loop circuit relies on a principle that underlies the operation of all oscillators. A field effect transistor is a current generator that is driven, without power being expended, by a voltage. When the polarities are suitably chosen (and maintained by dc sources not shown in Fig. 1.11), the current-voltage characteristic is of the form

$$i = gv - \beta v^3 \tag{1.56}$$

For small signals v, the transistor supplies a current i proportional to v. As the driving signal increases, a saturation effect occurs and the current increases more slowly than in the proportional region.

For a sinusoidal signal

$$v = V \sin \omega t \tag{1.57}$$

and if terms with pulsatance 3ω can be neglected, we obtain the linear relation

$$i = \left(g - \frac{3\beta}{4} V^2 \right) v \tag{1.58}$$

This is reasonable because the resonator tends to filter out pulsatances that are very different from the operating pulsatance ω_0.

A field effect transistor operating in sinusoidal regime is thus a current generator driven by a voltage, and its transconductance (the ratio i/v) decreases with increasing amplitude of the driving signal.

The oscillation-maintaining mechanism is now clear. The basic equation of the circuit is

$$C\frac{dv}{dt} + Gv + \frac{1}{L} \int v \, dt = \left(g - \frac{3}{4}\beta V^2 \right) v \tag{1.59}$$

which can also be written in the form

$$C\frac{dv}{dt} + \frac{1}{L} \int v \, dt = \left(g - \frac{3}{4}\beta V^2 - G \right) v \tag{1.60}$$

These expressions show that the transistor compensates the loss represented by R. On start up (small amplitude V), there is over-compensation, and the amplitude increases. Compensation begins when the amplitude reaches V_0, where

$$g - \frac{3}{4}\beta V_0^2 - G = 0 \tag{1.61}$$

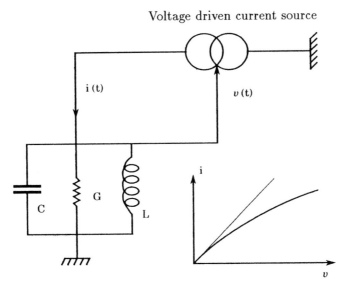

Voltage driven current source

Fig. 1.11 Principle used to maintain the oscillations of a resonator by means of an active circuit (the sources maintaining the polarity of the transistor are not shown).

The transistor then supplies, in each oscillation, the exact amount of energy necessary to maintain a constant amplitude. We have thus gone from the resonator to the oscillator with the help of the closed feedback loop.

1.5.3 The oscillator linewidth

The 'lossless' resonator (Section 1.4.1) delivers a sinusoidal signal, its spectrum is a Dirac pulse and its linewidth is zero (Section 1.3.1). Unfortunately, losses are unavoidable (Section 1.4.2) and the result is a nonzero linewidth $\Delta\omega$ related to the Q factor. On its own, the resonator can be used only during the coherence time $1/\Delta\omega$.

The oscillation of a resonator are maintained by compensating its losses, which results in higher Q and smaller linewidth. When the feedback loop is in place, the losses are compensated exactly in each oscillation. The Q is then infinite and the linewidth is zero. The oscillator spectrum is a Dirac pulse, which corresponds to a sinusoidal signal of constant amplitude (Section 1.5.2). We have thus seemingly realised a pure frequency signal.

Unfortunately, the current i produced by the transistor in the feedback loop is not the only signal that effects the resonator. There is also the background noise, a random signal from various sources, some of which are internal. A resistor is a source of noise because of the random motion of free electrons caused by thermal vibrations, and the transistor is also a source of noise because of the fluctuating flow of electrons across the source-drain junction. Other sources of noise are external to the oscillator and are due to unavoidable coupling to the environment. This creates fluctuations due to external temperature variations and mechanical vibrations transmitted by supports, which modify some of the oscillator parameters in a random manner.

The effect of some of these sources of noise, e.g., white noise, can be represented by a random current generator (Fig. 1.12) with power spectral density $A(F)$ $(cf.$ Section 1.2.3). The filtered noise in an interval $F, F+dF$ is then equal to power $A(F)dF$.

So long as the feedback loop is not closed, the phenomena are classical. The resonator transfer function $Z(F)$ [given by (1.55)] shows that the noise power in the interval $F, F + dF$ is

$$A\left(F\right)\left|Z\right|^2 \, dF = \frac{Z_0^2 A\left(F\right)}{1+4\left(\frac{F-f_0}{\Delta F}\right)^2}dF \qquad (1.62)$$

The output noise is filtered by the resonator.

The phenomena become more complicated if the loop is closed, which transforms the resonator into an oscillator. A sinusoidal signal is then generated and can be represented by a Dirac impulse at frequency f_0.

Once filtered by the resonator and the loop, the noise is reduced to its frequency components close to f_0 and is superimposed on the signal. The result is a Dirac line broadened at its base by noise (the noise level is low compared to the amplitude of the sinusoidal signal).

We note that an accurate study would also need to take into account the nonlinear characteristics of the oscillator, which cause frequency shifts.

The ideal oscillator delivering a pure sinusoidal signal is thus seen to be only a concept, since it would require either a lossless resonator (with an infinite Q) or noiseless electronics. The real shape of a line will be precisely defined in Chapters 3 and 5.

1.6 WHY BUILD OSCILLATORS AND CLOCKS?

An oscillator is a device that delivers a periodic signal with frequency, phase and amplitude known as accurately as possible. There are many classes of oscillator. Their stability depends to a large extent on the quality of their resonators and on the precautions taken to protect them from external perturbations and the influence of external sources of noise.

Oscillators (or equivalent devices) using atomic vibrations are the most stable currently available. The caesium atomic clock is in fact the primary standard of frequency and time.

Quartz oscillators cannot be considered as primary because their nominal frequency depends above all on the dimensions of the resonator crystal

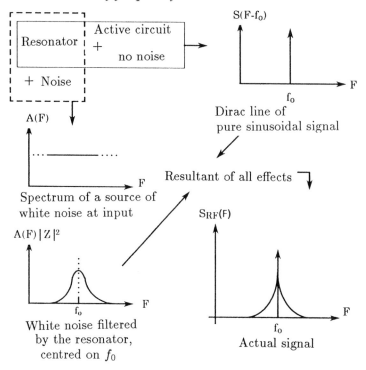

Fig. 1.12 The effect of a source of white noise on the spectrum of the oscillator signal.

i.e., directly on human intervention, so that its absolute value cannot, in principle, be determined in advance.

The range of applications is extensive. Oscillators supply the carrier frequencies of telecommunications networks. Accurate measurement of the propagation time of an electromagnetic wave is used in the localisation of moving objects, i.e., in radar. Measurement of the apparent frequency of a signal emitted by a moving object (the Doppler effect) is used in the measurement of its velocity.

These principles are also employed in navigation systems. Time must be known, and a set of clocks is used to establish a time scale. Many measuring instruments need a frequency reference. Many devices rely on the transformation of changes in a physical quantity into a frequency. This offers the highest resolution, because frequency can be measured with the highest precision.

Frequency and time measurements are therefore two complementary and inseparable aspects of oscillators and clocks.

1.6.1 Frequency measurement

Frequency is measured by an instrument consisting of an oscillator or a reference clock and an associated comparison system. The performance of the standard oscillator is described by a graph (Fig. 1.13) of the square root of the variance of frequency (its stability) as a function of the duration of measurement. This graph is different for different oscillators, and this determines the range of possible applications. Advances in metrology rely on the lowering of the level of the plateau, an increase in its length and improvements in the comparison techniques (there is no point in having a remarkable standard if the methods used to compare signals introduce large uncertainties).

Comparison systems will be discussed in Chapters 2 and 4. The quartz oscillator, which is an excellent secondary standard will be studied in Chapter 5. Finally, atomic clocks (primary caesium standard, hydrogen maser and so on) will be studied in Chapter 6. The current precision of frequency measurement is generally much higher than that of all other measured physical quantities. There are also cases where the present accuracy is at the limit, but does not satisfy the needs of experimentalists. This is the justification for the work of metrologists who try to perfect clocks and comparison techniques.

The peak of metrological precision is exploited when measured frequency is used to determine the wavelength of electromagnetic radiation. The relation

$$\lambda = \frac{c}{f} \tag{1.63}$$

has been raised to the status of fundamental metrological formula since the speed of light c was chosen, as a matter of convention, to be

$$c = 299\,792\,458 \text{ m/s}$$

Once this is accepted, λ can be measured with the same precision as frequency. These measurements are superior to interferometry which relies on the comparison of the wavelength λ with standard lines emitted by krypton. The above convention has relaunched research in very high resolution spectroscopy, especially in relation to the measurement of fundamental constants. In fact, everything is far from settled, since the frequencies of atomic oscillators, which are accurately known, often correspond to wavelength bands in which interferometry is difficult. We shall return to this problem in Chapter 4.

Certain physical laws are so securely established that they can be used to create a standard scale by relating it to a frequency. An example is provided by the superconducting diode based on the Josephson effect. In this device, pairs of electrons with charge $2e$ cross the Josephson junction as a result of the tunnel effect. If V is the potential difference across the junction, this transition causes an energy transfer of $2eV$ and the current in the junction oscillates with frequency

$$f = \frac{2e}{h}V \tag{1.64}$$

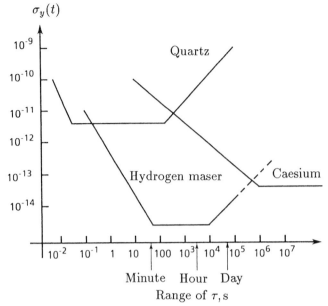

Fig. 1.13 The choice of a standard depends on the desired precision and the duration of measurement.

where h is Planck's constant.

Measurement of f provides information on the value of V, so that the Josephson junction can be the basis of a voltage standard. Since the January 1, 1990, the adopted result is (in GHz/V)

$$\frac{2e}{h} = 483\,597.9$$

Further advances are still needed in the measurement of fundamental constants such as e and h because their precision is still much lower than that of frequency measurement.

Yet another perspective is offered by the general theory of relativity which makes certain predictions for clocks in a gravitational field. For example, in the neighbourhood of the earth, an altitude change Δx causes the change in frequency

$$\frac{\Delta f}{f_0} = \frac{g}{c^2}\Delta x \approx 10^{-16}\Delta x \tag{1.65}$$

where Δx is in metres and g is the local value of the acceleration due to gravity.

We can envisage changing the altitude of a clock and thus measuring the gravitational field in the neighbourhood of the earth step by step in order to obtain the shape of the geoid. The general theory of relativity has been verified experimentally. One such experiment consisted of measuring changes in the frequency of a hydrogen clock travelling in a rocket on its ballistic trajectory. We can also envisage the detection of gravity waves by comparing two clocks a long distance apart. Such measurement require a relative precision of 10^{-18}. The proposal is to use a stabilised Michelson interferometer with linear dimensions of 3 km.

Telecommunications engineers use the analogue modulation of carrier waves. To increase the flow of information, a very large number of carriers separated by the band necessary for modulation, is sent down the same transmission line. When the frequencies fluctuate, the bands overlap, and this is not acceptable. A large number of frequencies in a pre-arranged pattern is therefore synthesised from a clock. Long-term precision of the order of relative precision of 10^{-11} is desirable.

Modulation by pulse and coding, which is the basis of digital communications, requires stable oscillators to sustain the pulsing; 10^{-11} is desirable and even internationally compulsory for digital networks (as recommended by *Comité Consultatif International Télégraphique et Téléphonique*).

There are numerous applications. Some receivers use the transduction of changes in a physical quantity into frequency changes, thus achieving high resolution, since frequency is measured with the highest accuracy among all other physical quantities.

1.6.2 Measurement of time intervals

The oldest application of clocks (in the most traditional sense of the word) is the measurement of the time interval between two events. This is done by counting the number of periods of a standard oscillator in a given interval. The overall accuracy depends on the uncertainty in the counting and in the period of the oscillations (i.e., in the frequency). Figure 1.13 illustrates the relation between the interval τ to be measured and the precision $\Delta f/f_0$ that can be attained (the variance depends on τ). It is advisable to specify the order of magnitude of the interval to be measured and the desired accuracy. Precision is degraded after a certain interval because of drift. Radar operators measure distances by determining the transit time of an electromagnetic pulse. High accuracy is required, in such cases, but only within a short interval of time.

A laser can be used to measure the distance between the Earth and the Moon to within a few dozen centimeters. Simultaneous measurement of the separation between several points can be used in triangulation to determine their relative position. This is the classical method used in navigation and in geodesy. In the GPS (global positioning system) and GLONASS systems, conceived by Americans and Russians, respectively, measurement is made of the intervals between the arrival of signals originating from satellites in geostationary orbits, which carry synchronized atomic clocks.

With four satellites, distances can be determined to within a few meters and the local time to a few nanoseconds.

The establishment of a time scale is the most fundamental application of clocks. A clock must function continuously, starting from an event chosen as the origin of time. This provides the ephemeris, marking each instant by its date, i.e., the time that has elapsed since the chosen origin. This problem is dominated by the long-term behaviour of the frequency standard and also by other scientific, technological and social factors, since the fixing of a date, and of the passing of time generally, play a considerable role in human affairs. We can leave a clock permanently switched on, count the periods and store the result in a memory after division by a known factor. Anyone can consult this memory and hence know the date. However, the clock fluctuates and drifts. The operation therefore requires the availability of a set of clocks, as identical as possible, which are constantly compared among themselves. The average result provides a time scale. Individual clocks fluctuate relative to this reference time, so that it is essential to have a programme of constant clock comparison. Clocks that drift in an abnormal manner can then be eliminated and only the most stable ones retained.

Every laboratory with a set of clocks can produce a time scale, but this is of interest only to those that live nearby. A time scale useful to all must be accessible from all points in space. The propagation of radiowaves, and a coding system allowing the insertion of information in a carrier wave, will play an important role in the dissemination of time. Everyone will be able to know 'the time' in the international reference laboratory and eventually relate the information provided by his local clocks to this basic scale. Clock comparisons wil therefore involve radio relays, cables, satellites and the transportation of clocks.

Numerous, and readily observed, phenomenona mark the passing of time. They include astronomical events (the alternation of day and night, the changing seasons, the crossing the meridian by stars and so on). Such observations permit the local realisation of a very elaborate astronomical ephemeris to which society is used to. This is, of course, the calendar. The drift of natural phenomena that is the basis for this *astronomical time* is greater than the drift of atomic clocks, so that today we adopt the atomic time. But we have to relate it, at each point on the globe (or in each space vehicle), to the astronomical time that everyone is used to. Because of the relative drift of these two scales, the conventions linking them have to be reviewed from time to time. Questions concerning time scales will be studied in Chapter 7. They involve the measurement of frequencies as well as many other fields of science and technology.

2

Radiofrequency measurements

2.1 FREQUENCY RANGES

The first to be considered is the radiofrequency range. It extends from very low frequencies to a few hundred megahertz. For convenience, we shall consider that it runs from 0 to 1 GHz. This range has been in use for the longest time, and a very high level of accuracy has been achieved within it. Digital counting techniques are particularly well suited to these frequencies, and so are analogue devices. They rely on the technology of lumped-parameter circuits. Frequency analysis and time analysis compete in this frequency range, so that it is important to link the results of these two types of measurement. The necessary techniques will be studied in this chapter and in Chapter 3.

The microwave range is next. It runs from a gigahertz to several hundred gigahertz. For convenience, we will consider that it covers the band between 1 GHz and 1 THz. Expressed in wavelengths, it extends from 30 cm (metre waves) to 300 μm (submillimetre waves). It is the range of circuits with distributed parameters; logic circuits circuits are of little use in this range. The essential problem is the translation of frequency to the radiofrequency range in which measurements are easier. It is therefore important to understand the effect of this down-conversion on the spectrum of the signal. In addition, the down-conversion relies on the availability of a reference oscillator with a similar frequency and a nonlinear mixing circuit. Metrology is constrained in this frequency range by the availability of a reference oscillator, synthesised from a standard oscillator. Chapter 4 will deal with this problem.

The third range extends to several hundred terahertz, i.e., down to wavelengths of a fraction of a micron, at which electromagnetic radiation is visible (400 to 800 THz). The principles of measurement are the same as for microwaves, i.e., down-conversion to the radiofrequency range with its well-established experimental techniques. Difficulties are encountered in the fabrication of nonlinear components capable of mixing two infrared lines (optical technology) and in heterodyning in the microwave range (technology of circuits with distributed parameters). Another problem involves intermediate standards. In optics we do not yet know how to synthesise a frequency mf_0/n from a standard f_0 (m and n being known integers).

In general, available oscillators are used together to produce frequencies that can be measured by classical methods. Measuring techniques and the results obtained are presented in Chapter 4 which brings together the microwave and radiofrequency ranges.

Figure 2.1 illustrates schematically the ideas introduced above. It is actually possible to link frequencies within a range extending from a fraction of a hertz to several hundred terahertz. The caesium clock, which will be discussed in Chapter 6, provides the reference frequency. Frequency synthesis techniques are readily used to provide secondary standards at the lower end of the frequency spectrum but difficulties arise at higher frequencies, say, up to 100 GHz. In practice, the internal circuits of the caesium clock provide a reference frequency of 5 MHz, and secondary standards are derived from it. There are now many lasers that provide fixed points in the optical spectrum. Multiplication of these fixed-point frequencies is possible with the aid of tunable lasers. However, it is still difficult to link the microwave spectrum (classical interactions between electrons and electromagnetic fields) and the optical spectrum (quantum mechanics and atomic phenomena). The boundary between them lies at the frequency at which photon energy hf is of the order of the thermal noise energy kT, i.e., at a few terahertz. We will return to this in Chapter 4.

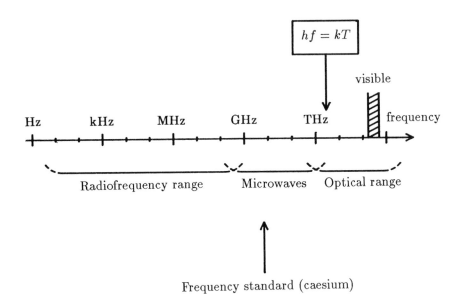

Fig. 2.1 The frequency scale.

2.2 THE RADIOFREQUENCY RANGE

The ideal signal without noise or drift is described by the sinusoidal function

$$v\left(t\right) = V_0 \sin\left(2\pi f_0 t\right) \tag{2.1}$$

where V_0 and f_0 are the amplitude and frequency of the signal, respectively. A real signal takes the form

$$v\left(t\right) = \left[V_0 + \varepsilon\left(t\right)\right] \sin\left[2\pi f_0 t + \Phi\left(t\right)\right] \tag{2.2}$$

where $\epsilon(t)$ and $\Phi(t)$ are random processes representing fluctuations in amplitude V_0 and the total phase, respectively, for a signal of nominal frequency f_0.

If we neglect amplitude fluctuations, which is generally valid for good quality oscillators, the real signal becomes

$$v\left(t\right) = V_0 \sin\left[2\pi f_0 t + \Phi\left(t\right)\right] \tag{2.3}$$

Fluctuations in the phase $\Phi(t)$ can be measured by demodulating the signal by mixing it with a reference signal of the same frequency in order to eliminate the deterministic component and extract the random component to be measured.

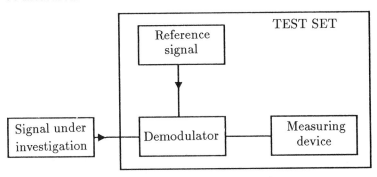

Fig. 2.2 Block diagram of a test set.

A test set for radiofrequency signals (Fig. 2.2) must therefore include a source of the reference signal, a demodulator of the signal under investigation and a noise detector.

2.2.1 The standards

We will distinguish between
- the primary standard which is directly related to the atomic standard

- secondary standards with long-term stability lower than that of the primary standard, but capable of presenting spectral quantities at fixed points

- tertiary standards derived from the above standards by multiplication, division, or synthesis.

Primary standard. The caesium clock is the primary standard providing the international standard of time and frequency. It offers the best long-term stability and reproducibility at present. By long-term we mean measuring times corresponding to integration times longer than 1000 seconds. Oscillator drift is measured relative to this standard. The relative frequency drift of oscillators is commonly quoted per day, per month, or per year (10^{-10} per day, for example).

Secondary standards. Quartz oscillators can be used as secondary standards in the short-term and very short-term ranges. The frequency delivered by these standards is often of 5 MHz or 10 MHz, but it can extend to a few hundred megahertz. Figure 2.3 compares the performance of caesium standards with an ultrastable quartz oscillator (USO).

The USO is better than the caesium clock in the short term, but drifts with time; it has to be reset periodically by comparison with the 5 or 10 MHz signal from a caesium standard or a signal emitted by a radiofrequency source whose carrier frequency is locked to the caesium standard, e.g., France-Inter (162 kHz), HBG (75 kHz), Loran C (100 kHz), Droitwich (200 kHz) and, nowadays, GPS (Global Positioning System).

There are also ultrastable oscillators producing higher frequencies, for example, 100 MHz, whose particular feature is very low phase noise above a few kilohertz. Figure 2.4 shows typical phase fluctuation densities of such oscillators. More recently, surface acoustic wave (SAW) resonators have been used in high-power oscillators that are stable at 500 MHz and offer very interesting characteristics.

The above reference oscillators can be compared by calculating the spectral density of relative frequency fluctuations, $S_y(F)$, each oscillator. This density is given by

$$S_y\left(F\right) = \frac{F^2}{f_0^2} S_\Phi\left(F\right) \tag{2.4}$$

where f_0 is the oscillator frequency and F is the Fourier frequency of the spectral density of frequency fluctuations (*cf.* Section 1.4).

Figure 2.5 shows a plot of three spectral densities. It is clear that no single secondary standard will cover the entire Fourier frequency range. The USO working at 5 MHz is the best up to 300 Hz carrier, whilst the 100 MHz oscillators take over above 300 Hz. The performance of 500 MHz oscillators is also given for the sake of completeness.

Consequently, if the signal under investigation has relative frequency fluctuations in zone 1 (Fig. 2.5), it can be assigned to the 5 MHz USO or

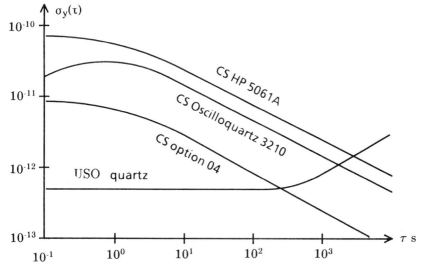

Fig. 2.3 Comparison of the medium-term relative frequency stability of a quartz USO with different caesium clocks

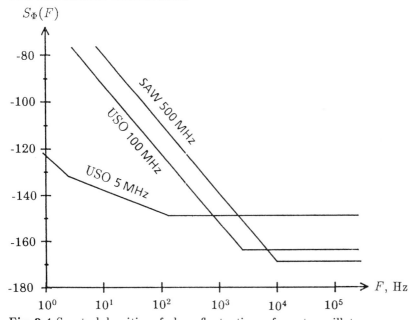

Fig. 2.4 Spectral densities of phase fluctuations of quartz oscillators.

the 100 MHz oscillator. In zone 2, however, only the USO will allow the measurement, and this also applies in zone 3 to the 100 MHz oscillator. A problem arises when the signal fluctuates in both zone 2 and zone 3. Metrology then has to rely on a 100 MHz oscillator linked to an USO with

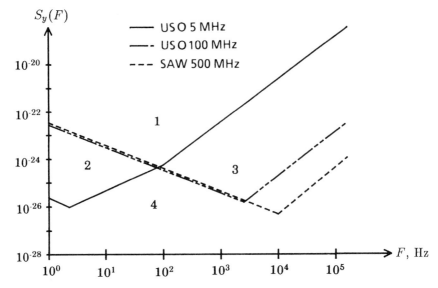

Fig. 2.5 Spectral densities of relative frequency fluctuations of quartz oscillators.

a passband of the order of 300 Hz (in the case of Fig. 2.5).

Tertiary standards. These are produced by multiplication or by division, or by what is commonly called frequency synthesis. The latter implements simultaneous frequency multiplication and division, as well as phase locked loops. These three processes will be described below.

(a) Frequency multiplication

Frequency multiplication is performed by exploiting the nonlinear properties of electronic devices which take a sinusoid and distort its shape or limit its amplitude in order to generate a nonsinusoidal signal containing the harmonics of the input signal. The required harmonic is then be selected by filtering.

The decomposition of different types of periodic signal into a Fourier series is illustrated in Fig. 2.6. The square-wave signal is produced by a flip-flop and the even harmonics are suppressed, at least in theory; in reality, fluctuations in bistable operation give rise to the appearance of even harmonics, but with much smaller amplitudes than those of odd harmonics.

A signal consisting of narrow pulses can be produced by a series of flip-flops or by digital processing of a sinusoidal signal followed by a monostable circuit. $k = 1/2$ takes us back to a square-wave signal, $k = 1/3$ suppresses harmonics 3,6, 9... and $k = 1/4$ suppresses harmonics 4, 8, 12 ... (k defines the relative width of the rectangular pulse).

A multiplier incorporating the step recovery diode (SRD) can be used to generate triangular pulse trains. In effect the diode generates a pulse of very short fixed duration (typically 100 psec) in each cycle. If the input signal has a frequency of 100 MHz, the magnitude of k for the $n = 70$ harmonic is then 0.01. This device therefore generates harmonics of relatively constant amplitude within the frequency range 100 MHz–7 GHz.

A signal consisting of sinusoidal arcs can be produced by transistor-driven multipliers (Fig. 2.7). The latter are usually symmetric in design. In fact, a push-pull circuit enables the generation of odd harmonics of the input signal while a push-push circuit delivers even harmonics. Harmonic selection is then carried out by the oscillating circuit that charges the transistors. This circuit must have a sufficiently high Q to suppress undesirable harmonics. With such devices, multiplication by a factor n is only possible if n factorises into primes greater than 7, so that the attenuation A of undesirable harmonics is at least 30 dB. Indeed the attenuation of a tuned circuit as a function of frequency deviation Δf is given by

$$A = \left|1 + j2Q\frac{\Delta f}{f}\right| \tag{2.5}$$

where f is the central frequency. The first undesirable harmonics occur at $2f_e$ where f_e is the input frequency. Since the central frequency is nf_e, we obtain

$$A = |1 + j4Q/n| \tag{2.6}$$

If we require an attenuation of ≥ 30 dB, i.e. a ratio of 30,

$$\frac{4Q}{n} \geq 30$$

so that

$$n \leq \frac{4Q}{30}$$

i.e., for $Q = 50$, we have $n \leq 7$.

A multiplier with $n = 75$ is produced by a cascade of two $n = 5$ multipliers and $n = 3$ multiplier. It is always preferable to put the high multiplication factors first because outputs that depend on the cut-off frequency of the transistors will then decrease with increasing frequency. These multipliers have the disadvantage that they have a fixed frequency and a very narrow bandwidth, the input signal passband being roughly that of the output stage divided by the multiplying factors of the preceeding stages. For example, for the $n = 75$ multiplier, 5 MHz input frequency and output-circuit Q of 50, the input passband is then only 50 kHz, i.e., 1% of the bandwidth.

(a) square wave

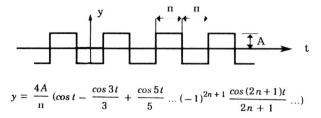

$$y = \frac{4A}{\pi}\left(\cos t - \frac{\cos 3t}{3} + \frac{\cos 5t}{5} \ldots (-1)^{2n+1}\frac{\cos(2n+1)t}{2n+1} \ldots\right)$$

(b) narrow rectangular pulses

$$y = A\left[(k + \frac{2}{\pi}\left(\sin k\pi \cos t + \frac{1}{2}\sin 2k\pi \cos 2t + \frac{1}{3}\sin 3k\pi \cos 3t \ldots + \frac{1}{n}\sin nk\pi \cos nt \ldots\right)\right.$$

(c) narrow triangular pulses

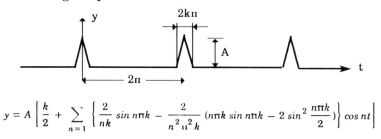

$$y = A\left|\frac{k}{2} + \sum_{n=1}\left\{\frac{2}{nk}\sin n\pi k - \frac{2}{n^2\pi^2 k}(n\pi k \sin n\pi k - 2\sin^2\frac{n\pi k}{2})\right\}\cos nt\right|$$

(d) sinusoidal arcs

$$y = \frac{A}{1 - \cos\theta/2}\left((\sin\frac{\theta}{2} - \frac{\theta}{2}\cos\frac{\theta}{2}) + (\frac{\theta}{2} - \sin\frac{\theta}{2}\cos\frac{\theta}{2})\cos t + \ldots\right.$$

$$\left.\ldots + (\frac{\sin(n+1)\vartheta/2}{n+1} + \frac{\sin(n-1)\vartheta/2}{n-1} - \frac{2\sin\vartheta/2\cos\vartheta/2}{n})\cos nt + \ldots\right)$$

Fig. 2.6 Decomposition of a periodic signal into a Fourier series (the fundamental frequency is such that $2\pi f = 1$).

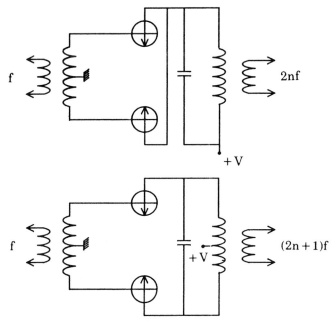

Fig. 2.7 Generation of a signal consisting of sinusoidal arcs.

Frequency multiplication amplifies phase noise, so that the multiplier input signal takes the form

$$v_i(t) = A_i \sin\left(2\pi f_i t + \Phi(t)\right) \qquad (2.7)$$

and the output signal v_0 can then be represented by

$$v_0(t) = A_0 \sin N\left[2\pi f_i t + \Phi(t)\right] = A_0 \sin\left[2\pi N f_i t + N\Phi(t)\right] \qquad (2.8)$$

which implies an amplification of the spectral phase density by a factor of N^2. In an ideal multiplier of order N, the phase noise is amplified with a gain (in dB) given by $G_{dB} = 20 \log N$.

(b) Frequency division

The frequency division relies on logic dividers that require a shaping circuit consisting (for a sinusoidal signal) of an amplifier, a smoothing circuit and a differentiator (if pulses are to be created). This solution is prefereable to an amplifier followed by a Schmitt trigger which is generally an additional source of noise. A pulse counter will then deliver one output pulse for N input pulses. This achieves division by N. The resulting signals have a better phase fluctuation spectral density than the input signal.

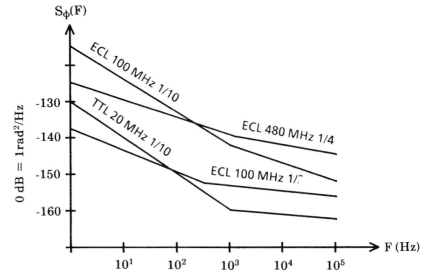

Fig. 2.8 Phase noise of different types of frequency divider.

Indeed, a divider is a multiplier of order of $1/N$. In principle, the noise reduction is by $G_{dB} = -20 \log N$.

The limiting factor is the intrinsic noise of the dividers. This is illustrated in Fig. 2.8.

(c) The phase locked loop

This is an essential circuit for frequency synthesis and phase comparison. Its principle of operation is illustrated in Fig. 2.9; details will be considered when we come to particular metrological applications.

The oscillator incorporates a variable capacitor (usually a varicap diode) which sets the frequency and is controlled by an external voltage v. The instantaneous frequency is a function of v and, if we confine our attention to small signals about a bias voltage, we have

$$f = \frac{1}{2\pi}\frac{d\Phi}{dt} = K\,v \tag{2.9}$$

where K is the frequency-voltage sensitivity of the oscillator. We therefore have

$$\Phi = 2\pi K \int_{t_0}^{t} v\,dt + \Phi_0 \tag{2.10}$$

This phase is compared with the phase Φ_1 of the reference signal, and the output signal of the comparator controls the voltage controlled oscillator

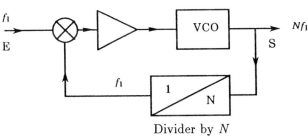

Fig. 2.9 Basic phase locked loop with N times multiplier.

(VCO). Thus

$$\frac{1}{2\pi}\frac{d\Phi}{dt} = K\mu(\Phi_1 - \Phi) \qquad (2.11)$$

where μ is the sensitivity of the phase comparator.

The result is that $\Phi = \Phi_1$ after a transitional period of the order of $1/2\pi K\mu$. The VCO is phase (and therefore frequency) locked to the reference signal with phase Φ_1.

The VCO produces frequency multiplication by means of a logic divider whose order N can be adjusted over a wide range (and is accurately known). In the course of this synthesis, the reference signal of frequency f_1 is compared with the VCO output divided by N (Fig. 2.9). The phase comparator controls the VCO and locks its frequency f until f_1 becomes equal to f_{VCO}/N.

Let E be the reference (or input) signal, S the signal from the locked VCO (or the output signal) and V the signal from the free-running VCO. We then have

$$S = \frac{NG_T}{1 + G_T}E + \frac{1}{1 + G_T}V \qquad (2.12)$$

where the total gain of the loop is given by

$$G_T = \mu K(p) G(p) / N \qquad (2.13)$$

In terms of phase fluctuation spectral densities, we have

$$S_{\Phi_{\text{VCO lock}}} = \left|\frac{NG_T}{1 + G_T}\right|^2 S_{\Phi_{\text{ref}}}(F) + \left|\frac{1}{1 + G_T}\right|^2 S_{\Phi_{\text{VCO free}}}(F) \qquad (2.14)$$

Two Fourier frequency domains appear: (1) the domain in which the gain G_T is greater than 1 and

$$S_{\Phi_{\text{VCO lock}}} \approx N^2 S_{\Phi_{\text{ref}}}(F) \qquad (2.15)$$

and (2) the domain in which G_T is less than 1 and

$$S_{\Phi_{\text{VCO lock}}} \approx S_{\Phi_{\text{VCO free}}}(F) \tag{2.16}$$

Different frequencies can be generated by varying the division factor N in steps of the input frequency. The working range of this device is set by the tuning range of the VCO, which can be up to an octave. These circuits are at the basis of all frequency synthesisers.

Frequency synthesisers constitute very useful sources of reference frequency. Indeed, any frequency can be generated with a resolution of the order of 1 Hz (0.01 or even 0.001 Hz in some of them). The long-term stability of the output signal follows the stability of the driver but, unfortunately, the synthesis itself generates phase noise depending on the operations performed in frequency generation.

The output frequency of most commercially available synthesisers is the result of subtractive mixing of two high frequencies synthesised by two multipliers fed with the same signal. For example, a synthesiser with working range of 1 to 160 MHz delivers a signal of 1 MHz by subtractive mixing of two frequencies greater than twice the maximum frequency and separated by 1 MHz. From the point of view of noise, if the reference frequency is of 5 MHz, this signal is multiplied by a factor of about 70 (350 divided by 5), i.e., there is a very significant amplification of noise (50 dB).

Better results are obtained by subjecting the VCO signal to a series of divisions by 2 (Fig. 2.10). This takes us back to the factor 1/5 of the optimum synthesiser generating 1 MHz from a 5-MHz driver. Noise still appears but is now divided by 160 (44 dB).

2.2.2 Phase comparator

This device receives two periodic signals with a phase difference $\Phi(t)$. It produces a signal that is a function of $\Phi(t)$ and performs phases comparison. There are two categories: phase comparators with sinusoidal response and those with linear response.

Sinusoidal signals are usually processed by phase comparators with sinusoidal response although it is possible to obtain logic signals from sinusoids and to use phase comparators with linear response. Logic signals are processed by linear phase comparators utilising logic circuits. In any case, each type is used for small signals near a working point for which the response may be considered to be linearised.

Phase comparator with sinusoidal response. A two-diode phase comparator consists of two voltage sensors connected in parallel. It receives two signals E_1 and E_2. When a signal is applied to only one input, the output voltage is zero if the circuit is well balanced. When E_1 and E_2 have equal frequencies but different phases, a dc output E_s appears and is a function of the phase difference between the two input signals (Fig. 2.11).

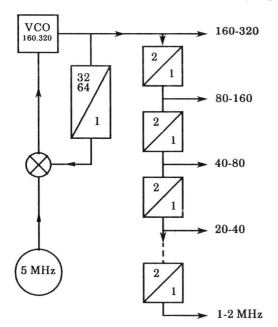

Fig. 2.10 Synthesiser with dividers.

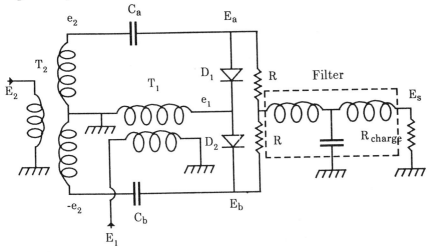

Fig. 2.11 Phase comparator with two diodes.

Suppose that

$$E_1 = E \sin \omega t \tag{2.17}$$

$$E_2 = E \sin (\omega t + \Phi) \tag{2.18}$$

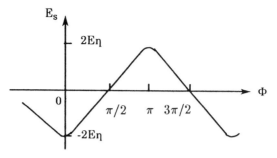

Fig. 2.12 Output voltage of the phase comparator as the function of the phase difference Φ.

The output of the 1:1 transformer T_1 is then

$$e_1 = E \sin \omega t \qquad (2.19)$$

and output of the centre-tapped 2:1 transformer T_2 is

$$e_2 = E \sin (\omega t + \Phi) \qquad (2.20)$$

and

$$-e_2 = -E \sin (\omega t + \Phi) \qquad (2.21)$$

A current flows through diodes D_1 and D_2 only when the voltages e_a and e_b applied to item are

$$e_a = e_1 - e_2 = 2E \cos \frac{2\omega t + \Phi}{2} \sin \frac{\Phi}{2} > 0 \qquad (2.22)$$

$$e_b = e_1 + e_2 = 2E \sin \frac{2\omega t + \Phi}{2} \cos \frac{\Phi}{2} > 0 \qquad (2.23)$$

The dc components E_a and E_b across the capacitors C_a and C_b are

$$E_a = 2E \sin \frac{\Phi}{2} \qquad 0 < \Phi < 2\pi \qquad (2.24)$$

$$E_b = 2E \cos \frac{\Phi}{2} \qquad 0 < \Phi < \pi$$

$$E_b' = -2E \cos \frac{\Phi}{2} \qquad \pi < \Phi < 2\pi \qquad (2.25)$$

Bearing in mind the direction of the diodes and the attenuation in the resistors $[\eta = R_{\text{load}}/(R_{\text{load}} + R)]$, the voltage E_s (Fig. 2.12) across the load is

$$E_s = \eta (E_a - E_b) = 2E\eta \left[\sin \frac{\Phi}{2} - \cos \frac{\Phi}{2} \right] \quad \text{for } 0 < \Phi < \pi$$

and

$$E_s = 2E\eta \left[\sin \frac{\Phi}{2} + \cos \frac{\Phi}{2} \right] \qquad \text{for } \pi < \Phi < 2\pi \qquad (2.26)$$

In square-law operation, the response of the phase comparator is practically linear. The sensitivity μ (in V/rad) is given by

$$\mu = \eta E \sqrt{2} \tag{2.27}$$

Near $\Phi = 0$ or π, the output voltage E_s is then proportional to E and we have an amplitude detector.

It is also possible to construct a four-diode phase comparator, called a double balanced mixer (Fig. 2.13.)

The characteristics of the four-diode phase comparator are identical, apart from a factor of 2, with those of the two-diode phase comparator. Its output being higher, the former is preferred.

Phase comparator with linear response. Consider two square-wave signals $E_1(t)$ and $E_2(t)$ differing in phase by Φ. These signals drive an exclusive-or gate (Fig. 2.14).

The width of the output pulses $S(t)$ is a function of the dephasing between E_1 and E_2. If the dc component of $S(t)$ is extracted by a low pass filter, a triangular signal is obtained at the filter output

$$V_s = \frac{A\Phi}{\pi} \quad \text{for} \quad 0 < \phi < \pi \tag{2.28}$$

$$V_s = \frac{A}{\pi}(2\pi - \Phi) \quad \text{for} \quad \pi < \Phi < 2\pi \tag{2.29}$$

where A represents logic level 1.

This type of triangular characteristic is very useful in phase locked loops as it does depend on the direction of dephasing of the locked oscillator. For a given voltage, there are actually two values of the phase: one is unstable and the other stable.

A phase comparator with linear characteristic in the range $0 - 2\pi$ can be built according to the schematic of Fig. 2.15.

The monostable circuits switch on the rising fronts of the input signals E_1 and E_2. The narrower the pulses, the nearer we get to the limits 0 and 2π. These pulses drive the SET and RESET inputs of a flip-flop, and the output can be either Q or $\overline{Q}$. It is therefore be possible to measure E_1 relative to E_2 or *vice versa*.

When the dc component is removed by a low pass filter, the characteristic shown on Fig. 2.15 is obtained.

If the two periodic signals have the same mean frequency, Φ is a constant quantity (apart from fluctuations), but if the mean frequencies are slightly different (by Δf), Φ becomes a function of time. Its variation during a time T is $2\pi\Delta fT$. If the duration corresponding to a variation of 2π is measured, the interval $\Delta f/f$, which is equal to $1/fT$, can be determined. This is, therefore, a way of measuring a frequency interval.

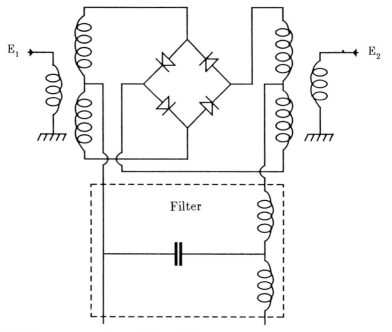

Fig. 2.13 Phase comparator with four diodes.

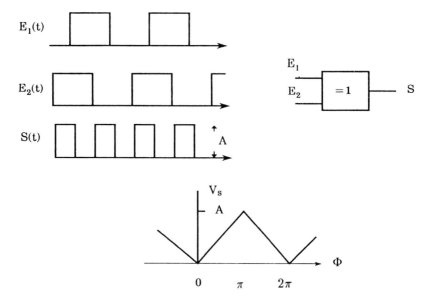

Fig. 2.14 Phase comparator with linear response in the interval $(0, \pi)$.

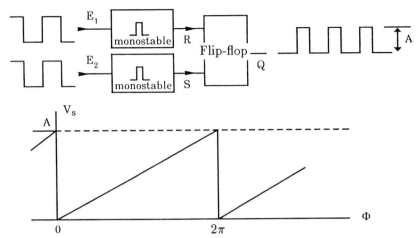

Fig. 2.15 Phase meter with linear response in the interval $(0, 2\pi)$.

2.2.3 Measuring devices

The measuring devices used in the time domain are different from those in the spectral domain. They are frequency counters and spectrum analysers, respectively.

(a) Classical frequency counter

This electronic device counts, during the open time T of a gate, the number of zero crossings by the signal in either positive or negative direction after shaping by a trigger (Fig. 2.16). The resolution is therefore of ± 1 count.

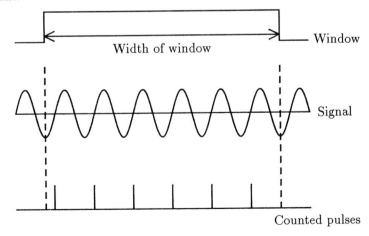

Fig. 2.16 Operation of the classical counter.

The relative resolution (Fig.2.18a) for a given counting time is therefore a function of the measured frequency.

(b) Reciprocal counter

The reciprocal counter measures the period of the signal and displays the frequency. It is actually a double counter (Fig. 2.17). One counter counts the number N of signal periods when the gate is open and the other counts the number M of periods of a clock. This gives the measured time

$$NT = MT_0 \qquad (2.30)$$

and hence

$$T = \frac{MT_0}{N} \quad \text{or} \quad F = \frac{N}{MT_0} \qquad (2.31)$$

The resolution of such counters is plus or minus the clock period.

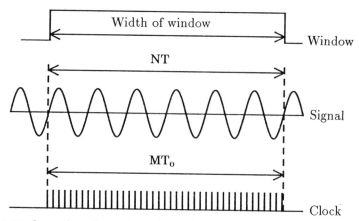

Fig. 2.17 Operation of a reciprocal counter.

The resolution (Fig. 2.18b) is now a function of only the measuring time and no longer of the relative frequency and the measuring time. For a clock operating at 500 MHz, the resolution is $2 \times 10^{-9}/\tau$.

(c) The window spectrum analyser

A spectrum analyser is a device that displays on a screen (or otherwise) the energy distribution among the different Fourier components of a signal.

The analogue window analyser is mostly aimed at radiofrequency analysis.

The digital fast Fourier transform analyser (FFT) is a low frequency analyser (maximum 100 kHz). The signal to be tested is translated to an

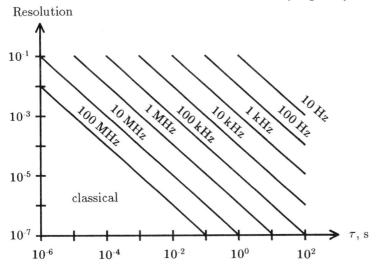

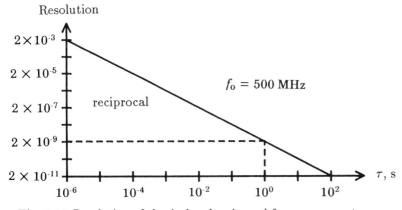

Fig. 2.18 Resolution of classical and reciprocal frequency counters.

intermediate frequency with the aid of a local oscillator which is swept in frequency within $\pm f$, so that the components of the signal within the range $\pm f$ can be examined. The analyser window is thus determined by the width of the corresponding intermediate-frequency filter.

Most commercial analysers have a minimum analyser window of 1 kHz and occasionally down to 100 Hz. These analysers do not therefore allow measurements to be made near the carrier. Moreover, the intrinsic noise of the local oscillator is significant and does not usually allow the low-noise performance of quality radiofrequency sources to be fully exploited.

These analysers are nevertheless very useful in determining the amplitudes of harmonics (or sub-harmonics in the case of signals produced by multiplication). Their sensitivity is of the order of $-100\,\mathrm{dBm}$ with a 10 kHz passband. The sensitivity is a function of the window width.

(d) Fast Fourier transform analyser (FFT)

The Fourier transform of a signal $s(t)$, which allows the quantitative identification of the different harmonics of $s(t)$, is

$$\tilde{S}_e\,(F) = \int_{-\infty}^{\infty} s(t)\exp(-j2\pi Ft)dt \qquad (2.32)$$

Spectrum analysers being digital devices, operating by sampling the function $s(t)$, the spectral sampling density $S_e(F)$ is given by

$$\tilde{S}_e\,(F_k) = \sum_{i=0}^{N-1} s\,(t_i)\,e^{-j2\pi F_k t_i}\,(t_{i+1} - t_i) \qquad k = 0,1,...N-1 \qquad (2.33)$$

This expression shows that if we have N samples of the signal and we need the amplitudes of N harmonics, we have to perform N^2 multiplications. A microprocessor that takes 7 ms per multiplication needs approximately 7 s to perform a 1024-point analysis.

The fast Fourier transform algorithm developed by Cooley and Tukey in 1965 has reduced the number of operations to $N\log_2 N$. In the last example, the computation time is thus reduced by a factor of 200.

The dynamic performance of these spectrum analysers is a function of the capacity of the digital-to-analogue converter that samples the input signal (16 bits = 96 dB, 14 bits = 84 dB, 12 bits = 72 dB and so on). The resolution depends on the scale and can reach -116 dBV. Phase spectral density measurements at -170 dB/rad^2 need an interface whose gain will have to allow, in view of the phase-voltage transfer function of the phase comparator used (typically 0.3 V/rad), a voltage greater than -116 dBV in order to be detectable.

2.3 MEASUREMENTS IN THE SPECTRAL DOMAIN

2.3.1 Characterisation of a signal of the same frequency as that of the reference signal

Demodulation circuit. This method requires a voltage controlled reference signal of the same frequency as the signal to be tested. The reference oscillator is locked to the signal to be tested as illustrated in Fig. 2.19.

If the phase comparator is linear, we can use the Laplace formalism.

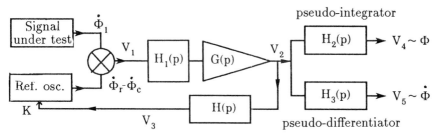

Fig. 2.19 Comparison of two oscillators of the same frequency.

In the feedback loop, $\dot{\Phi}_1(p)$ is the angular frequency fluctuation of the oscillator to be tested, $\dot{\Phi}_r(p)$ is the angular frequency fluctuation of the reference oscillator, $\dot{\Phi}_c(p)$ is the fluctuation correction applied to the reference oscillator, μ is the sensitivity of the phase comparator in volts per radian, $H_1(p)$ is the transfer function of the low-pass filter, $G(p)$ is the amplifier gain, $H(p)$ is the transfer function of the feedback circuit, K the sensitivity of the reference oscillator to the control voltage (radians per second per volt) and $p = j2\pi F$ where F is the frequency under analysis.

If the signals driving the balanced mixer are in quadrature, the signal V_1 is proportional to the phase fluctuation between the two input signals (the primitive of the frequency fluctuations), which can be written as

$$V_1(p) = \frac{\mu}{p}\left(\dot{\Phi}_1 - \dot{\Phi}_r - \dot{\Phi}_c\right) \tag{2.34}$$

Since the main function of the low-pass filter is to remove the carrier, we will assume that $H_1(p) = 1$ at low frequencies. The output voltage V_2 of the amplifier is then given by

$$V_2(p) = \frac{\mu G(p)}{p}\left(\dot{\Phi}_1 - \dot{\Phi}_r - \dot{\Phi}_c\right) \tag{2.35}$$

and hence

$$V_3(p) = \frac{\mu H(p) G(p)}{p}\left(\dot{\Phi}_1 - \dot{\Phi}_r - \dot{\Phi}_c\right) \tag{2.36}$$

The correction applied to the reference oscillator is

$$\dot{\Phi}_c = K V_3 \tag{2.37}$$

which leads to

$$V_2 = \frac{\mu G(p)}{p + \mu K G(p) H(p)}\left(\dot{\Phi}_1 - \dot{\Phi}_r\right) = \frac{\mu G(p)}{1 + (\mu/p)K G(p) H(p)}(\dot{\Phi}_1 - \dot{\Phi}_r) \tag{2.38}$$

Two cases must now be considered. The first corresponds to a feedback loop gain $(\mu/p)KG(p)H(p) \gg 1$ in which case

$$V_2 = \frac{\left(\dot{\Phi}_1 - \dot{\Phi}_r\right)}{KH(p)} \tag{2.39}$$

The second corresponds to a small loop gain for which

$$V_2 = \mu G(p)(\Phi_1 - \Phi_r) \tag{2.40}$$

Thus, by suitably choosing G, μ and K, we obtain an output voltage proportional to the frequency or phase fluctuations.

The random signal V_1, which has zero mean (it is frequency-locked to the reference oscillator), is sent to the FFT spectrum analyser, which generates the signal spectral density as a function of Fourier frequency.

Since the fluctuations $\dot{\Phi}_1$ and $\dot{\Phi}_r$ are uncorrelated, the power spectral densities are added when the spectral analysis of the difference $\dot{\Phi}_1 - \dot{\Phi}_r$ is performed. The result is that the spectrum analyser responds to V_2 by displaying the spectral density proportional to the sum of the spectral densities of the frequency fluctuations $\dot{\Phi}_1$ and $\dot{\Phi}_r$ of each one of the oscillator (if the loop gain is high) or the spectral density proportional to the sum of the spectral densities of the phase fluctuations Φ_1 and Φ_r of each one of the oscillators, (if the loop gain is low).

*Determination of loop transfer function.*In practice, G, K and μ can be assumed constant within the LF band used in the spectral analysis. The loop gain is then $F_c H(p)/p$, where $F_c = GK\mu$ is the feedback cutoff frequency.

To determine F_c we inject white noise, under the control of the reference oscillator (Fig. 2.19) through a high-value resistor. The signal output V_2 is then practically determined by this noise

$$\dot{\Phi} = KV(p) \tag{2.41}$$

which gives

$$V_2 = \frac{\mu G}{p + \mu KG} KV = \frac{1}{1 + p/p_c} V \tag{2.42}$$

where $p_c = \mu KG$. In terms of spectral density, we have

$$S_{V_2}(F) = \frac{1}{1 + (F/F_c)^2} S_0 \tag{2.43}$$

where S_0 is the spectral density of white noise.

To determine the feedback cutoff frequency, it is sufficient to measure the frequency F_c for which the noise output of the circuit is less by $3\,\mathrm{dB}$ than its value at very low frequencies.

Correction by a pseudo-integrator. To obtain a signal proportional to the phase fluctuations only, i.e., to access the phase-fluctuation spectral density, we have to correct the response of the feedback loop by an external filter (Fig. 2.20). This can be done with the help of a pseudo-integrator. Let $H_2(p)$ be its transfer function. It must be such that

$$V_4 = H_2(p) \, V_2 \tag{2.44}$$

where

$$V_4 = A\left[\phi_1(p) - \phi_{\text{ref}}(p)\right] \tag{2.45}$$

within the frequency band under analysis. This filter must therefore have a cutoff frequency equal to the feedback cutoff and a limited gain near the low-frequency end, i.e., a second cutoff frequency at, say, 1 Hz. Its transfer function is

$$H_2(p) = H_0 \frac{1 + T_1 p}{1 + T_2 p} \tag{2.46}$$

where $T_1 = (2\pi F_c)^{-1}$, $T_2 = (2\pi F_2)^{-1}$ and $F_2 = 1\,\text{Hz}$. The circuit with this transfer function is shown in Fig. 2.20.

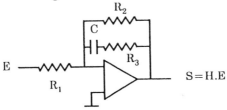

$$H = \frac{S}{E} = \frac{R_2}{R_1} \frac{1 + R_3 C p}{1 + (R_3 + R_2) C p} \simeq \frac{R_2}{R_1} \frac{1 + T_1 p}{1 + T_2 p}$$

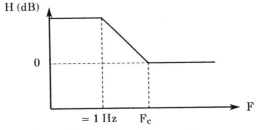

Fig. 2.20 Diagram and transfer function of a pseudo-integrator filter.

Correction by a pseudo-differentiator. A similar procedure can be used to produce a signal proportional to the frequency fluctuations, except that this time we employ a pseudo-differentiator (Fig. 2.21) with the transfer function

$$V_5 = H_3(p) \, V_2 \tag{2.47}$$

$$V_5 = B\left[\Phi(p) - \Phi_{\text{ref}}(p)\right] \tag{2.48}$$

within the frequency band under analysis.

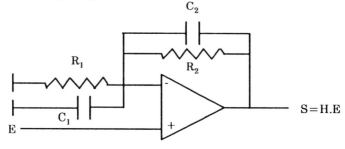

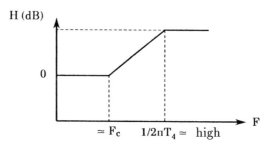

$$H = \frac{S}{E} = \frac{R_2 + R_1}{R_1} \; \frac{1 + R_2 R_1 \, (C_1 + C_2) \, p}{1 + R_2 C_2 \, p} \simeq \frac{1 + T_3 p}{1 + T_4 p}$$

Fig. 2.21 Diagram and transfer function of a pseudo-differentiating filter.

This pseudo-differentiator corrects the response to frequencies above the feedback cutoff. The circuit must work up to the frequency chosen by the user. Its transfer function is

$$H_3 = H_0 \frac{1 + T_3 p}{1 + T_4 p} \tag{2.49}$$

where

$$T_3 = (2\pi F_c)^{-1} \quad \text{and} \quad T_4 \le T_3$$

Several circuits have this transfer function; the chosen circuit offers the distinctive feature of having the differentiating capacitor grounded at one end.

Calibration of the test set. The test set can be calibrated in two ways. We can calculate the overall gain of the system from the known gain G and the measured sensitivity μ of the balanced mixer. This measurement is carried out using a known phase step, or by determining the slope of the mixing signal after the low-pass filter, with the system on open circuit (the slope of Fig. 2.12).

Another method is to apply to the reference oscillator a sinusoidal signal of frequency F_m and known amplitude V_m, which produces a frequency shift KV_m. This gives rise to a modulation line in the frequency-fluctuation spectral density, whose amplitude is a measure of the frequency shift produced; at the same time, the phase-fluctuation spectral density acquires a line whose amplitude is a measure of the phase shift $\Delta\Phi = KV_m/F_m$.

Noise within the test set. The test-set noise can be examined with the help of the circuit shown in Figure 2.22. The two inputs are driven by the same signal, but with a 90° phase difference introduced between the two lines, as shown. The mixer then produces an output voltage close to zero, under the same conditions as during measurement.

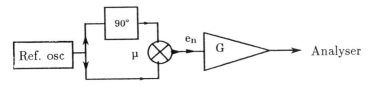

Fig. 2.22 Determination of intrinsic noise by a phase bridge.

The absolute detection limit is set by the input noise e_n of the low-noise amplifier which acts as phase noise during measurement. This 'equivalent phase noise' must obviously depend on the sensitivity μ of the mixer.

The detectable phase-noise spectral density is given by

$$S_{\Phi_{limit}} = -20\log\left(\frac{e_n}{\mu}\right) \tag{2.50}$$

It is a function of the sensitivity μ of the mixer and, hence, of its input threshold levels.

Figure 2.24 shows noise level limit in the phase bridge (including the amplifier) as a function of the mixer sensitivity. This limit is measurable in most cases and is lower than the noise of the reference oscillator. The effective limit in measurement is the noise of the latter.

2.3.2 Characterisation of a signal containing a harmonic or subharmonic frequency of the reference signal

The principle here is the same as in the last Section, except that frequency translation is necessary before the signals are introduced into the phase

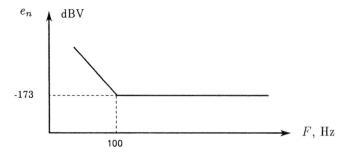

Fig. 2.23 Amplifier noise (relative to the input).

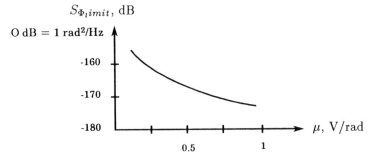

Fig. 2.24 Noise in the phase bridge.

comparator. Thus, either the frequency of the oscillator to be tested is acted upon or, the frequency of the reference oscillator is modified by division or multiplication.

Modification of the reference oscillator frequency. Here a multiplier or a divider, depending on the frequency of the signal to be tested, is used in the feedback loop (Fig. 2.25), so that error signal is of the form

$$V(p) = \frac{\mu GN}{p + \mu KGN} \left[\frac{\dot{\Phi}_1}{N} - \dot{\Phi}_r \right] \qquad (2.51)$$

where $\dot{\Phi}_1$ is the frequency fluctuation of the signal to be measured at frequency Nf and $\dot{\Phi}_r$ is the frequency fluctuation of the reference signal at frequency f. The cutoff frequency is μKGN. As before, two cases have to be considered:

$p \ll \mu KGN$, in which case

$$V(p) = \frac{1}{K} \left[\frac{\dot{\Phi}_1}{N} - \dot{\Phi}_r \right] \quad \text{and} \quad S_\Phi(f) = \frac{S_{\dot{\Phi}_1}(F)}{K^2} \qquad (2.52)$$

The signal frequency fluctuation, translated to the frequency of the reference oscillator, are measured.

$p \gg \mu KGN$, in which case

$$V\left(p\right) = \mu G\left[\Phi_1 - N\Phi_r\right] \quad \text{and} \quad S_\Phi\left(F\right) = \mu^2 G^2\left[S_{\Phi_1}\left(F\right) + N^2 S_{\Phi_r}\left(F\right)\right]$$
$$(2.53)$$

The measured phase fluctuations are the phase fluctuations of the signal under test plus the phase fluctuations of the reference signal, translated to the frequency Nf.

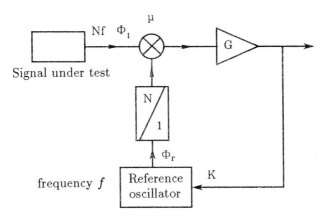

Fig. 2.25 Multiplication or division of the reference signal frequency.

Modification of the frequency of the signal under test. The oscillator frequency to be measured is now multiplied or divided to make it equal to that of the reference oscillator (Fig. 2.26).

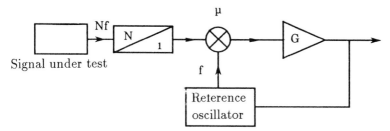

Fig. 2.26 Multiplication or division of the frequency of the signal to be measured.

The error signal is given by

$$V\left(p\right) = \frac{\mu G}{p + \mu KG}\left(\frac{\dot{\Phi}_1}{N} - \dot{\Phi}_r\right)$$
$$(2.54)$$

so that,

$$S_{\dot{\Phi}}\left(F\right) = \frac{1}{K}\left[\frac{S_{\dot{\Phi}_1}\left(F\right)}{N^2} + S_{\dot{\Phi}_r}\left(F\right)\right] \quad p \ll \mu KG$$

(2.55)

$$S_\Phi\left(F\right) = \mu^2 G^2 \left[\frac{S_{\Phi_1}\left(F\right)}{N^2} + S_{\Phi_r}\left(F\right)\right] \quad p \gg \mu KG$$

The fluctuations of the oscillator under test, translated to the reference oscillator frequency, are measured inside and outside the pass band.

Special case. When $N = 1/5, 1/3, 3$ or 5, it is possible to use the mixer as a harmonic mixer (Fig. 2.27).

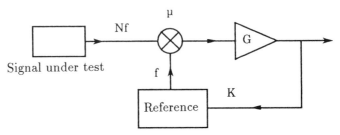

Fig. 2.27 Harmonic mixer arrangement.

In this case the mixer behaves directly as a divider or a multiplier of the loop gain.

We note that the sensitivity of the mixer is a function of the order of the odd harmonic present:

$$N = 1/5 \quad \text{or} \quad N = 5 \rightarrow \mu_5 = \mu_1/5$$

$$N = 1/3 \quad \text{or} \quad N = 3 \rightarrow \mu_3 = \mu_1/3$$

where μ_1 is the sensitivity of the mixer for $N = 1$.

2.3.3 Characterisation of a signal of any frequency

The simplest (and obvious) solution is to use a synthesiser driven by a reference oscillator (but bearing in mind the remarks made in Section 2.5.5 about the two types of synthesiser).

(a) Type I synthesiser

This synthesiser produces a signal whose phase fluctuation spectral density is independent of the generated frequency. It is therefore interesting to divide the synthesiser output frequency and use it within a range near

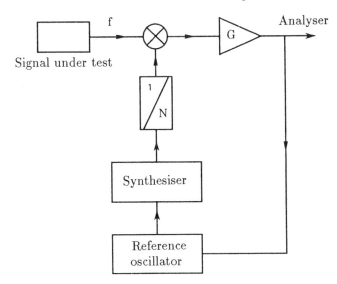

Fig. 2.28 Type I synthesiser.

the maximum output frequency in which the spectral density of the relative frequency fluctuations is best defined. The principle of this type of measurement is illustrated schematically in Fig. 2.28.

The first step is to determine accurately the loop transfer function. Indeed, in some synthesisers, the internal driver is not replaced by an external driver, but is instead locked to the external signal, thus producing a second phase loop within the measurement loop. This often leads to a second-order phase loop that produces a peak in the neighbourhood of the cutoff frequency, which interferes with measurements.

These measurements have to be corrected by the transfer function obtained by injecting white noise into the low-noise amplifier. The expressions giving the spectral density of phase and frequency are the same as those of Section 2.3.2.

(b) Type II synthesiser

We saw earlier that this synthesiser produces a signal that is generated by division of a high-frequency signal. The spectral density of the relative frequency fluctuations is then constant whatever the output frequency. It is therefore unnecessary to use dividers when the effective working frequency lies within the operating range of the synthesiser. The principle of this measurement is illustrated in Fig. 2.29.

It is sufficient to measure the feedback cutoff frequency in order to determine the two ranges in which the expressions for the phase and frequency

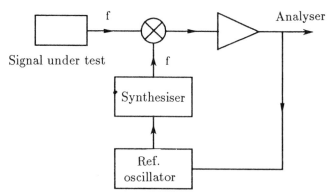

Fig. 2.29 Type II synthesiser.

fluctuation spectral densities are valid. This takes us back to the general case.

(c) The synthesis method

In this method, frequency translation is used to reduce the measurement of any frequency to the examination of a harmonic frequency of the reference oscillator. The principle is illustrated in Fig. 2.30. The divided frequency of the signal from the synthesiser is added to or subtracted from that of the signal under test, thus producing an integer-value harmonic of the reference oscillator. The desired frequency is selected by a reasonably narrow-band filter which rejects undesirable frequencies. This brings us back to the measurement of a harmonic frequency of the reference oscillator.

This method is particularly useful for frequencies close to a round value (e.g., 5.020600 MHz). The synthesiser can then be used at 2.06 MHz, the output frequency divided by 100 and then shifted to 5 MHz by mixing.

It is thus clear from the different diagrams shown above that the metrology of frequency requires a certain number of frequency multipliers, frequency dividers, filters and synthesisers for ranges in which there are no universal reference frequencies.

2.3.4 Cross-spectral or cross-correlation method

The phase or frequency spectral density is related to the correlation function. We can therefore place a correlator at the output of the circuit producing the signal V_2 (Fig. 2.19). This simple case is not as useful as it might seem because it forces us to calculate the spectral density by FFT, and we find ourselves back in the previous situation. In particular, the result obtained is the sum of the individual spectra of the two oscillators. In

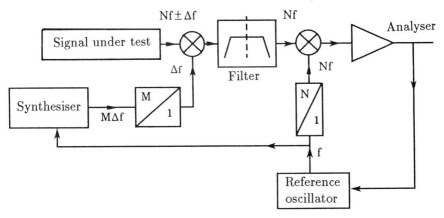

Fig. 2.30 Comparison of harmonic frequencies by synthesis.

contrast, the cross-correlation method removes this limitation and allows us to obtain the spectrum (or the stability) of only the oscillator under test.

The method employs three oscillators, two of which are locked to the third as shown in Fig. 2.31.

The estimated cross-correlation function of the two signals $V_{ab}(t)$ and $V_{bc}(t)$ is

$$C_{V_{ab} V_{bc}}(\tau) = \frac{K_1 K_2}{T} \int_{-T/2}^{T/2} [\Phi_a(t) - \Phi_b(t)][\Phi_b(t-\tau) - \Phi_c(t-\tau)]\, dt$$

(2.56)

where T is to the integration time and τ is the delay between the two paths.

Products of the form $\Phi_a(t) \times \Phi_b(t-\tau)$, where a and b are not equal, are zero on average if the fluctuations $\Phi_a(t), \Phi_b(t), \Phi_c(t)$ are uncorrelated. The only term that remains is $\Phi_b(t)\Phi_b(t-\tau)$ and the cross-correlation function then takes the form

$$C_{V_{ab} V_{bc}}(\tau) = \frac{K_1 K_2}{T} \int_{-T/2}^{T/2} [\Phi_b(t)\, \Phi_b(t-\tau)]\, dt$$

(2.57)

which is the phase cross-correlation function of oscillator b. If we know this autocorrelation function, we can (1) obtain time-domain information by calculating the variance $\sigma^2(\tau)$ of this oscillator and (2) obtain the phase fluctuation spectral density, which is none other than the Fourier transform of the autocorrelation function (Section 1.3.1).

It is thus possible to determine the phase fluctuation spectral density of a single oscillator. This is then equivalent to the triangulation method in which three oscillators are compared two by two.

We have, so far, ignored the additive noise in the two lines, such as the amplifier noise (which is relatively low). This currently limits to $-167\,\mathrm{dB}$

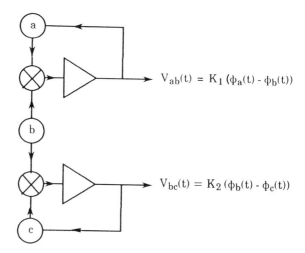

$$V_{ab}(t) = K_1 (\phi_a(t) - \phi_b(t))$$

$$V_{bc}(t) = K_2 (\phi_b(t) - \phi_c(t))$$

Fig. 2.31 Comparison of three oscillators.

the minimum level of classical measurements. This uncorrelated noise is also substantially reduced by the triangulation technique.

This can be verified by applying to the correlator the outputs of the two amplifiers, their inputs being shunted by the 50-Ω output impedance of the balanced mixers. This measures the two autocorrelation functions $C_{V_{ab}V_{ab}}(\tau)$ and $C_{V_{bc}V_{bc}}(\tau)$ at the outputs of the two amplifiers and the cross-correlation function $C_{V_{ab}V_{bc}}(\tau)$. The resulting spectra are obtained by Fourier transforms. The plateau of the spectrum deduced from the cross-correlation lies at about -180 dB (Fig.2.32).

We therefore conclude that these methods allow spectral purity measurements with resolution in the region of -180 dB, starting at 2000 Hz from the carrier. There is no other method at present that allows this.

2.4 TIME-DOMAIN MEASUREMENTS

Time-domain measurements are easier than spectral domain measurements. Indeed, there is no need for the reference signal at the exact frequency of the tested signal as there is in frequency-domain measurements. A relatively low-frequency beat between the signal under test and the reference signal suffices for signal characterisation.

Moreover, certain circuits developed for frequency-domain measurements can be adapted for measurements in the time domain. We will consider a number of methods that rely on the comparison of the unknown signal with a single reference or with two such signals in the most sophisticated systems that offer higher stability.

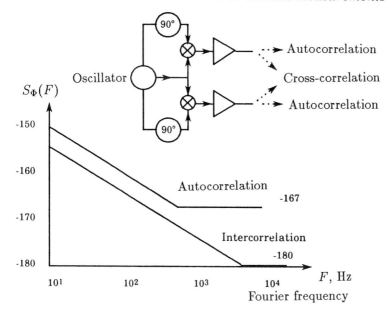

Fig. 2.32 The resolutions of the cross-correlation and autocorrelation methods.

2.4.1 The counter method

The signal is sent to a reciprocal counter (Section 2.2.3) in which logic circuits perform the automatic display of frequency. The counter is connected to a calculator which evaluates the variance $\sigma^2(\tau)$ (defined in Chapter 1; detailed calculations will be discussed later). The Allan variance is usually calculated (Chapter 3). The counting gate window T is determined by N periods of the reference oscillator of frequency f.

Long-term measurements are acceptable only if the counter is itself driven by a stable reference signal, i.e., a 5 or 10 MHz signal derived from a caesium clock or a rubidium clock checked periodically against a caesium clock (see Section 2.2.1).

The resolution of this device is theoretically $2 \times 10^{-9}/\tau$ (if the internal clock of the counter is running at 500 MHz), where τ is the duration of measurement. We note that counter switching errors depend on the slope of the signal under test and the noise that accompanies it.

The measurement error depends on the quality of the driver which can be either a temperature controlled crystal oscillator (TCXO, with a drift of $\pm 5 \times 10^{-8}$ per day) or an USO (drift of 2×10^{-10} per day). In ten days the TCXO can vary by $\pm 5 \times 10^{-7}$ which is equivalent to ± 5 Hz (with the driver running at 10 MHz). A measured frequency of 200 MHz will have an error of $\pm 5 \times 10^{-7}$ which is equivalent to ± 100 Hz, showing that several of the displayed digits are often not significant.

2.4.2 The use of a mixer and a counter

The apparent resolution of the counter can be raised by mixing the signal under investigation with a reference signal of slightly different frequency in such a way that the resulting beat signal appears in the region of, say, 1 kHz. This can be done by using a nonlinear component, followed by a low-pass filter.

The beat frequency is counted with the same resolution as before, i.e., that is $2 \times 10^{-9}/\tau$ if the signal entering the counter input has a sufficient slope ($1 \, V/\mu s$), which means that beat signal has to be amplified with a gain of about 1 000.

The accuracy of frequency measurement is considerably improved in the ratio of the effective measured frequency (beat frequency of 1 kHz) to that of the signal (say, 50 MHz).

2.4.3 The use of a difference multiplier

The difference multiplier takes two neighbouring frequencies f_1 and f_2 and produces a frequency f_3 given by

$$f_3 = f_1 + N \, (f_1 - f_2) = f_1 + N \Delta f \tag{2.58}$$

where

$$\Delta f = \left[\Delta f_0 + \dot{\Phi}_1 \, (t) + \dot{\Phi}_2 \, (t) \right] \tag{2.59}$$

The difference $f_3 - f_1$ is measured by counting as described above. The accuracy of Δf is increased by a factor of N. Various types of difference multiplier are available commercially. They consist of devices incorporating multipliers or phase loop systems based on dividers.

Multipliers. These devices multiply the frequency f_1 to be measured by ten and the reference frequency f_2 by 9. Subtractive mixing then gives the frequency

$$f_s = 10f_1 - 9f_2 = f_2 + 10 \, (f_1 - f_2) = f_2 + 10\Delta f \tag{2.60}$$

By using n identical devices in series, we then achieve multiplication by 10^n (Fig. 2.33).

This device would operate perfectly if the frequency multipliers were perfect and capable of delivering the single output frequency Nf. Actually, the spectral analysis of the output signal reveals appreciable side bands at frequencies $(N \pm 1)f$, which means that the output signals from the $\times 9$ and $\times 10$ multipliers contain three frequency components $8f_2, 9f_2, 10f_2$ and $9f_1, 10f_1, 11f_1$, respectively, which cannot be completely eliminated by filtering.

The mixer output will therefore contain the components

$$9f_2 + 9\Delta f - 8f_2 = f_2 + 9\Delta f$$

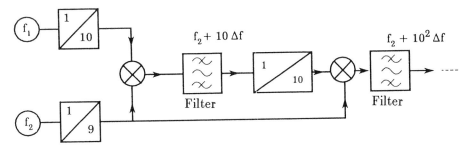

Fig. 2.33 Difference multiplier.

$$10f_2 + 10\Delta f - 9f_2 = f_2 + 10\Delta f$$
$$11f_2 + 11\Delta f - 10f_2 = f_2 + 11\Delta f$$

The frequency components $f_2 + 9\Delta f$ and $f_2 + 11\Delta f$ have low amplitudes in comparison with the wanted component $f_2 + 10\Delta f$, but we must not forget that the following $(n-1)$ stages amplify the amplitude of these parasitic modulation lines by the factor of $20 \log(10^{n-1})$.

The device thus supplies a signal with the required amplified difference $N\Delta f$, amplified frequency noise and parasitic signal $m \cos 2\pi\Delta ft$.

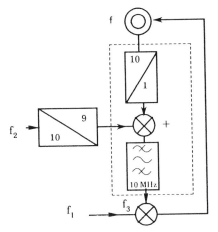

Fig. 2.34 Difference multiplier with division and phase locked loop.

Dividers (Fig. 2.34). Here the mixer receives the frequency $f/10$ from a VCO and a frequency equal to $9/10$ of the reference frequency f_2. A filter selects the sum frequency

$$f_3 = f/10 + (9/10) f_2$$

and a phase comparator receives f_3 and f_1. The output signal controls a VCO. When it is phase locked, the frequency f_3 becomes equal to f_1 and $f_1 = f_2 + \Delta f$ becomes $f = f_2 + 10\Delta f$. Multiplication by 10^n is achieved by using n successive states.

Difference-multiplier noise. The difference multiplier not only supplies the amplified difference between two frequencies f_1 and f_2, but it is also a source of frequency noise.

Its intrinsic noise can be measured by applying the reference signal to the two inputs simultaneously and calculating the output fluctuations. The noise in commercial and laboratory difference multipliers is shown in Fig. 2.35.

These devices are particularly useful when one wishes to track the short-term operation of oscillators and to measure their medium-term stability. The only inconvenience is that only integer frequencies can be used at the input (5 MHz or 10 MHz ± 10 Hz).

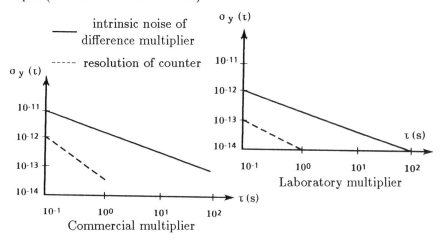

Fig. 2.35 Noise in the difference multipliers.

2.4.4 Systems using several oscillators

Here we have a transposition of the methods of Section 2.34. for the determination of variance.

Methods employing two mixers. A pair of oscillators of the same frequency f_0 and a synthesiser delivering a frequency f_s are employed. A schematic diagram illustrating the underlying principle is shown in Fig. 2.36.

The two beating signals f_b and f'_b are amplified, and pulses are generated on each zero crossing. These pulses drive a counter acting as an interval meter. The time intervals measured in this manner fluctuate in step with the fluctuations of oscillators O_1 and O_2, but the synthesiser fluctuations are eliminated by subtraction.

Measurement by triangulation. This method involves simultaneous measurement using pairs of oscillators. The principle of the method is illustrated in Fig. 2.37. It yields three variances, each representing relative fluctuations of two of the three oscillators. The relative variances $(\sigma_{ab}^2, \sigma_{ac}^2, \sigma_{bc}^2)$ are related (Section 3.3.4) to the individual variances of the oscillators $(\sigma_a^2, \sigma_b^2, \sigma_c^2)$ assumed uncorrelated. The three measurements $(\sigma_{ab}^2, \sigma_{ac}^2, \sigma_{bc}^2)$ enable us to calculate the three unknowns $(\sigma_a^2, \sigma_b^2, \sigma_c^2)$, and this provides us an absolute measurement of each variance. A fairly large number of samples is obviously required for this calculation.

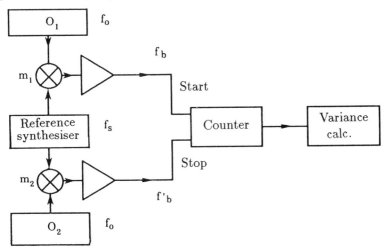

Fig. 2.36 Comparison of two mixers.

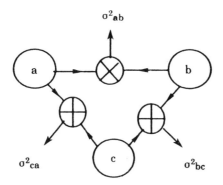

Fig. 2.37 Measurement by triangulation.

Figure 2.38 shows a measuring device employing three difference multipliers and three counters. In a demonstration, the three oscillators were deliberately degraded with separate white noise. Each oscillator was first tested individually relative to a higher-stability reference. This was followed by simultaneous triangulation. Figure 2.39 illustrates the results.

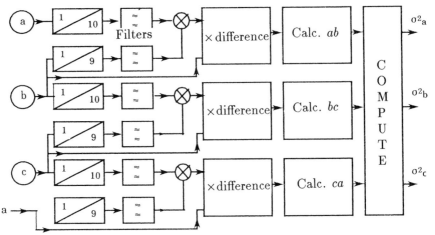

Fig. 2.38 Triangulation with difference multipliers.

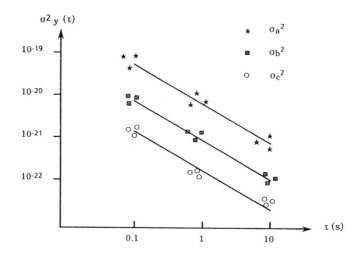

σ_a^2	1205	957	1150
σ_b^2	121	133	114
σ_c^2	8	10	11

simultaneous measurements

σ_a^2	1172	1084	1061	1089	1051	1167
σ_b^2	154	242	29	3	213	97
σ_c^2	-11	-117	100	124	-84	47

non-simultaneous measurements

Fig. 2.39 Characterisation of three oscillators by triangulation. The tables show the importance of simultaneous measurements.

The straight lines represent variances measured individually by comparing each perturbed oscillator with a higher-quality reference. The experimental values were found by computation from three series of measurements by the above triangulation procedure (points). There is, clearly, good agreement between the triangulation measurements and the individual measurements. We emphasise the importance of simultaneous measurements. The table included in Fig. 2.39 speaks for itself. We note that the values of σ_b^2 and σ_c^2 obtained by nonsimultaneous measurements, exhibit considerable scatter (which can give rise to spurious results such as negative variance), whilst simultaneous measurements produce consistent results.

3

Experimental results and their interpretation

3.1 POWER SPECTRAL DENSITY OF PHASE AND FREQUENCY FLUCTUATIONS

3.1.1 Summary of results

The methods detailed in Chapter 2 give the mean frequency $f_0 = \omega_0/2\pi$ of an oscillator and the spectral densities of phase and relative frequency fluctuations, $S_\Phi(F)$ and $S_y(F)$, which we will refer to as *low-frequency spectra* for short. The relation between these densities is

$$S_y(F) = \frac{4\pi^2 F^2}{\omega_0^2} S_\Phi(F) \tag{3.1}$$

where $y = (1/2\pi)/(\dot{\Phi}/f_0)$, the oscillator signal is given by

$$v = V_0 \sin[\omega_0 t + \Phi(t)] \tag{3.2}$$

and amplitude fluctuations are assumed to be negligible. Figure 3.1 shows a few examples of the results obtained. The frequency axis has a logarithmic scale and the ordinates are plotted in dB/Hz since the spectral densities are normalised to the total power of the oscillator signal. The frequency spectrum is usually well described by a five-parameter model of the form

$$S_y(F) = \frac{h_{-2}}{F^2} + \frac{h_{-1}}{F} + h_0 + h_1 F + h_2 F^2 + ... \tag{3.3}$$

In fact, the spectrum is confined to the frequency band in which the analysis is performed. This band runs from a minimum frequency F_m to a maximum frequency F_M.

The phase-fluctuation spectrum can also be described by the same model spectrum:

$$S_\Phi(F) = \frac{\omega_0^2}{4\pi^2 F^2} \left[\frac{h_{-2}}{F^2} + \frac{h_{-1}}{F} + h_0 + h_1 F + h_2 F^2 + ... \right] \tag{3.4}$$

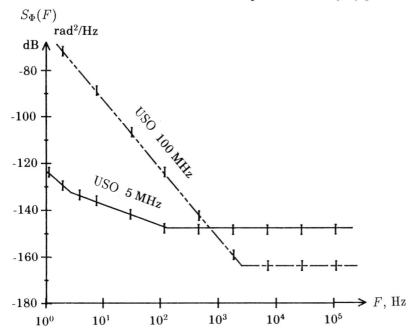

Fig. 3.1 Examples of oscillator phase spectra.

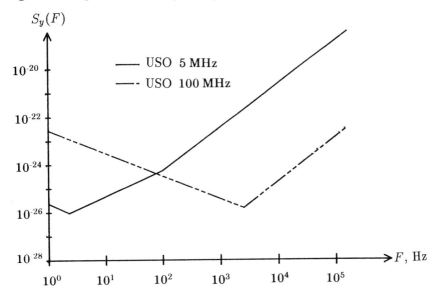

Fig. 3.2 Relative-frequency spectra of the oscillators of Fig. 3.1.

Figure 3.2 illustrates the agreement between these models and experimental results.

The prime purpose of such measurements is to compare the oscillators with one another and to verify the theoretical models of these oscillators. The experimental curves do not allow a realistic extrapolation for F_m tending to zero for which the spectra S_Φ and S_y seem to tend to infinity. Many of the difficulties of the theory that will be presented in this chapter derive from this phenomenon. It might be thought that advances in experimental methods would allow a better determination of the shape of the curve, so that a reasonable extrapolation could be made to small values of F. Unfortunately this is virtually impossible, as we are about to show.

Indeed, the spectral analysis of a quantity involves observing it through a very narrow filter ΔF, centred on the frequency F under investigation, and measuring the mean power delivered.

As we approach the carrier $(F \to 0)$, the necessary duration of measurement increases. However, ΔF is obviously smaller than $2F$, since the frequencies are positive, so that we have to measure smaller and smaller values because the window filter is narrower. It is tempting to push the spectral analysis to 10^{-4} Hz, which needs a filter with a pass band of the order of 10^{-5} Hz and a measuring time of at least one day, since measurement on a sinusoidal signal presupposes a duration of at least one period. Spectral analysis at very low frequencies is physically limited by the very long time that it demands.

3.1.2 Different types of LF spectra

The autocorrelation function. The behaviour of phase and frequency variables in the frequency domain can be described in terms of the power spectral density which is related to the autocorrelation function $\Gamma(\tau)$ by the Fourier transform:

$$\Gamma(\tau) = \int_{-\infty}^{+\infty} S^B(F) \exp(j2\pi F\tau) \, dF \tag{3.5}$$

and, conversely,

$$S^B(F) = \int_{-\infty}^{+\infty} \Gamma(\tau) \exp(-j2\pi F\tau) \, d\tau \tag{3.6}$$

This is the theorem of Wiener and Khintchine. For signals that are real functions of time, $\Gamma(\tau)$ is a real even function, so that $S^B(F)$ is also real and even:

$$S^B(F) = 2 \int_0^{+\infty} \Gamma(\tau) \cos 2\pi F\tau \, d\tau \tag{3.7}$$

In these expressions, $S^B(F)$ contains both positive and negative frequencies and is the mathematical or two-sided spectrum. It is readily obtained

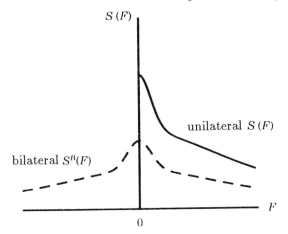

$S(F)$

unilateral $S(F)$

bilateral $S^B(F)$

0

F

Fig. 3.3 Two-sided spectrum $S^B(F)$ and one-sided spectrum $S(F)$. The relation between them is $S(F) = 2S^B(F)$.

by transposition, relative to zero frequency, of the spectrum of a signal centered on the frequency f_0. This transposition can be carried out by phase detection. Since negative frequencies cannot be distinguished experimentally from positive frequencies, the one-sided spectrum is obtained by folding the mathematical spectrum on itself (Fig. 3.3).

We are thus led to the following definition of the physical or one-sided spectrum $S(F)$:

$$S(F) = 2S^B(F) \quad \text{for} \quad 0 \le F < \infty \quad \text{and} \quad S(F) = 0 \quad \text{for} \quad F < 0 \quad (3.8)$$

Substitution in (3.7) then gives

$$S(F) = 4 \int_0^\infty \Gamma(\tau) \cos 2\pi F\tau \, d\tau \tag{3.9}$$

The inverse expression obtained from (3.5) can be written as

$$\Gamma(\tau) = 2 \int_0^\infty S^B(F) \cos 2\pi F\tau \, dF \tag{3.10}$$

or as

$$\Gamma(\tau) = \int_0^\infty S(F) \cos 2\pi F\tau \, dF \tag{3.11}$$

This integral is sometimes impossible to evaluate for the models used to represent experimental results because the integral diverges at $F = 0$ or for $F \to \infty$.

We shall proceed step by step, using the different terms of the model. A very simple case corresponds to

$$S(F) = h_0 \qquad 0 < F < F_M \tag{3.12}$$

This is white noise with a limited spectrum. Interpolation towards $F = 0$ does not then present any risk. We have

$$\Gamma(\tau) = h_0 F_M \left(\frac{\sin 2\pi F_M \tau}{2\pi F_M \tau} \right) \tag{3.13}$$

which is shown in Fig. 3.4. When F_M is very large, the correlation function tends to a Dirac pulse, and the noise has no memory. On the other hand, white noise with a limited spectrum does have some memory, i.e., of the order of $\frac{1}{2} F_M$.

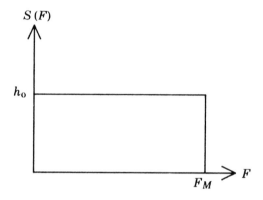

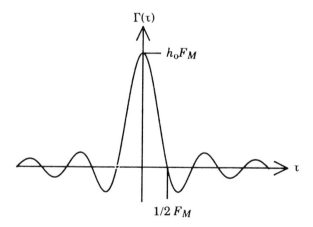

Fig. 3.4 The Fourier transform of white spectrum at F_M.

It is clear that the curve representing S_Φ can never be compared to a rectangle; the climb towards high values, for small values of F, is always found to occur. A model of the correlation function that can be considered is

$$\Gamma(\tau) = \Gamma_0 e^{-\tau/\tau_0} \qquad (3.14)$$

which represents a memory phenomenon that vanishes exponentially with the delay τ. The corresponding spectrum is

$$S(F) = 4 \int_0^\infty \Gamma(\tau) \cos 2\pi F \tau d\tau = \frac{4\Gamma_0 \tau_0}{1 + 4\pi^2 F^2 \tau_0^2} \qquad (3.15)$$

and is called a Lorentzian. By superimposing finite white noise with large F_M on noise of the above type with a very long τ_0 of the order of $1/F_m$, we get closer to the experimental result. Indeed, the Lorentzian is identical to a $1/F^2$ law with the exception of the region of very small values of F (Fig. 3.5).

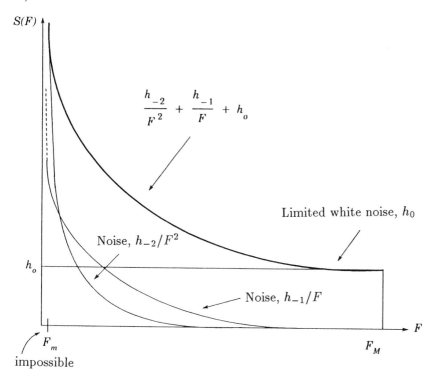

Fig. 3.5 Graphs of $h_0, h_{-1} F^{-1}, h_{-2} F^{-2}$ and of their sum.

The F^2 noise will be called the *random walk noise*. It is represented by the coefficient h_{-2} in the expression for $S(F)$. It can be described as the result of filtering of very short random pulses by a very narrow pass band filter. This filtering produces a response with a very long decay time, which continues until a second pulse arrives and abruptly modifies the state of the device at a random time.

It is much harder to find models of correlation functions which give a $1/F$ spectrum. Nevertheless, experiment always yields this law for part of the spectrum (the term h_{-1} in the phase spectrum). It is therefore essential to add to the white noise and the random-walk noise a further noise, which is called $1/F$ noise and is characterised by the coefficient h_{-1} in $S(F)$.

White noise can be regarded as a random phenomenon without memory, whereas $1/F^2$ noise retains its memory as long as a new pulse does not arrive and does not introduce a perturbation. We can imagine that $1/F$ noise is the result of an intermediate case, i.e., noise with memory, but with a decay time τ_0 that can have all the possible values from the largest observable (of the order of $1/F_M$) downwards. The spectrum can be approximated by a superposition of Lorentzian. The outcome of this superposition depends on the distribution of the time constants τ_0. The simple assumption of uniform probability leads to

$$
\begin{aligned}
S(F) &= \int_{t_m}^{t_M} \frac{S_0}{1 + 4\pi^2 F^2 \tau_0^2} d\tau_0 \\
&= \frac{S_0}{2\pi F} \left[\arctan 2\pi F \tau_M - \arctan 2\pi F \tau_m \right]
\end{aligned}
\tag{3.16}
$$

When F is contained within the band between $F_m \sim 1/\tau_M$ and $F_M \sim 1/\tau_m$, and is not too close to the limits, we can approximate the first arctan by $\pi/2$ and the second by the argument $2\pi F/F_M$. We then get

$$
S(F) = \frac{S_0}{4F} - \frac{S_0}{F_M}
\tag{3.17}
$$

which represents effectively a power spectrum inversely proportional to F.

Two conclusions can be drawn from the above discussion. First, we have to develop measuring instruments capable of exploring the spectrum as close as possible to the carrier. Theoretical models more sophisticated than the correlation function have to be developed in order to avoid computing difficulties for $F \to 0$. It is also necessary to take into consideration the different factors responsible for phase fluctuations. We cannot explain everything in terms of a single phenomenon. How many phenomena are needed? What are they? Are they independent or are they correlated? Is their origin in the physics of the components or in the operation of the circuit? These are the questions that will have to be addressed later.

3.2 TRUE VARIANCE: THE SEARCH FOR ACCURACY

3.2.1 Description of a random variable

Let $X(t)$ be a random variable, e.g., phase fluctuation $\Phi(t)$ or angular frequency fluctuation $\dot{\Phi}(t)$. Instantaneous measurements of $X(t)$ that can follow all such variations are out of the question. On the other hand, we can easily measure the mean of X between t and $t + \tau$:

$$\overline{X(t, \tau)} = \frac{1}{\tau} \int_t^{t+\tau} X(t) dt \qquad (3.18)$$

We can also replace $X(t)$ in the interval $t, t + \tau$ by a horizontal straight segment with ordinate $X(t, \tau)$. To retain the maximum amount of information, we recalculate a new mean as soon as the previous measurement is over. We therefore put $t = n\tau$, so that the variable $X(t)$ is replaced by a sequence X_n that depends on τ. We thus represent $X(t)$ by the discontinuous curve illustrated in Fig. 3.6. The representation is better for smaller τ, but the measuring techniques impose a lower limit on τ.

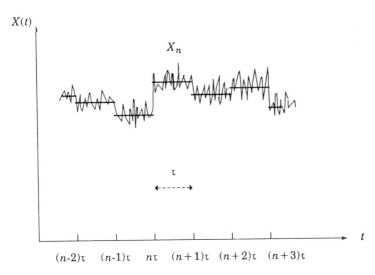

Fig. 3.6 Representation of the signal $X(t)$ by successive mean values X_n.

To make use of these data, we introduce the probability that X_n lies between X and $X + dX$. Starting with N samples and a corresponding time interval $N\tau$, we count the number of samples dN that have the value X_n within the given interval. To the first approximation, this probability is dN/N. We then repeat the operation over succesively longer time intervals $N\tau$, and we define the probability $P(X)$ as the limit of dN/N as N tends to infinity. We assume stationarity, which will have to be verified later. By integrating over all values of X, we obtain the normalisation condition

$$\int_{-\infty}^{+\infty} P(X) dX = 1 \qquad (3.19)$$

since summation of all the dN yields N. The mean of X is given by

$$\langle X \rangle = \int_{-\infty}^{+\infty} X P(X) \, dX \qquad (3.20)$$

which is the expectation value of X. We can then calculate the mean square deviation

$$\langle \Delta X^2 \rangle = \int_{-\infty}^{+\infty} \left(X - \bar{X} \right)^2 P(X) \, dX \qquad (3.21)$$

where we assume that the integral converges, in which case, the signal is said to be stationary in the second order.

The mean value of a function $F(X)$ is usually given by

$$\langle F(X) \rangle = \int_{-\infty}^{+\infty} F(X) \, P(X) \, dX \qquad (3.22)$$

It is also often useful to calculate the n-th order moment of X or of $X - \bar{X}$, which corresponds to $F(X) = X^n$ or to $F(X) = (X - \bar{X})^n$.

All these are clear definitions, but they do not readily yield correct results in practice because we cannot have an infinite number of samples (or the equivalent infinite duration). Indeed, we have to assume that a limit exists and make measurements using a finite number of samples. The resulting variances are estimates, but are readily computed (Section 3.3). We also note the unrealistic nature of the previous assumption for a system such as an oscillator that drifts and wears out as it operates. Stationary operation of infinite duration is impossible.

In order to place the theoretical definitions on a firm footing, we now imagine a very large number of identical oscillators. At a given instant t, and for a given duration τ, we measure the mean (3.18) for each of the oscillators as a function of time. Let X_p be the set of measurements made at a given instant on the different oscillators (Fig. 3.7). The data are used to calculate the mean, the mean square deviation and so on. We are thus dealing with the statistics of a very large number of identical objects, and obtain the spatial or ensemble means. We thus see that there are two procedures. In one, we sample a given oscillator output many times and evaluate the mean and, in the other, we sample each of a large number of identical oscillators only once, and evaluate the mean. The former is the time mean and the latter the ensemble mean. When the two means are the same, the phenomenon is said to be ergodic.

The hypotheses of invariance and ergodicity are absolutely necessary for a rigorous theory. They enable us to perform calculations and to interpret measurements. If the results are confirmed by experience, the hypotheses are verified. Actually, rigorous theory is possible only in exceptional cases (Section 3.2.3.) and we will have to make do with approximate calculations,

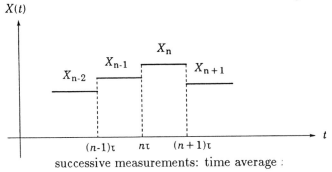

successive measurements: time average :

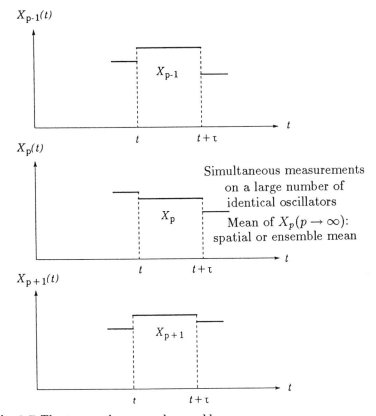

Simultaneous measurements
on a large number of
identical oscillators

Mean of $X_p(p \to \infty)$:
spatial or ensemble mean

Fig. 3.7 The temporal mean and ensemble mean.

e.g., means calculated for a finite number of objects. This was noted in previous chapters when we compared two identical oscillators and found that their fluctuations had the same statistical properties.

3.2.2 True variance, correlation function and spectral density

Suppose we compare two oscillators by measuring their variance under similar conditions. It does not matter much whether this is the true variance or an approximate variance. To compare the measurements with oscillator theory, we use the variance to calculate the correlation function and then the spectrum of the random variable $\Phi(t)$.

Neglecting amplitude fluctuations, the oscillator signal can now be described by

$$v(t) = V_0 \sin\left[\omega_0 t + \Phi(t)\right] \tag{3.23}$$

Let us now consider the random variable y that characterises instantaneous frequency fluctuations and can be written in the form

$$y = \frac{1}{\omega_0}\dot{\Phi} \tag{3.24}$$

It follows that

$$\overline{y(t,\tau)} = \frac{1}{\tau}\int_t^{t+\tau} y(t)\,dt = \frac{1}{\omega_0\tau}\left[\Phi(t+\tau) - \Phi(t)\right] \tag{3.25}$$

We now assume that the true mean is zero and introduce the true variance of $\bar{y}(t,\tau)$ by means of the expression

$$
I^2(\tau) = <(y(t,\tau))^2> =
$$
$$
\frac{1}{\omega_0^2\tau^2}\left\{\left\langle\left[\Phi(t+\tau)\right]^2\right\rangle + \left\langle\left[\Phi(t)\right]^2\right\rangle - 2\left\langle\Phi(t+\tau)\Phi(t)\right\rangle\right\} \tag{3.26}
$$

where the means are evaluated for an an infinite number of samples. If we assume that the system is stationary and ergodic, the mean square of $\Phi(t)$ is independent of t, so that

$$I^2(\tau) = \frac{2}{\omega_0^2\tau^2}\left[\Gamma_\Phi(0) - \Gamma_\Phi(\tau)\right] \tag{3.27}$$

where we have introduced the correlation function $\Gamma_\Phi(\tau)$ of the variable $\Phi(t)$. Combining these expressions with the Wiener-Khintchine theorem, we obtain

$$I^2(\tau) = \int_0^\infty S_\Phi(F)\frac{\sin^2 \pi F\tau}{\pi^2 f_0^2\tau^2}\,dF \tag{3.28}$$

If we now use (3.1) to introduce the spectrum $S_y(F)$ of the relative frequency fluctuations, we obtain the equivalent formula

$$I^2(\tau) = \int_0^\infty S_y(F)\left(\frac{\sin \pi F\tau}{\pi F\tau}\right)^2\,dF \tag{3.29}$$

Although this formula is rigorous, it is unfortunately not very useful. Indeed, when $S_y(F)$ refers to $1/F$ or $1/F^2$ noise, the integral diverges because the quantity

$$K^2(F) = \left(\frac{\sin \pi F \tau}{\pi F \tau}\right)^2 \tag{3.30}$$

tends to unity at low frequencies. Hence it follows that the true variance cannot always be calculated from measurements of $S_y(F)$. For white frequency noise, we have

$$S_y(F) = h_0 \qquad \Gamma^2(\tau) = \frac{h_0}{2\tau} \tag{3.31}$$

The true variance is therefore inversely proportional to τ.

We shall see in Section 3.3 how this method of calculation can be extended to the determination of the spectrum by using approximate variances. This then provides the basis for an objective comparison of the two approximate measures, one which employs time analysis to obtain the variance and the other performs a frequency analysis to find the spectrum.

3.2.3 An example of the evaluation of true variance

Consider a random event, whose probability is independent of previous events of the same type. The probability that this event will occur in a time interval dt is proportional to dt and will be represented by λdt. The interval dt is infinitesimal, so that it can contain only one event Actually, the probability of two independent events occuring during dt is $(\lambda dt)^2$, which is negligible in the approximation considered here. The probability that *no* event will take place in dt is simply $(1 - \lambda dt)$.

What then is the probability $P_n(t)$ that n events will occurr in a time t? It consists of two terms, namely,

- the probability of n events in time $(t - dt)$, multiplied by the probability of having no events in time dt, i.e.,

$$P_n(t - dt)(1 - \lambda dt) \tag{3.32}$$

and
- the probability of $n - 1$ events in time $(t - dt)$, multiplied by the probability of one event in time dt, i.e.,

$$P_{n-1}(t - dt) \lambda dt \tag{3.33}$$

The interval dt can contain only one event, so that there are no other possibilities, and the total probability is the sum of the above two:

$$P_n(t) = (1 - \lambda dt) P_n(t - dt) + \lambda dt P_{n-1}(t - dt) \tag{3.34}$$

or

$$\frac{dP_n}{dt} + \lambda P_n = \lambda P_{n-1} \qquad (3.35)$$

This is a simple differential equation whose solution yields

$$P_n(\tau) = \frac{(\lambda\tau)^n}{n!} e^{-\lambda\tau} \qquad (3.36)$$

Once we have a rigorous expression for the probability, we can calculate the mean and the variance for this special case. The true mean is then

$$\langle n \rangle = \lim_{N \to \infty} \sum_{n=0}^{n=N} n P_n(\tau) = \lambda\tau \qquad (3.37)$$

where we have used a property of the series in powers of $e^{\lambda\tau}$. Similarly,

$$\langle \Delta n^2 \rangle = \langle (n - \bar{n})^2 \rangle = \lim_{N \to \infty} \sum_{n=0}^{n=N} (n - \lambda\tau)^2 P_n(\tau) = \lambda\tau \qquad (3.38)$$

so that, in this special case, the true variance is

$$I^2(\tau) = \frac{\langle \Delta n^2 \rangle}{\langle n \rangle^2} = \frac{1}{\lambda\tau} \qquad (3.39)$$

This is a Poissonian process consisting of the random succession of independent events. There is therefore no memory, which corresponds to white noise. The agreement with Section 3.2.2 is therefore satisfactory and, in this case, $\lambda = 2/h_0$.

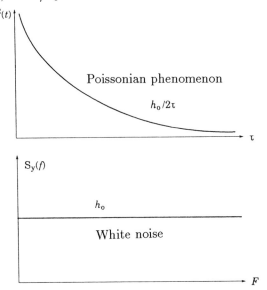

Fig. 3.8 Two variances and spectrum of a Poissonian phenomenon.

We could now advance towards a theoretical interpretation if the experimental results were as simple as those illustrated in Figure 3.8. Unfortunately, this is not always the case, and we will repeat the variance calculation, using a more pragmatic approach in order to get to approximate quantities that are readily calculated and are particularly convenient in linking time and frequency measurements.

3.3 THE CONVENIENT VARIANCE: THE SEARCH FOR EFFICACY

3.3.1 The Allan variance

The true variance is a limit obtained by considering an infinite number of measurements. It cannot be measured directly. Nor is it calculable, except in special cases such as the Poissonian process; even then, problems remain when the variance has to be used at at very low frequencies ($F \rightarrow 0$) or for very long measurement times ($\tau \rightarrow \infty$).

As part of a search for pragmatic methods, consider a set of successive frequency measurements $\overline{f}_1, \overline{f}_2, \overline{f}_3 ... \overline{f}_n$, performed on an oscillator in a time interval τ. We have a twofold objective. On the one hand, to characterise the oscillator we need an approximate variance that does not involve very long or very strenuous calculations. On the other hand, the variance must be readily related to other characteristics of the same oscillator, e.g., its phase spectrum, exactly as measured (we have to avoid the problem of divergence for $S \rightarrow 0$).

Assuming that the phenomenon is stationary, the mean frequency deduced from a finite number of measurements is

$$f_0 = \frac{1}{N} \sum_{i=1}^{N} \overline{f}_i \qquad (3.40)$$

where N is a reasonably large number. This is obviously meaningless if there is appreciable long-term drift. It is therefore advisable to correct primary measurements if they indicate that drift may be present (cf. Section 3.3.2). It is essential to bear in mind the simplifying assumptions that have been made (in our case, the assumption of stationarity) in order to achieve reasonable results in practice.

We shall now consider relative frequency fluctuations, using the notation

$$\overline{y_n} = \frac{\overline{f}_n - f_0}{f_0} \qquad (3.41)$$

We begin our calculation of variance by considering the mean square deviation of N consecutive values:

$$P_N^2 (\tau) = \frac{1}{N} \sum_{n=1}^{N} \left(\overline{y_n} - \frac{1}{N} \sum_{i=1}^{N} \overline{y_i} \right)^2 \qquad (3.42)$$

This is a fast calculation when N is small, which suggests a procedure for the the variance. It also implies that the variance obtained in this manner is itself a random quantity. We can therefore try to improve the result obtained by repeating the operation several times and calculating the mean. The calculation is performed for as large a number of measurements as possible, hoping to approach the true mean (which corresponds to an infinite number of samples).

Actually, if we perform this operation using expression (3.42) for the simple case of white noise, we find that the relative difference between $< P_N^2(\tau) >$ and $I^2(\tau)$ is proportional to $1/N$.

If we adopt the slightly different definition

$$P_N^2 (\tau) = \frac{1}{N-1} \sum_{n=1}^{N} \left(\overline{y_n} - \frac{1}{N} \sum_{i=1}^{N} \overline{y_i} \right)^2 \qquad (3.43)$$

we can eliminate the difference between this and the true variance for all N.

Encouraged by these results, we now consider P_N^2 for the smallest value of N ($N = 2$) and calculate the mean of $P_2^2(\tau)$ for the largest possible numbers of values, hoping to achieve stationarity. In this way we obtain

$$P_2^2 (\tau) = \frac{1}{2} (\overline{y}_1 - \overline{y}_2)^2 \qquad (3.44)$$

and the ensemble mean becomes

$$\sigma_y^2 (\tau) = \frac{1}{2} \left\langle (\overline{y}_1 - \overline{y}_2)^2 \right\rangle \qquad (3.45)$$

This is the Allan or two-sample variance. If the frequency noise is white, it is approximately the same as the true variance and is simple to calculate (the mean of the squares of two consecutive deviations). It is very quickly apparent whether this mean tends reasonably to a limit. It has been proposed that the Allan variance should always be used to characterise an oscillator, except for special cases in which the need to identify some properties of the signal justifies the use of other variances. We shall return to this later.

Rigour, which is absolutely essential in metrology, requires greater precision. The frequency samples $f_1, f_2, \ldots$ measured in time intervals τ with a period τ_0, can be adjacent or not, as illustrated in Figure 3.9. The way in which the samples are obtained has an influence on the result. In the general case, it is therefore necessary to specify the values of N, τ_0 and τ in any variance calculation. This will be indicated by writing $\sigma_y^2(N, \tau_0, \tau)$. By definition, the Allan variance corresponds to $N = 2$ and $\tau_0 = \tau$. This is written as $\sigma_y^2(2, \tau, \tau)$ or, more simply, $\sigma_y^2(\tau)$.

This variance can be related to the frequency spectrum $S_y(F)$ or the phase spectrum. We saw earlier that replacing the instantaneous value of the signal by its mean over τ, at time $t = n\tau$, was equivalent to replacing the spectrum $C(F)$ signal by the quantity

$$C(F) \frac{\sin \pi F \tau}{\pi F \tau} \exp(-j2\pi n F \tau) \qquad (3.46)$$

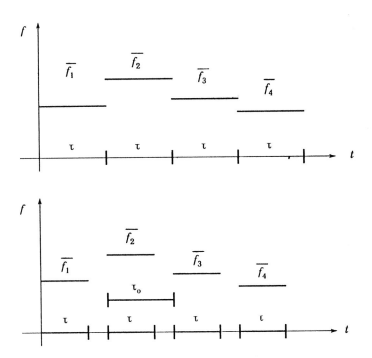

Fig. 3.9 Adjacent ($\tau_0 = \tau$, no dead time) and nonadjacent ($\tau_0 \neq \tau$, dead time $\tau_0 - \tau$) samples.

In the case of the Allan variance, we have to calculate $y_{n+1} - y_n$ using the Fourier transform:

$$
\overline{y_{n+1}} - \overline{y_n} = \int_{-\infty}^{+\infty} C(F) \frac{\sin \pi F \tau}{\pi F \tau} \{ \exp\left[-j\pi F (n+1) \tau \right]
$$

$$
- \exp\left[-j\pi n F \tau \right] \} \exp\left(j2\pi F \tau \right) dF
$$

$$
= \int_{-\infty}^{+\infty} -C(F) 2j \frac{\sin^2 \pi F \tau}{\pi F \tau} \exp\left[-j2\pi F (n+1/2) \tau \right] dF \qquad (3.47)
$$

This leads to the coefficient,

$$
K_A^2 (F) = \frac{4 \sin^4 \pi F \tau}{(\pi F \tau)^2} \qquad (3.48)
$$

which should be compared with that of the true variance, given by expression (3.29). The formula connecting the Allan variance and the spectrum $S_y(F)$ is

$$
\sigma_y^2 (\tau) = \frac{1}{2} \left\langle (\bar{y}_{n+1} - \bar{y}_n)^2 \right\rangle = \int_0^\infty S_y (F) \frac{2 \sin^4 \pi F \tau}{(\pi F \tau)^2} dF \qquad (3.49)
$$

This relationship is very interesting because it represents the equivalent of 'noise filtering'. The $1/F^2$ in $S_y(F)$ will not cause a divergence. We can then consider the comparison of time-domain measurements, described by $\sigma_y^2(\tau)$, and frequency-domain measurements, described by $S_y(F)$. For example, in order to go from the realistic representation of the low frequency noise by a law such as (3.4) to the Allan variance, we use the formulae given in the table below and derived from equation (3.49), in which M is the upper cutoff frequency of the noise (these relationships are valid for $2\pi F_M \tau \gg 1$)

Term in $S_y(F)$	Designation	Allen variance $\sigma_y^2(\tau)$
$h_2 F^{-2}$	freq. random walk	$h_{-2}(2/3)\pi^2 \tau$
$f_{-1} F^{-1}$	$1/f$ noise (freq.)	$h_{-1}2\log 2$
h_0	freq. white noise	$h_0/2\tau$
$h_1 F (F < F_M)$	$1/f$ noise (phase)	$\frac{h_1}{4\pi^2}(1 + 3\log 2\pi F_M)/\tau^2$
$h_2 F^2 (F < F_M)$	phase white noise	$h_2 \frac{3}{4\pi^2} F_M /\tau^2$

3.3.2 Difficulties arising from the long-term drifts

The above table illustrates one of the difficulties encountered in the statistics of time-domain measurement. Indeed, the term proportional to τ in

$\sigma_y^2(\tau)$, which represents the increase in the variance for very long measuring times, can be interpreted as the consequence of a frequency random walk ($1/F^2$ noise). But the slope and continuous drift of the oscillator frequency, which we have ignored in the calculation of variance on the grounds of stationarity, have analogous consequences.

To describe these difficult problems more precisely, consider a slow drift of oscillator frequency in the absence of fluctuations. This can be described by

$$v = V_0 \sin\left(\omega_0 t + \alpha t^2\right) \tag{3.50}$$

The instantaneous pulsatance can then be written as

$$\omega\left(t\right) = \omega_0 + \alpha t \tag{3.51}$$

and

$$\overline{y_n} = \frac{\alpha n \tau}{\omega_0} \tag{3.52}$$

so that

$$\sigma_y^2\left(\tau\right) = \frac{1}{2}\left(\bar{y}_{n+1} - \bar{y}_n\right)^2 \approx \frac{\alpha^2 \tau^2}{2\omega_0^2} \tag{3.53}$$

If the variation of $\sigma_y^2(\tau)$ over long measuring times is proportional to τ^2 we can interpret this as a linear frequency drift. The first series of measurements will give us this term. We then recalculate the Allan variance for successively smaller differences $y_{n+1} - y_n$.

There is no satisfactory theoretical model to describe the drift of the oscillator. Difficulties of principle are very important, but a slow drift of a sinusoidal fluctuation of very low frequency can be described, at least at the beginning, in the same manner ($\sin x \approx x$). A very long observation is required to confirm that the drift is irreversible or that a fluctuation comes and goes. It is as difficult to measure $\sigma_y^2(\tau)$ over a very long duration (τ of the order of hour) as it is to determine $S_y(F)$ at very low frequencies ($F \approx 10^{-4}$ Hz). We can equally say that the spectrum, as defined above, has no meaning for a signal that is not stationary. This is the case with a drifting of the frequency.

3.3.3 Hadamard variance

The Allan variance is a typical example of an effective statistical tool that leads to results that are approximate, but specific and usable. The Hadamard variance is a more sophisticated example based on the same principle. It was proposed in 1971 by R.A. Baugh as a convenient method of spectral analysis, but it is actually based on very general results established in the last century by J.Hadamard.

To progress further towards an improved variance, we recall the expression given by (3.49). It describes the filtering of frequency fluctuations by a filter for which the modulus of the transfer function is

$$|K(F)| = \frac{\sqrt{2}\sin^2 \pi F\tau}{\pi F\tau} \tag{3.54}$$

The corresponding impulse response $D(t)$ is given by the Fourier transform of this transfer function. This case is illustrated in Fig.3.10 and is described by

$$\begin{aligned} D(t) &= -1/\tau\sqrt{2} & -\tau < t < 0 \\ D(t) &= 1/\tau\sqrt{2} & 0 < t < \tau \\ D(t) &= 0 & |t| > \tau \end{aligned} \tag{3.55}$$

The origin of time is unimportant since the transfer function is determined by its modulus; a shift of the origin of time would modify only the argument. This is so because the basic function $\bar{y}_{n+1} - \bar{y}_n$ used to calculate the variance involves two consecutive measurements of duration τ, taken with positive and negative signs, respectively. We can also envisage the calculation of variance from a more complicated basic function corresponding to the impulse response shown in Fig. 3.11.

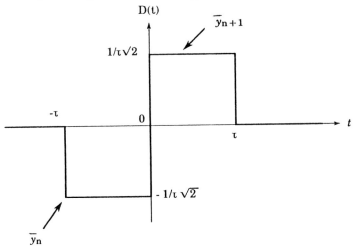

Fig. 3.10 Equivalent impulse response for the Allan variance.

We now introduce the dead time $\tau_0 - \tau$ and we alternate the N values obtained by consecutive measurements. The transfer function is the Fourier transform of $D(t)$. The result is

$$|K_H(F)| = \left|\frac{\sin \pi F\tau}{\pi F\tau}\right| \left|\frac{\sin N\pi F\tau_0}{\cos \pi F\tau_0}\right| \tag{3.56}$$

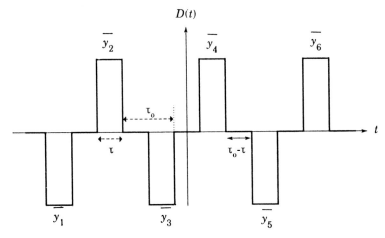

Fig. 3.11 Sampling over a time interval τ with periodicity τ_0 and dead time $\tau_0 - \tau$.

and is shown in Fig. 3.12 for even N – the only interesting case. It is a narrow pass band filter centered on the frequency

$$F_0 = \frac{1}{2\tau_0} \tag{3.57}$$

with the pass band given by

$$\Delta F = \frac{1}{N\tau_0} \tag{3.58}$$

The secondary ripples are inconvenient, but a correct choice of the ratio τ_0/τ will reduce the most important of them. The figure illustrates the most interesting case for which $\tau_0 = 3\tau/2$.

To obtain the corresponding variance, we compare the two impulse responses. The basic term $(\bar{y}_{n+1} - \bar{y}_n)^2$ gives the Allan variance corresponding to Fig. 3.10, and it is then clear that the term corresponding to Fig. 3.11 is

$$P_n^2 = (\bar{y}_6 - \bar{y}_5 + \bar{y}_4 - \bar{y}_3 + \bar{y}_2 - \bar{y}_1)^2 \tag{3.59}$$

If necessary, the definition can be refined by introducing weighting factors for each sample. The Hadamard variance is then the mean of the values of P_n^2 over the largest possible number of measurements:

$$\sigma_H^2 (N, \tau_0, \tau) = \langle P_N^2 \rangle \tag{3.60}$$

To relate the low-frequency spectrum to this variance, we can proceed as in the case of the Allan variance. There result is an analogous formula with the new filtering function:

$$\sigma_H^2 (N, \tau_0, \tau) = \int_0^{\infty} S_y (F) \left(\frac{\sin \pi F \tau}{\pi F \tau} \right)^2 \left(\frac{\sin N \pi F \tau_0}{\cos \pi F \tau_0} \right)^2 dF \qquad (3.61)$$

The advantage of this variance is clear. Just like the Allan variance, it eliminates the difficulties at very low frequencies for which $S_y(F)$ does not seem to have an easily definable limit. If we ignore the secondary ripples and compare the filter to a narrow rectangle (Fig. 3.12), we obtain the approximate expression

$$\sigma_H^2 (N, \tau_0, \tau) \approx S_y \left(\frac{1}{2\tau_0} \right) \left| K_H \left(\frac{1}{2\tau_0} \right) \right|^2 \frac{1}{2 N \tau_0} \qquad (3.62)$$

from which

$$S_y (F_0) \approx 0.73 \frac{\tau_0}{N} \sigma_H^2 (N, \tau_0, \tau) \qquad (3.63)$$

where $F_0 = 1/2\tau_0$.

We therefore have a process that allows us to calculate the low-frequency spectrum from the variance obtained by counting. All this involves prolonged measurements. A realistic value of $S_y(F)$ near $F = 0.01$ Hz can be reached if τ_0 is of the order of 200 sec. To obtain meaningful statistics, N has to be large and the measurements have to be repeated many times. Patience and experimental skill are needed to ensure that the measuring circuit does not drift. But it is the only method that permits the exploration of the LF spectrum near very low frequencies.

3.3.4 A few other useful practical procedures

Boileau and Picinbono have proposed a variance employing samples of three consecutive measurements $(\tau_0 = \tau)$. The basic expression is

$$P_3^2 = \frac{1}{9} (\bar{\gamma}_{n+2} - 2\bar{\gamma}_{n+1} - \bar{\gamma}_n)^2 \qquad (3.64)$$

and the mean of this is evaluated over the largest possible number of consecutive measurements. This can be related to variance

$$\sigma_p^3 (3, \tau, \tau) = \langle P_3^2 \rangle \qquad (3.65)$$

and the spectrum $S_y(F)$ by a method similar to the methods used previously. The result is

$$\sigma_p^2 = \frac{16}{9} \int_0^{\infty} S_y (F) \frac{\sin^2 \pi F \tau}{(\pi F \tau)^2} dF \qquad (3.66)$$

$$K_H(F)^2 = \left(\frac{sin\,(\pi F\tau)}{\pi F\tau} \right)^2 \left(\frac{sin\,(N\pi F\tau_o)}{cos\,(\pi F\tau_o)} \right)^2$$

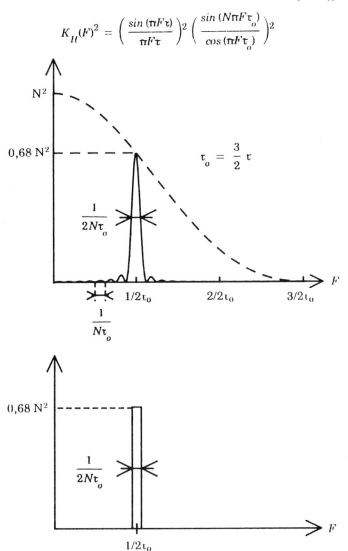

Fig. 3.12 Exact and equivalent transfer functions for the Hadamard variance.

This shows that components of $S_y(F)$ that are very important at very low frequencies will not cause a divergence (even if it is necessary to introduce F^{-4} noise to account for the very significant increase near the carrier).

One advantage of the three-sample variance is that it eliminates most of the very low frequency components of noise. It is therefore used to remove the signal drift, which then brings out more clearly the presence of the random-walk frequency noise (in $\sqrt{\tau}$) and the level of the $1/F$ noise.

In the time domain, variance is equivalent to the autocorrelation function, and true variance can be expressed directly in terms of it [equation (3.26)]. We can also compare two random signals by calculating their mutual correlation function

$$\Gamma_{ab}(\tau) = \lim_{\Theta \to \infty} \int_{-\Theta/2}^{+\Theta/2} v_a(t)\, v_b(t+\tau)\, dt \qquad (3.67)$$

If the signals are totally uncorrelated, we have $\Gamma_{ab} = 0$. This is the case of two independent sources of noise. However, if the signals come from the *same* source via different circuits, they are correlated and $\Gamma_{ab} \neq 0$. Approximate calculations of variance are related to those of covariance. If we have two series of measurements $\bar{y}_1^a, \bar{y}_2^a, ... \bar{y}_n^a$ and $\bar{y}_1^b, \bar{y}_2^b, ... \bar{y}_n^b$, we can calculate the covariance

$$\sum_{ab}^{2} = \frac{1}{2}\left\langle (\bar{y}_{n+1}^a - \bar{y}_n^a)(\bar{y}_{n+1}^b - \bar{y}_n^b) \right\rangle \qquad (3.68)$$

which is zero if the measurements $\bar{y}_n^a$ and $\bar{y}_n^b$ are uncorrelated.

For example, consider the circuit with two mixers and three oscillators (Fig. 3.13) described in Chapter 2. The quantities $\bar{y}_n^a, \bar{y}_n^b$ and $\bar{y}_n^c$ represent the relative frequency fluctuations of each of the three oscillators.

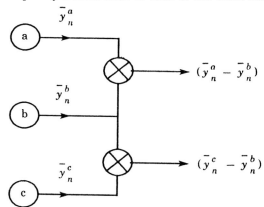

Fig. 3.13 Comparison of relative frequency fluctuations using the calculated covariance.

The output signals of the mixers therefore contain the fluctuations $(\bar{y}_n^a - \bar{y}_n^b)$ and $(\bar{y}_n^c - \bar{y}_n^c)$. Their covariance is given by

$$\Sigma^2 = \frac{1}{2}\left\langle [(\bar{y}_{n+1}^a - \bar{y}_{n+1}^b) - (\bar{y}_n^a - \bar{y}_n^b)]\,[(\bar{y}_{n+1}^c - \bar{y}_{n+1}^b) - (\bar{y}_n^c - \bar{y}_n^b)] \right\rangle \qquad (3.69)$$

which can be rewritten as

$$\Sigma^2 = \sum_{ac}^{2} - \sum_{bc}^{2} - \sum_{ab}^{2} + \sigma_b^2 \qquad (3.70)$$

where σ_b^2 is the Allan variance of oscillator b. The assumed independence of the oscillators leads to

$$\sum\nolimits_{ac}^2 = \sum\nolimits_{bc}^2 = \sum\nolimits_{ab}^2 = 0 \tag{3.71}$$

and therefore

$$\Sigma^2 = \sigma_b^2 \tag{3.72}$$

This method thus provides a basis for the characterisation of the common oscillator by discarding the influence of the fluctuations of the two auxiliary oscillators. It is completely equivalent to the mutualcorrelation method.

3.4 THE ORIGINS OF SIGNAL FLUCTUATIONS

3.4.1 Complexity of the problem

We have just described an analysis of the oscillator signal. The three previous Sections were concerned with the development of the concepts introduced in the first part of Chapter 1 (Sections 1.1 – 1.3). We now continue our study in the same spirit, and proceed to the analysis of the structure of the oscillator and the origin of its noise.

An oscillator modelled by an electrical circuit is a deterministic system. Its behaviour is described by differential equations and is perfectly known as soon as the initial conditions are given. However, the *observed* signal is subject to fluctuations; the determinism is blurred; and after an interval of a few coherence times (Section 1.3), the memory of the initial conditions is lost. The signal becomes random and disorder sets in. The simplest explanation is to admit that we have ignored some sources of noise in the modelling of the oscillator. This noise can come from internal sources (resistors, transistors and so on) or it may be the consequence of external noise acting on the circuit (external vibrations causing the fluctuations in a particular parameter). We shall pursue this line of thought in an attempt to describe these phenomena by completing the basic picture. It is a complex adventure that will be continued in Chapter 5; research is far from finished in this field.

There is another possible explanation. We can put it in the form of a question: are we certain that a circuit described by differential equations always leads to a solution determined simply by the initial conditions? Whatever we do, the initial values are always subject to a very slight uncertainty. This is not important if this uncertainty intervenes only in the brief transient state that is followed by a steady state. This is the case of linear circuits. Unfortunately, real oscillators are not linear. When this distinction between transient and steady states is no longer possible, the situation becomes very complex. The transient state can last for a very long time, or it can last indefinitely, so that the steady state does not emerge. Worse still, this state can be chaotic, i.e., the difference between two solutions corresponding to quasi-identical initial conditions may increase in such a way that it is impossible to predict the final value, and an infinitesimal variation in the initial value may cause a large jump in the final value. The behavior of the nonlinear system thus becomes apparently disordered. Such chaotic phenomena are likely to occur in nonlinear systems described by more than two variables of state. They are very sensitive to the parameter values and may or may not occur for very small changes in the parameters values. The resulting chaos cannot therefore be taken as a general explanation of oscillator noise if only because fluctuations are always present even in linear circuits, and for only two variables of state. However, all this points in the direction of a more sophisticated theory that will need to be used when the more classical explanations that we are going to consider do not lead to a solution. We will return to this in Chapter 5.

3.4.2 Noise and the oscillator: a simple scheme

We introduced a very simple oscillator in Section 1.5.2. It consists of a resonator with natural frequency f_0 and very high Q, locked to an amplifier with real gain G. Near resonance, the transfer function $T(f)$ of the resonator is adequately represented by

$$T(f) = \frac{T_0}{1 + 2jQ(f - f_0)/f_0} \qquad (3.73)$$

Locking occurs at frequency f_0 when $GT = 1$. Amplitude stabilisation is due to the nonlinearity, and the gain is then a function of signal amplitude. In practice, the gain decreases as the amplitude increases (because of saturation). When Q is very large, T decreases very rapidly as soon as we move away from f_0. Since G varies very little with frequency, locking cannot take place at neighbouring frequencies. On the one hand, the gain is insufficient to compensate for losses and, on the other hand, T introduces a phase difference other than 2π. For $f = f_0$, the nonlinear circuit is an oscillator delivering a sinusoidal signal that can be characterised by its output power P.

However, when the circuit receives signals of frequency near f_0, its responses add to the f_0 line due to the sinusoidal signal. These additional inputs are due to noise originating in the circuit components (resistors and trsnsistors) and noise picked up from outside by the elements of the circuit. We must therefore complete our description of the system by combining all the sources of noise in an effective source at the input, a technique well known to the amplifier specialist. This is illustrated in Fig. 3.14.

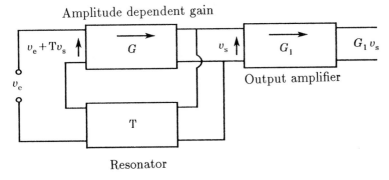

Fig. 3.14 Schematic diagram of an oscillator.

The input v_e is introduced together with an effective noise generator representing all the real sources of noise, both internal and external. The output v_s of the loop that generates the oscillation feeds a second amplifier that produces the usable signal. We have to be aware of the approximation underlying all this. For $f = f_0$, and only for this value, we are in a band of zero width that noise cannot enter and we have large-signal nonlinear operation. Elsewhere, in a narrow but nonzero band, we are in the small-signal linear regime, and the classical formalism of harmonic analysis is valid.

The transfer function v_s/v_e of the oscillator circuit is given by

$$\frac{v_s}{v_e} = \frac{G}{1 - GT} = \frac{G\left(1 + 2jQ\frac{F}{f_0}\right)}{1 - GT_0 + 2jQ\frac{F}{f_0}} \qquad (3.74)$$

where

$$F = f - f_0 \qquad (3.75)$$

which, for $F \neq 0$, reduces to

$$\frac{v_s}{v_e} = \frac{1}{T_0}\left(1 + \frac{f_0}{2jQF}\right) \qquad (3.76)$$

when the loop is locked. The frequency $f - f_0$ at which the spectral line is the Dirac impulse

$$S_{RF}(f) = P\delta(f - f_0) \qquad (3.77)$$

has to be distinguished from the neighbouring frequencies for which the broad spectrum is described by a hyperbolic law. Within a given band Δf, the power output depends on the noise that enters this band via the input v_e as in any linear circuit. Let $A(F)$ be the noise power density equivalent to all the noise sources included in v_e. The noise power output ΔP_B within the band ΔF is

$$\Delta P_B = \left| \frac{v_s}{v_e} \right|^2 A(f)\, \Delta F \tag{3.78}$$

so that

$$\frac{1}{P} \frac{\Delta P_B}{\Delta F} = \frac{A(f_0 + F)}{PT_0^2} \left(1 + \frac{f_0^2}{4Q^2 F^2} \right) \tag{3.79}$$

The power spectrum is shown in Fig. 3.15 and is the superposition of a Dirac line on the a broad spectrum. From this we deduce that the sum of the phase fluctuation spectrum S_Φ and the amplitude fluctuation spectrum S_ϵ is given by

$$S_\Phi(F) + S_\epsilon(F) = 2\frac{A(f_0 + F)}{PT_0^2} \left(1 + \frac{f_0^2}{4Q^2 F^2} \right) \tag{3.80}$$

and, since amplitude fluctuations will be neglected, we thus obtain the first approximation to the phase spectrum. It is interesting to note that this is again the superposition of white phase noise and $1/F^2$ noise (white frequency noise) noted at the beginning of this chapter.

When A is constant within a frequency band close to f_0, i.e., if the resultant effective noise at the input, both internal and external, is white, we can rewrite (3.80) in the notation of Section 3.1:

$$S_\Phi(F) = \frac{\omega_0^2}{4\pi^2} \left(h_2 + \frac{h_0}{F^2} \right) \tag{3.81}$$

so that, for example, the Allan variance becomes

$$\sigma_y^2(\tau) = \frac{3h_2 F_M}{4\pi^2} \frac{1}{\tau^2} + \frac{h_0}{2\tau} \tag{3.82}$$

For very small F, the phase spectrum is proportional to $1/F^2$ and is represented by a straight line with the slope of -2 when plotted in terms of logarithmic coordinates. For large F, the spectrum is horizontal and the two straight lines intersect at $f = f_0/2Q$, which is equal to one half of the resonator bandwidth (Fig. 3.16).

Actually, these results are incomplete because they ignore the ever present $1/F$ noise. Since the sinusoidal oscillation (say, the carrier) is always there, the noise that appears near the carrier frequency, e.g., at $f_0 + F$, can also appear at low frequency in resistors or transistors or in the resonator. This low frequency noise modulates directly the elements of the oscillator loop, which determine its frequency. The modulation side lines created in this way can be readily calculated on the realistic assumption of a small modulation factor.

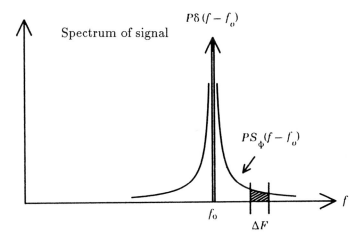

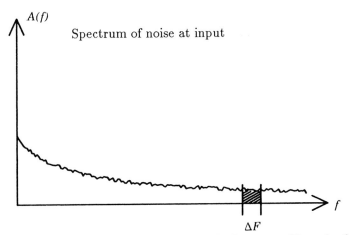

Fig. 3.15 RF spectrum of the output signal of an an oscillator in the presence of noise noise.

The main results obtained on LF noise in resistors, transistors and so on are summarised in Appendix I. Such studies began around 1935; they are far from finished and many of the results are not as yet fully interpreted. However, on the whole, noise in physical systems is reasonably well described by

$$A(F) = A_0 + \frac{A_1}{F} \qquad (3.83)$$

This is a superposition of white noise and $1/F$ noise.

When this is translated to the neighbourhood of the carrier, we obtain

$$S_\Phi(F) = \frac{1}{P} a_0 \left(1 + \frac{F_1}{F}\right)\left(1 + \frac{f_0^2}{4Q^2 F^2}\right) \qquad (3.84)$$

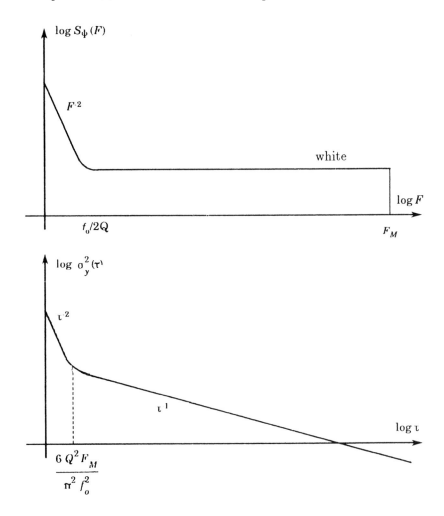

Fig. 3.16 Phase spectrum and Allan variance of an oscillator with white noise.

When the coefficient F_1 is greater than $f_0/2Q$, the spectrum plotted on a logarithmic scale has a linear region of slope -1, as shown in Fig. 3.17. The spectrum contains the $1/F^3$ term and hence some $1/F$ frequency noise.

We could continue this analysis by introducing the contribution of the output amplifier, which would add some white noise and introduce a cutoff frequency in the oscillator noise which it amplifies. The circuit can contain other resonators in addition to the locked resonator that produces the oscillation. This is the case of the maser. We shall discuss this in Chapter 6 and will find that additional filtering will be present.

This simplistic theory can be used to interpret some quite sophisticated

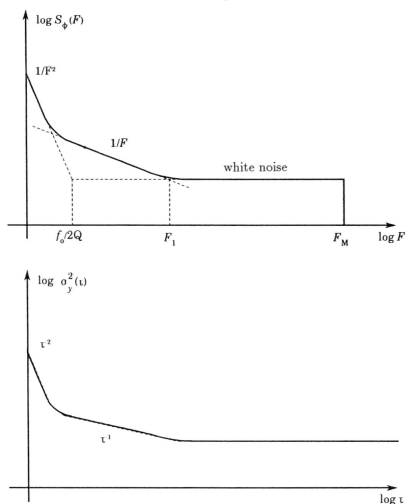

Fig. 3.17 Phase spectrum and Allan variance in the presence of $1/F$ noise (the τ^{-1} region is actually quite rare).

measurements, but in Chapter 5 we will introduce a somewhat more advanced nonlinear theory. The agreement between theory and experiment is reasonable, at least qualitatively. However, little if anything can be concluded at the quantitative level, since we do not know how to determine a_0 and F_1 from known circuit components. It is convenient to set $a_0 = kT$ which brings in the effective input noise temperature. Since the values of T calculated from spectral measurements are well above 300 K, we cannot draw any reasonable conclusions about the origin of this noise.

If we determine the Allan variance from the above simplistic theory, we obtain

$$\sigma_y^2(\tau) = \frac{a_0}{4\pi^2 P f_0^2} \left(3F_M + F_1 + 3F_1 \log 2\pi F_M \tau\right) \frac{1}{\tau^2} + \frac{a_0}{8Q^2 P} \frac{1}{\tau} + \frac{2a_0 F_1 \log 2}{4Q^2 P}$$

$$(3.85)$$

The essential point here is the presence of a plateau on which $\sigma_y^2(\tau)$ reaches its minimum value and remains constant as long as the drift does not intervene. The plateau is related to F_1 and to the $1/F$ noise. The plateau allows us to exploit ultimate oscillator accuracy, as indicated in Chapter 1.

This appears to be a universal phenomenon. It is important to note that the $1/F$ noise is still poorly known, especially in so far as its origin is concerned, but is nevertheless fundamental in the performance of the oscillator.

4

Frequency measurement above 1 GHz

4.1 EXTENSION TO HIGHER FREQUENCIES

4.1.1 The radiofrequency domain ($< 1\,\text{GHz}$)

The frequency spectrum can be divided into 'domains' by more or less well defined fixed points. Each of them has associated with it particular measuring techniques that require the appropriate secondary standards. Some of these fixed points are linked to technological advances, others originate in basic scientific principles. We begin by recalling the techniques available for frequencies below $1\,\text{GHz}$ for which the methods of Chapter 2 are appropriate.

Until the end of the nineteenth century, mechanical clocks were the only oscillators that could be used as standards. The frequencies were of the order of the hertz, this limit being fixed by the then available technology for the fabrication of balances, wheels and escapement mechanisms. Oscillators working at higher frequencies were available, but their operation was poorly understood, which meant that they could not be used as standards. They were musical instruments such as the whistle, the flute, the vibrating chord and so on. The whistle can be used to generate up to several dozen kilohertz, but the frequency of the other instruments, readily identified by the pitch of the emitted sound, is not unique. The invention of the triode in 1905, led to the construction of oscillators that could be used as standards for frequencies up to about a megahertz. This *leap forward* in the direction of higher frequencies was a consequence of the replacement of the escapement mechanism (a metal piece, weighing a few grams, operating by contact and subject to friction and wear) by the electronic tube (electrons are the lightest of particles whose interaction at a distance is conveyed by the electromagnetic field).

The triode valve was first used with resonators consisting of coils and capacitors, whose properties were well known. Oscillators of this kind can be described by the model of Chapter 1 (Fig.1.11). The frequency is determined by the self-inductance L and the capacitance C. Values of C as small as 100 pF and of L as small as $1\,\mu\text{F}$ are technologically possible and give $f_0 = 1/2\pi\sqrt{LC} \approx 16\,\text{MHz}$. The triode functions very well at such frequencies. On the other hand, in order to use the electrical oscillator invented around 1912 as a primary standard, we need to calculate the self-inductance and capacitance, and this we cannot do with the necessary precision. It is even difficult to use it as a secondary standard, because L and C have to be kept constant (there are various losses, the coil is heated by the current flowing through it and so on).

Let us now consider lower frequencies. LC resonators working at, say, 500 Hz are not easy to construct, but this is not important, since we can produce *beats* between two oscillators with frequencies f and $f + F$ and choose f to lie in a range in which generation is easy (500 kHz, for example). The frequency of the second oscillator can then be tuned within a limited range by means of a varactor. By mixing the two signals in a nonlinear circuit, we can produce composite frequencies, including the difference between the two frequencies. A low-pass filter is then employed to isolate the lowest of these frequencies, i.e., the required frequency F. The spectrum can be covered in this way between the audiofrequency range and frequencies of the order of 10 MHz.

The electronics involved in this arrangement can operate at very high frequencies. It relies on electrons (mass m, charge e) accelerated by the voltage V between the heated filament (cathode) that emits them and the plate (anode) that captures them. The filament and plate are a distance d apart, and the electron transit time is of the order of

$$\tau = \frac{d}{v} \approx \frac{d\sqrt{m}}{\sqrt{2eV}} \approx 1.7 \times 10^{-6} \frac{d}{\sqrt{V}} \qquad (4.1)$$

For the system to operate satisfactorily, the electrons have to be bunched by an electrode called a grid, so that they reach the plate together at the right time and use up their energy maintaining the oscillation. The bunching and acceleration process must be complete within the transit time τ at frequency f, so that τ must be small compared with $1/f$. The system functions so long as

$$f < \frac{1}{\tau} = 6 \times 10^5 \frac{\sqrt{V}}{d} \qquad (4.2)$$

For $V = 100\,\text{V}$ and $d = 1\,\text{mm}$, the frequency is 600 MHZ.

Transistors have now replaced triodes, but the above analysis remains valid, except that the dimensions are much smaller. Frequencies in the gigahertz range have become accessible. The necessary resonators must therefore operate at these frequencies, they must be amenable to theoretical calculation and they must resist environmental damage and wear. Resonators that are not too damped can be used as passive frequency standards, provided the measuring time is less than the coherence time. This provides us with frequency standards, but obviously not clocks. Musical instruments can act as resonators of this kind. Vibrating rods (in the tuning fork) plates (bells, cymbals) and membranes (drums) are useful standards that fix the frequency during the coherence time. Elasticity theory, developed since the beginning of the nineteenth century, enables us to model such deformable resonators.

We now know how to use objects with small linear dimensions and well-chosen deformations to construct elastic resonators operating up to the gigahertz range and relying on surface waves. They are essentially multimode resonators of the Fabry-Perot type (*cf.* Section 4.2.4). The important quantity is the wavelength $\lambda = V/f$ of the mechanical vibrations, where V is the speed of the elastic waves. This speed is not precisely known and is subject to numerous disturbing factors; it is of the order of a few kilometers per second, i.e., 1 kHz for a wavelength of the order of a few microns, which is feasible in a Fabry-Perot resonator with very small linear dimensions. Electrical resonators of this kind are made from piezoelectric materials, which allows direct coupling to the associated electronics. We shall return to this in Chapter 5, where we shall consider in detail oscillators incorporating piezoelectric resonators that can cover the entire radiofrequency range. They provide remarkable secondary standards.

4.1.2 Microwave and optical frequencies (> 1 GHz)

Above 1 GHz, the resonator takes the form of an electromagnetic cavity with metal walls. The energy trapped in the cavity is periodically shared between its magnetic and electric components. The resonator thus resembles the pendulum whose total energy is shared periodically between its kinetic and potential components. The important quantity is the electromagnetic wavelength $\lambda = c/f$. There are two essential differences as compared with elastic resonators, but the mathematical models are similar. Since c is of the order of 3×10^8 m/s, the resonators can operate at 100 GHz with cavities having linear dimensions of a few milimetres. The multimode Fabry-Perot cavity can reach frequencies of several hundred terahertz (Section 4.2.4). At the same time, it is not really possible to work with cavities having dimensions greater than a few dozen centimeters, this process cannot be used beyond ~ 1 GHz ($\lambda \sim 30$ cm). On the other hand, c is a fundamental constant whose value is fixed by convention to be 299 792 458 m/s. The resonance frequency of a cavity of simple shape can be calculated, so that we can have a resonator that, at least theoretically, will cover the entire frequency spectrum if we choose the dimensions correctly. Finally, we need reflecting walls, which means that matter must be considered to be continuous, which is not valid when the wavelength is of the order of the interatomic separation. The upper limit of the working frequency of such cavities is therefore in the X-ray range ($\lambda = 10^{-3}\,\mu$m, i.e.,

$f = 300\,000$ THz), which has not been reached by frequency measurements as yet.

Actually, further fixed points appear as we continue our examination. The transistor is a discrete component that can operate at higher frequencies if we reduce its linear dimensions. This becomes technologically more and more difficult, and is accompanied by a reduction in ouput power. A few dozen gigahertz seems to be the current frequency limit. The alternative is to use a beam of accelerated electrons that interact with the wave in the cavity. The cavity must then have a suitable shape, which becomes more and more difficult to calculate, and there are always several resonant modes for synchronisation purposes. Oscillators that employ a cavity and an oscillation-maintaining electron beam are the microwave tubes. Frequencies of 200 GHz can thus be reached at the cost of substantial technological advances. We shall return to these systems in Chapter 5. They can be used as standards, but only with an auxiliary device providing frequency stabilisation. Moreover, they can generally oscillate in several modes, which complicates their use in metrology.

Einstein showed in 1917 that there was a process whereby a naturally absorbing medium could be transformed into an amplifying medium. This process is called stimulated emission. We shall return to it in Section 4.4.1 and again in Chapter 6. The technique of population inversion, which is used to produce amplification, is sophisticated, but is now readily manageable. By placing an amplifying medium of this kind in a resonant cavity, we can compensate cavity losses and produce an oscillator called the *maser* (in the microwave range) and the *laser* (at optical frequencies). For this we need a medium that absorbs at the frequency at which we wish to work (i.e., the resonance frequency of the cavity) and we need some method for inverting the population in order to achieve amplification. This is relatively easy in the optical ($0.4\,\mu$m $< \lambda < 0.8\,\mu$m) and near infrared (up to $\lambda = 10\,\mu m$) ranges that stretch from 30 to 750 THz and include laser oscillators whose frequency can be measured with the help of stabilising devices (*cf*. Section 4.4). Much greater sophistication is necessary in the ultraviolet spectrum where 1 000 THz can be reached with difficulty. As we move into the far infrared ($\lambda = 300\,\mu$m), the devices become more and more difficult to construct. We now know how to build an oscillator that works at $300\,\mu$m (1 THz), but this level of technological performance is difficult to achieve.

In practice, there is a gap between 300 GHz and 1 THz. It separates the limit of performance of electron beams (i.e., classical mechanics) from the limit of performance of the laser (i.e., quantum mechanics). The origin of this is very fundamental. At very high frequencies, the energy hf of a photon is large compared with kT, i.e., the thermal noise level that sets the lower limit for all energy measurements. The readily identifiable photon then plays a very important role, and the oscillator depends on quantum physics. In contrast, at low frequencies, the individual photon has so little energy that it is practically undetectable. A very large number of photons is then required to produce a phenomenon that can be detected by measuring instruments, and the energy flow becomes effectively continuous, i.e., we have the case of the classical oscillator. The dividing point lies at the frequency

$$f = \frac{kT}{h} \approx 2 \times 10^{10} T \tag{4.3}$$

i.e., approximately 1 THz at the temperature of liquid nitrogen. This is the region of the spectrum in which the individual quanta begin to manifest themselves, but the quantum numbers are high and the phenomena are complicated. However, the corresponding metrological problems have not as yet arisen because of the lack of oscillators that qualify.

4.1.3 Metrological constraints in different frequency domains

Figure 4.1 summarises the discussion given in the last Section. We now turn to an analysis of the associated metrological problems. In Chapter 2, we dealt with frequency measurements in the radio-frequency domain, which we assumed to stretch between a few hertz and a few gigahertz. This is the band in which lumped-parameter circuits predominate (a quartz resonator being considered as a discrete component). It is also the frequency band in which flip-flop circuits operate correctly and allow the counting of oscillations, as required in time-domain analysis. These circuits include transistors, and electrons must have enough time to trigger the flip-flop during their transit (the principles of this are discussed in Section 4.1.1). These counting circuits perform frequency division and work with nonlinear multiplying circuits. They carry out the synthesis of frequencies in the form mf_0/n, where f_0 is a known standard frequency. Finally, a phase locked loop controlled by a varactor ensures that the mean frequencies are equal. That way we not only measure the frequencies, but also frequency differences.

Let us now turn to the microwave range, which technically begins at around 800 or 900 MHz and stretches up to 100 or 150 GHz. As we enter the gigahertz range, the metrological techniques gradually become different from those presented in Chapter 2. The essential difference derives from the fact that stable counting circuits cease to work below a few dozen nanoseconds, so that there is no point in considering oscillation counting at 10 GHz. Direct time-domain analysis is then no longer possible. The generation of intermediate standards mf_0/n by synthesis is therefore a very sophisticated matter. On the other hand, we can use computable passive circuits, exemplified by the balanced bridge, to measure frequency and frequency difference. Frequency analysis using passive circuits is a relatively new technique that measures the wavelength by comparing it

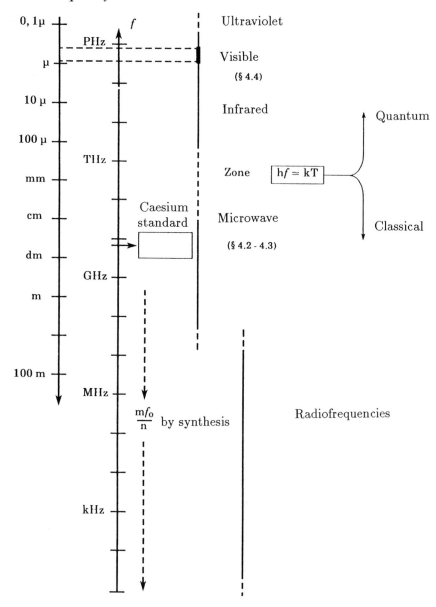

Fig. 4.1 Frequency scales.

with the length of the circuit. Since the velocity c is fixed, this yields the frequency as well. Passive methods will be studied in Section 4.2.

Of course, the mixing of signals with frequencies f_1 and f_2 in a nonlinear component that produces the composite frequencies $m f_1 \pm n f_2$ is still a possible method for changing the frequency. Diodes provide the necessary nonlinearity and a filter can ensure the selection of the required frequency. This brings us down from the microwave to the radio-frequency range in which we can look at beat frequencies, using all the techniques of Chapter 2. However, this approach relies on the availability of a stable auxiliary source, whose frequency differs from that of the oscillator under test by an amount corresponding to the radiofrequency range in which measurements are easier. We therefore have to construct frequency-stabilised oscillators and provide them with suitable connecting networks. The first difficulty to overcome concerns the structure of circuits that must deal simultaneously with 500 THz and 5 MHz. The second difficulty is that we then need nonlinear components that will work at these frequencies.

The calibration ranges linking fixed points on the frequency scale will be studied in Section 4.3. The techniques do not differ markedly from those of Chapter 2 and, under certain conditions, frequency synthesis remains possible. In Section 4.4 we will discuss the operation of these circuits in the visible and infrared ranges. The associated networks require some novel nonlinear components.

4.2 MEASUREMENTS USING A PASSIVE CIRCUIT

4.2.1 Variable-length wavemeter

The impedance Z of a shorted line of characteristic impedance Z_0 and length l, between points a and a', is given by

$$Z = Z_0 \tanh (\alpha + j\beta) l \qquad (4.4)$$

where α is the attenuation coefficient of the line and β is the propagation constant (Fig. 4.2).

In the neighbourhood of a frequency f_0 such that

$$\beta (f_0) = \frac{n\pi}{l} \qquad (4.5)$$

where n is an integer, we have

$$Z = Z_0 a l \left[1 + j \frac{f_0}{a} \left(\frac{d\beta}{df} \right)_0 \frac{f - f_0}{f_0} \right] \qquad (4.6)$$

If al is small compared with unity (low-loss line), the section of the line behaves as a resonant circuit with

$$Q = \frac{f_0}{2a} \left(\frac{d\beta}{df} \right)_0 \qquad (4.7)$$

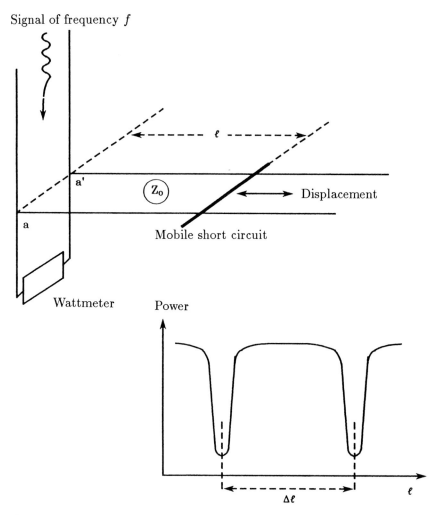

Fig. 4.2 Frequency measurement using a variable-length line.

Placed as a shunt across the main line, this device is practically a short circuit between a and a' at the frequency f_0, and has little effect on neighbouring frequencies if Q is high (Fig.4.2). The length l can be varied by moving the short circuit with a micrometre screw, and the minimum of transmitted power can be found. If we know the law relating β to the frequency, we can thus measure the frequency.

The propagation constant β can be calculated for lines of simple geometry, for example, an empty cylindrical guide of radius R, for which

$$\beta^2 = \frac{4\pi^2 f^2}{c^2} - \frac{K^2}{R^2} \tag{4.8}$$

where K is a numerical factor that depends on the propagation mode (it will be defined more precisely later). The distance Δl between two short circuit positions giving consecutive transmitted-power minima is therefore related to frequency by

$$f = c\sqrt{\frac{1}{(\Delta l)^2} + \frac{K^2}{4\pi^2 R^2}} \qquad (4.9)$$

It would seem therefore that an absolute measurement can be envisaged.

In reality, it is very difficult to construct a system based on the simple model just described. The factor K must be accurately known, but which factor? To avoid multimode operation, we choose it so that only one mode can propagate within the frequency band studied. The symmetry of a transverse electric mode TE (the electric field has no longitudinal component) is such that we can take

$$0.6 < \frac{fR}{c} < 1.1 \qquad (4.10)$$

since β^2 must be positive. Only the TE_{01} mode will then be present, and ite can be shown that $K = 3.8$. It will still be necessary to avoid the TM mode and ensure axial symmetry. All will depend on the manner in which the measurement line is coupled to the main line, i.e., on how the junction aa', represented by the simple contact in Fig. 4.2, is constructed. Finally, the positions of the minima must be accurately located. The width of a resonance peaks is inversely proportional to Q, and calculations show that Q is of the order of δ/λ_g where $\lambda_g = 2\pi/\beta$ is the line wavelength and δ is the skin depth of the cavity wall. Values of Q of the order of 40 000 are possible at frequencies of a few gigahertz. Other difficulties arise with the design of the mobile short circuit and the connection to the supply line. Finally, perturbations due to mechanical and thermal vibrations must be suppressed. Absolute frequency measurement using a variable-length cavity is therefore not a simple matter. The arrangement is, however, very convenient for relative measurements since the micrometer screw can be calibrated and will allow direct reading of frequency to within a few parts in 10^{-4}.

The passive method just described could be translated to the radio-electric range, since it relies on the properties of a resonant LCR circuit with variable capacitance. However, at low frequencies, computable capacitances are very difficult to fabricate. It is even difficult to eliminate the parasitic effects of contacts and casing. Nor is it easy to produce high-grade coils with high Q. The bridge of Fig. 4.3 can be used in relative frequency measurements in the RF range to within a few parts in 10^{-4}. A condenser can be used to balance out the signal $v(t)$ across the bridge terminals. If the frequency fluctuates, the output will not stay at zero. The bridge is therefore a frequency-to-amplitude converter in the neighbourhood of the reference (but tunable) frequency. By translating the bridge technique to the microwave domain, we can perform spectral analysis of the noise of microwave oscillators, which we shall now consider.

4.2.2 Spectral analysis using passive circuits at microwave frequencies

The first method employs a frequency-to-phase converter and a phase detector. The former tackles frequency fluctuations and the latter is usually a balanced mixer. The complete unit constitutes a frequency discriminator. The passive component is either a cavity or a delay line. The second method brings together frequency-to-amplitude conversion and an amplitude detector. Once again, this a frequency discriminator a device generally consists (in the microwave domain) of a resonant cavity and a magic T ensuring amplitude detection.

Figure 4.3 llustrates the principle of the delay-line discriminator. The incoming signal is divided into two parts, one of which is sent via the delay line and the other into the phase shifter that produces a phase difference of $\lambda/4$ across the input terminals of the microwave mixer. The latter then operates as a phase detector (*cf.* Section 2.2.4).

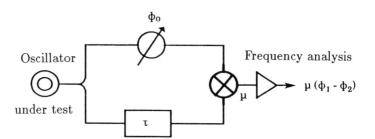

Fig. 4.3 Delay-line frequency discriminator.

We can now examine the behaviour of the system for a sinusoidally modulated source frequency, in which case the signal is

$$v(t) = V_0 \cos\left[2\pi f_0 t - \frac{\Delta f}{F} \cos 2\pi F t\right] \quad f = f_0 + \Delta f \sin 2\pi F t \quad (4.11)$$

where Δf and F are the amplitude of the frequency excursion and the modulating frequency, respectively. The phase difference $\Delta\Phi(t)$ across the terminals of the mixer is then given by

$$\Delta\Phi(t) = -2\pi f_0 \tau + \frac{2\Delta f}{F} \sin(\pi F \tau) \sin\left[2\pi F\left(t - \frac{\tau}{2}\right)\right] - \Phi_0 \quad (4.12)$$

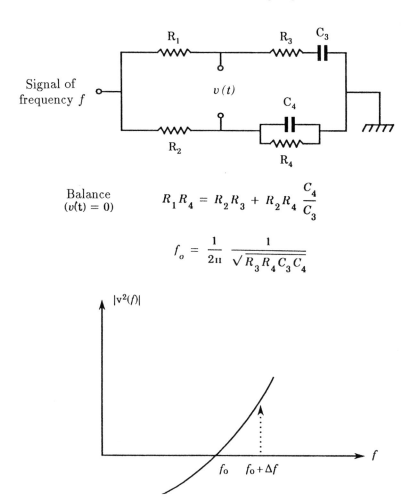

Balance
$(v(t) = 0)$
$$R_1 R_4 = R_2 R_3 + R_2 R_4 \frac{C_4}{C_3}$$

$$f_o = \frac{1}{2\pi} \frac{1}{\sqrt{R_3 R_4 C_3 C_4}}$$

Fig. 4.4 A frequency measurement bridge for the LF spectrum (Wien bridge).

where Φ_0 is the phase difference introduced by the phase shifter. We choose it so that

$$-2\pi f_0 \tau - \Phi_0 = \pi/2 \qquad (4.13)$$

After filtering, we obtain the following response across the output of the phase detector:

$$\Delta V = 2\mu \frac{\Delta f}{F} \sin \pi F \tau = 2\pi\mu\tau \frac{\sin \pi F \tau}{\pi F \tau} \Delta f \qquad (4.14)$$

where μ is the sensitivity (in volts per radian) of the detector. From these formulae we deduce the relationship between the spectrum $S_{\Delta V}(F)$ of the output signal of the phase detector and the frequency spectrum $S_{\Delta f}(F)$ of the signal under examination:

$$S_{\Delta V}(F) = (2\pi\mu\tau)^2 \frac{\sin^2 \pi F\tau}{(\pi F\tau)^2} S_{\Delta f}(F) \tag{4.15}$$

A low-pass filter represented by the function $(\sin \pi F\tau/\pi F\tau)^2$ is introduced, even though the delay line is actually a wide-band device. This filter is present because the discriminator detects frequency fluctuations by looking at the phase variations in terms of the delay τ. Its efficacy is therefore restricted to modulating frequencies lower than $1/2\pi\tau$. We can therefore write

$$\Delta V = 2\pi\mu\tau\Delta f = K_d\Delta f \qquad S_y(F) = \frac{1}{f_0^2 K_d^2} S_{\Delta V}(F) \tag{4.16}$$

where $K_d = 2\pi\mu\tau$ (volts per hertz).

A resonator can be used as a dispersive system producing frequency-to-phase conversion. The arrangement is similar to that described above, with a resonant cavity replacing the delay line. For example, one can use a cavity with transmission coefficient $S_{21}(F)$ given by

$$S_{21}(f) = \frac{2\sqrt{a_1 a_2}}{1 + a_1 + a_2 + jX} \qquad X = 2Q_0 \frac{f - f_0}{f_0} \tag{4.17}$$

where Q_0 is the intrinsic Q of the cavity and α_1 and α_2 are the coupling factors between the cavity and the input and output connections. If we introduce the load Q, and use a symmetrically coupled cavity $\alpha_1 = \alpha_2 = \alpha$, we obtain

$$Q = \frac{Q_0}{1 + 2\alpha} \qquad S_{21}(f) = \frac{S_{21}(f_0)}{1 + 2jQ(f - f_0)/f_0} \tag{4.18}$$

The dephasing introduced by the resonator is given by

$$\Phi_{21} = -\arctan \frac{2Q(f - f_0)}{f_0} \tag{4.19}$$

In the immediate neighbourhood of the resonance we obtain a linear variation of the phase ψ_{21} with frequency. Hence, a frequency fluctuation Δf corresponds to a phase change

$$\Delta\Phi = \frac{2Q}{f_0}\Delta f \tag{4.20}$$

Phase detection performed in the same way as before gives the output signal

$$\Delta V = \mu \frac{2Q}{f_0} \Delta f \tag{4.21}$$

The same result is obtained by introducing a cavity-equivalent delay $\tau = (d\phi/2\pi df)_{f_0} = Q/\pi f_0$, which leads to

$$\Delta V = 2\pi \mu \tau \Delta f = \mu \frac{2Q}{f_0} \Delta f = K_d \Delta f \tag{4.22}$$

In this way we obtain the frequency spectrum

$$S_{\Delta f} F = \frac{1}{K_d^2} S_{\Delta V} (F) \tag{4.23}$$

This result is valid provided the phase varies linearly with frequency, i.e., in practice, within the system pass band. Outside the pass band, the side bands corresponding to frequency noise are filtered off. The system then operates as a simple phase detector with sensitivity μ.

$$\Delta V = \mu \Delta \Phi \qquad S_{\Delta V} (F) = \mu^2 S_{\Delta \Phi} (F) \tag{4.24}$$

By tuning the cavity to the frequency of the input carrier, we ensure that the amplitude fluctuations associated with the frequency-to-amplitude transformation in the resonator appear only in the second order in Δf. They are also rejected by the phase detector. Similar results are obtained for a reflecting cavity

$$S_{21}(f) = \frac{1 + 2jQ_0 (f - f_0)/f_0}{1 + \alpha + 2jQ_0 (f - f_0)/f_0} \qquad \Phi_{21} = \frac{2Q}{f_0} \alpha (f - f_0) \tag{4.25}$$

Frequency-to-amplitude conversion can be achieved with an analogous device. The method relies on the transformation of the oscillator frequency noise into amplitude noise that is then detected by a classical device. The amplitude noise of the oscillator under test can be measured in a similar way. The circuit (Fig.4.5) now includes two arms, one of which contains a resonator and a circulator, which together work as a filter for the carrier. Only the side bands of noise are allowed to remain. The carrier is reinjected with a suitable phase, using the signal from the second arm. The process can be easily visualised with the help of the vector representation of the frequency-modulated wave (Fig.4.5). The FM carrier is suppressed and reinjected with a phase shift of $\pi/2$. This gives the diagram for an amplitude-modulated wave. Amplitude detection is often performed by a microwave T consisting of two detecting diodes whose output is fed to

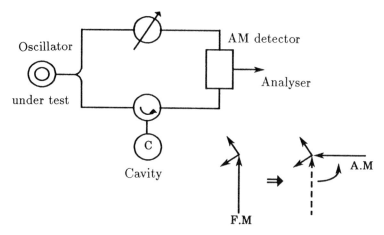

Fig. 4.5 Discriminator using the frequency-amplitude conversion.

a differential amplifier that removes the amplitude noise of the oscillator under test.

The operation of this device can be described in terms of the overall transmission coefficient S_{21} of the circulator and cavity

$$S_{21}(f) = \frac{\alpha - 1 - 2jQ_0\,(f - f_0)/f_0}{\alpha + 1 + 2jQ_0\,(f - f_0)/f_0} \tag{4.26}$$

When $\alpha = 1$ and $f \approx f_0$, we have

$$S_{21} \approx -jQ_0 \frac{f - f_0}{f_0} \tag{4.27}$$

A frequency-modulated sinusoidal input signal can be expressed in the form of equation (4.11), which for small Δf leads to

$$v_e\,(t) = V_0 \cos 2\pi f_0 t + \frac{V_0 \Delta f}{2F}\,[\sin 2\pi\,(f_0 + F)\,t + \sin 2\pi\,(f_0 - F)\,t] \tag{4.28}$$

The first term (of frequency f_0) gives zero response since $|S_{12}| = 0$ for $f = f_0$.

The output signal is therefore given by

$$v_e\,(t) = \frac{V_0 \Delta f Q_0}{2f_0}\,[\cos 2\pi\,(f_0 + F)\,t - \cos 2\pi\,(f_0 - F)\,t] \tag{4.29}$$

where we have used the transfer function given by (4.27). By adding a $\pi/2$ shifted carrier, i.e., $V_0 \sin 2\pi f_0 t$, we obtain an amplitude modulated signal. Linear detection then gives

$$\Delta V = \frac{k\,V_0 Q_0}{f_0}\Delta f = K_d \Delta f \tag{4.30}$$

where k is a numerical factor characterising the detector operation.

The spectrum associated with the frequency fluctuation is

$$S_{\Delta V}(F) = k^2 V_0^2 \left(\frac{Q_0}{f_0}\right)^2 S_{\Delta f}(F) \tag{4.31}$$

i.e., an expression similar to that obtained for the discriminator using frequency-to-phase conversion.

The sensitivity of systems using discriminators is measured by the ratio $\Delta V / \Delta f$ of the signal obtained by detection and the signal being analyzed. The ratio is equal to $2\mu\pi\tau$ for the delay line system, $2Q\mu/f_0$ for the frequency-to-phase conversion discriminator and $kV_0 Q_0/f_0$ for the frequency-to-amplitude conversion discriminator of the previous sections. The sensitivity of the passive measuring system is thus seen to increase with τ and Q. On the other hand, an increase in these quantities produces an undesirable reduction in the pass band of the system. Moreover, the above results are valid provided the frequencies involved in the analysis lie within the pass band of the system. In practice, we use only a fraction of the latter (a quarter or even a fifth) and a compromise between sensitivity and pass band must be sought. The sensitivity of the system increases with μ and kV_0, which implies high power levels (if possible, of the order of a few milliwatts) at the detector inputs. When the discriminators employ transmitting or reflecting cavities, and when the input signal power is low, cavity losses must be low, which implies optimisation of the product $|S_{21}|\Phi_{21}$. It can be shown that the optimum coupling coefficient is $\alpha = 1/2$ and $\alpha = 1$ for transmission and reflection, respectively.

The ultimate sensitivity of a measuring stage is measured by the minimum phase or frequency noise that can be detected by the system. This quantity is the translation (in terms of phase or frequency noise) of the total noise of the system by the factor K_d. We make a distinction between the actual residual noise of the measuring system and the total noise. The former is due to sources of noise internal to the measuring system, to the exclusion of the noise introduced by the standard, which is a passive standard, a resonator or a delay line. Frequency fluctuations in the standard are hand included in the expression for the total noise.

The residual noise of a passive measuring stage can be measured by replacing the dispersive element (delay line or cavity) by a nondispersive section of the line. This measurement requires a standard oscillator of high spectral purity, and the symmetry of the two circuit arms is used to reject the phase fluctuations of the standard oscillator. This symmetry can be broken by introducing a phase shifter into one of the arms that ensure that the signals at the detector inputs are in quadrature. This phase shifter produces very little dispersion at high frequencies, and close to 30 dB rejection of standard-oscillator noise can be achieved. Calibration of the equipment, prior to measurement of any oscillator noise, requires a knowledge of the sensitivity expressed in terms of K_d. The latter is obtained by injecting into the spectrum of the oscillator under test a line with a known modulation factor $\Delta f/F$ determined by the spectrum analyser from the relative heights of the side lines and the carrier.

The spectrum of the minimum detectable frequency noise $S_{\Delta F}^{min}$ of passive measuring stages is generally related to the noise of the detector and the low-noise amplifier that follows it. A long noise plateau corresponding to the additive noise of the output preamplifier, and a region of flicker noise associated with the active elements of the detector operating at high power levels, are obtained. This noise can be reduced by using Schottky diodes. Figure 4.6 shows the minimum detectable phase and frequency noise spectra of passive measuring stages operating in the X band.

The noise of the actual resonators which of appears in the expression for the global noise of the passive measuring system has been measured using a symmetric circuit containing two resonators as identical as possible in each arm. A very high quality oscillator is used as excitation source (Fig.4.7). The phase fluctuations detected can be related back to the frequency noise of the resonator, if we know the frequency-phase characteristic of the latter. The results of measurements carried out on dielectric resonators are given in Fig. 4.8. The frequency noise is negligible compared to the residual noise of the measuring stage using these same resonators as dispersive components, and this occurs in all the ranges of spectral analysis used (1 Hz − 100 GHz).Figure 4.9 gives a few examples of phase spectra of oscillators operating in the X band (10 GHz).

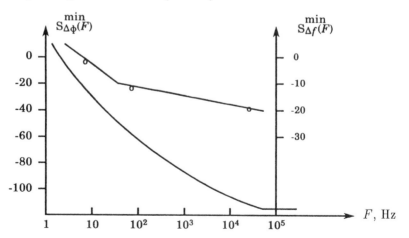

Fig. 4.6 Minimum detectable phase and frequency noise.

4.2.3 Spectral analysis in the optical domain: the Michelson interferometer

Interferometers have been used in the visible spectrum for two centuries.

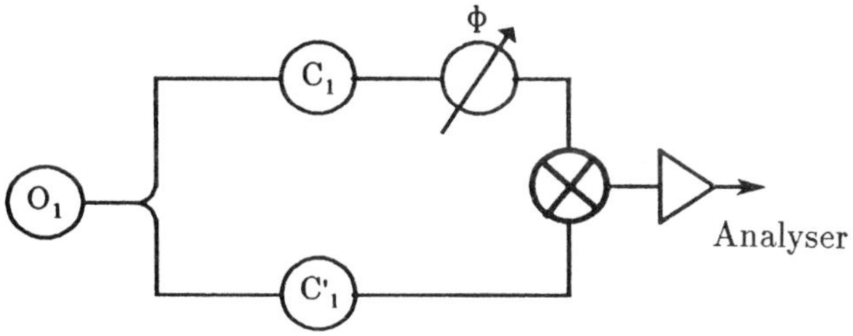

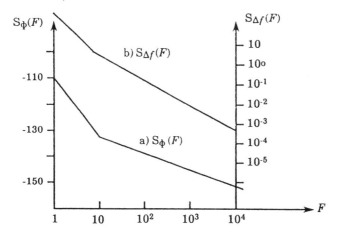

Fig. 4.7 Measurement of characteristic noise in cavities and resonators (C_1, C_1' are identical cavities).

Fig. 4.8 (a) Phase noise in two identical dielectric resonators, (b) frequency noise in a dielectric resonator at 9.1 GHz.

They compare the wavelength with a physical standard of length. The velocity of light being a fundamental constant, interferometers can also be used to measure frequency, so long as they operate in vacuum. The principle of the Michelson interferometer (Fig. 4.10) is simple. The incident wave $v(t)$ is split into two beams by a semi-transparent mirror. One beam is reflected by mirror A and the other by mirror B. The beams recombine on their return to the semi-reflecting mirror and are collected by a photodetector that measures the mean power of the sum wave, which can be written in the form

$$P = \frac{1}{\vartheta} \int_0^{\vartheta} \left[v(t) + v\left(t - 2\frac{l_1 - l_2}{c}\right) \right]^2 dt \qquad (4.32)$$

since the delay of one wave relative to the other is equal to twice the difference between the interferometer arm lengths, divided by c. The integration

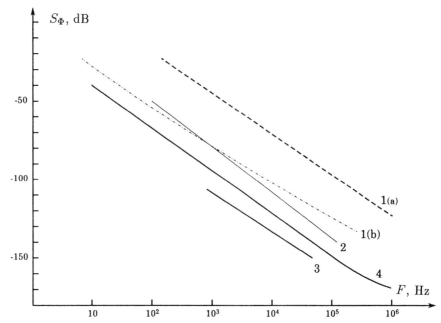

Fig. 4.9 1– DRO ($f_0 = 9.2\,\text{GHz}, \text{P} = 20\,\text{mW}$), $1(a)$ – free, $1(b)$ – locked to 1 GHz quartz, 2–low-power reflex klystron, 3–double cavity klystron, 4–theoretical curve $S_\Phi(F)$ from (3.84) with $a_0 = FKT, F = 3, P = 0.03\text{W}, f_0 = 10\,\text{GHz}, Q = 3000, F_1 = 106\,\text{Hz}$.

time ϑ must be as long as possible; in practice a few thousand periods are sufficient. This gives

$$P = \frac{1}{2}P_0 + \Gamma\left(2\frac{l_1 - l_2}{c}\right) \qquad (4.33)$$

where Γ is the autocorrelation function of $v(t)$ and P_0 is the average power of the primary beam. When the wave is generated by a sinusoidal oscillator of frequency f, we have

$$\Gamma\left(2\frac{l_1 - l_2}{c}\right) = \frac{1}{2}P_0\cos 4\pi f\frac{l_1 - l_2}{c} = \frac{1}{2}P_0\cos 2\pi\frac{x}{\lambda} \qquad (4.34)$$

If we vary the delay by moving one of the mirrors, the observed power will vary sinusoidally (Fig. 4.11). The same power is recorded after the displacement $\Delta x = c/f$, equivalent to scanning one fringe.

We can measure the wavelength by counting the number N of fringes crossing the field of view for a displacement L of one of the mirrors. This gives

$$f = \frac{c}{\lambda} = \frac{cN}{2L} \qquad (4.35)$$

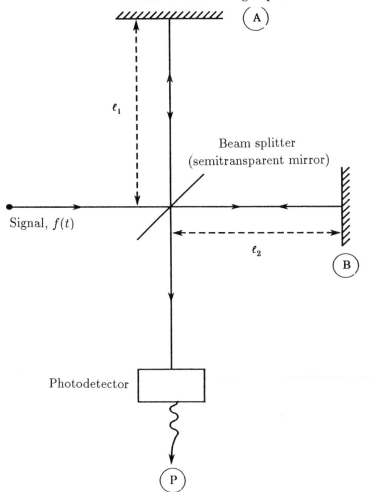

Fig. 4.10 The Michelson interferometer.

Wavelengths can be compared by making two successive measurements for the same displacement and using two sources. At optical frequencies, a displacement of a few millimetres corresponds to a few thousand fringes so that accurate measurement is possible. The method can be extended to the microwave range, but the displacement is then only a few wavelengths and the accuracy of fringe measurements is limited. This takes us back to the arrangement of Section 4.2.1. In the limit, a displacement of one fringe is sufficient, which is equivalent to metre waves ($f \approx 300\,\text{MHz}$). However, to obtain an accurate measurement, we have to determine the beginning and the end of the wave as accurately as possible, and this can only be done if the the correlation curve is a true sinusoid.

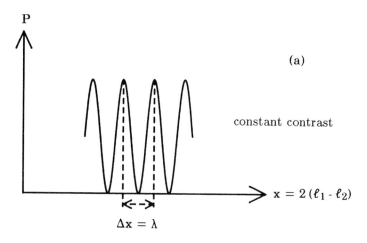

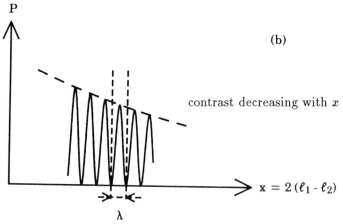

Fig. 4.11 Power collected by the interferometer photodetector for (a) a sinusoidal signal (infinitely narrow spectrum), (b) a signal with a broadened spectrum.

Actually, the variation of power with $x = 2(l_1 - l_2)$ is *not* a true sinusoid. Figure 4.11 shows the power curve obtained for a light source in the form of a discharge lamp. A displacement of a few thousand fringes reveals that

$$\Gamma = \frac{1}{2} P_0 \exp\left(-x/x_0\right) \cos 2\pi \frac{x}{\lambda} \qquad (4.36)$$

i.e., visibility declines with increasing path difference. As explained in Chapter 1, this allows the description of the source by a power spectrum centered on f. The width of the spectrum is

$$\Delta f = \frac{1}{\tau} = \frac{c}{x_0} \qquad (4.37)$$

where x_0 is the coherence length. In practice, if we use discharge tubes, x_0 is of the order of 1 m. Δf is then of the order of 100 MHz and the relative width of the spectrum is $\Delta f / f = 10^{-6}$. Interferometry thus yields both frequency and coherence time τ. Reasonable accuracy can be achieved for a few thousand fringes, which means that the method must be confined to classical sources in the visible or infrared domain for which the coherence time is short.

The above method yields directly the spectrum at high frequencies. Unfortunately, it is not effective in the microwave range, and even less so in the RF band. A displacement of several thousand fringes is inconceivable in the spectral ranges in which oscillators are available with coherence time (theoretically infinite) limited only by noise. The relative line width $\Delta f/f$, estimated by the methods outlined in previous chapters, is of the order of 10^{-12}. To observe a significant reduction in Γ (say 10 %) and thus determine τ, it would be necessary to count 10^{11} fringes, the fringe separation being a few decimetres (at 300 MHz). This is quite impossible, so that in the spectral domain in which classical oscillators are available (up to 100 GHz) interferometric measurements cannot be used to obtain Δf and it is very difficult to determine the frequency. On the other hand, these methods of measurements work well in the visible and infrared ranges, both for frequency and for the coherence time. However, in this domain, conventional sources do not emit continuous vibrations and the coherence time is short. Such light sources are not oscillators: they emit wave packets of short duration.

4.2.4 The optical cavity spectrum

The wavelength is now of the order of a micron and we cannot make single-mode resonant cavities because the distance between the walls would have to be of the order of one micron. The solution is a multimode cavity in which resonance occurs when a stationary regime is established with an integral number of half-waves between the walls. This number is obviously very high (10^6 for walls 0.25 m apart in the visible range). This leads to interesting properties and also to considerable practical difficulties.

These resonant cavities were invented in 1900 by Fabry and Perot and were used by them in multiple-beam interferometers. We shall now consider a simple transmission-line model of an oscillator (Fig. 4.12).

A line with a characteristic impedance Z_0 transmits waves propagating in opposite directions. The voltage v and current i are given by

$$v = A \exp -\alpha x \exp \left[j \left(2\pi f t - \beta x \right) \right] + B \exp \alpha x \exp \left[j \left(2\pi f t + \beta x \right) \right]$$

$$(4.38)$$

$$i = \frac{1}{Z_0} \left[A \exp \left(-\alpha x \right) \exp j (2\pi f t - \beta x) - B \exp \left(\alpha x \right) \exp j \left(2\pi f t + \beta x \right) \right]$$

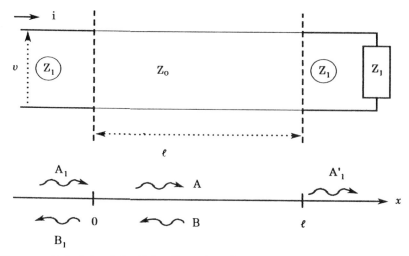

Fig. 4.12 Principle of the Fabry-Perot interferometer.

where α is the attenuation factor (in Np/m) and $\beta = 2\pi/\lambda$ is the propagation constant. Consider a segment of the line of length l. Its extremities are connected to lines of characteristic impedance Z_1. One of these lines supplies the resonator and the other is terminated with impedance Z_1, which prevents all reflections. Since the voltages and currents must be continuous across the two extremities, we have

$$A_1 + B_1 = A + B$$

$$\frac{1}{Z_1}(A_1 - B_1) = \frac{1}{Z_0}(A - B)$$

$$A \exp(-\gamma l) + B \exp(\gamma l) = A'_1 \qquad (4.39)$$

$$\frac{1}{Z_0}[A \exp(-\gamma l) - B \exp(\gamma l)] = \frac{1}{Z_1}A'_1$$

where $\alpha + j\beta = \gamma$

Hence the transmission coefficient is given by

$$|T|^2 = \left|\frac{A'_1}{A_1}\right|^2 = \frac{4|t|^2}{[\exp(\alpha l) - |\rho|^2 \exp(-\alpha l)]^2 + 4|\rho|^2 \sin^2 2\pi f l/V}$$

$$t = \frac{2Z_1 Z_0}{Z_1 + Z_0} \qquad \rho = \frac{Z_1 - Z_0}{Z_1 + Z_0} \qquad \beta = \frac{2\pi}{\lambda} = \frac{2f}{V} \qquad (4.40)$$

where V is the phase velocity.

We thus obtain the multimode resonator illustrated in Fig. 4.13. The resonance frequencies are

$$f_n = \frac{nV}{2l} \qquad (4.41)$$

A resonator, 1 m long, in which electromagnetic waves propagate with $V = 3 \times 10^8$ m/s, has an infinite series of resonance frequencies, spaced at 150 MHz. If we take $n = 3 \times 10^6$, we find a resonance at 450 THz (and many others). In the neighbourhood of resonance,

$$|T|^2 = \frac{4\,|t|^2}{[\exp{(\alpha l)} - |\rho|^2 \exp{(-\alpha l)}]^2} \frac{1}{1 + 4Q^2(f - f_0)^2/f_0^2} \tag{4.42}$$

which is similar to the classical resonance curve. The Q factor is given by

$$Q = \frac{n\pi\,|\rho|}{\exp{(\alpha l)} - |\rho|^2 \exp{(-\alpha l)}} \tag{4.43}$$

If αl is small, the medium has low losses and $|\rho|$ is close to unity; Z_1 and Z_0 then have to be very different and we find that

$$Q \approx \frac{n\pi}{1 - |\rho|^2 + 2\alpha l} \tag{4.44}$$

For $1 - |\rho^2| = 10^{-4}$ and $\alpha l = 10^{-4}$, this gives $Q = 10^{10}$. The line width at 450 THz is smaller by a factor of 3 000 than the separation.

Physics allows us to achieve this order of performance only within narrow frequency bands. Exceptional values of $|\rho|$, practically equal to unity, and small values of αl, can be obtained in the neighborhood of the frequency f_0 of a strong narrow line. As we move away from f_0, $|\rho|$ decreases, $\alpha \ell$ increases and the peaks become smaller and broader (Fig.4.13). In this way, we obtain a useable, but probably still multimode, resonator. The simplest arrangement consists of two plane parallel mirrors carrying multilayer insulating coatings with slightly different layer impedances Z_1 and Z_1'. The layer thicknesses can be chosen so that $|\rho| \approx 1$ within a narrow frequency band.

Since the layer thickness is of the order of $\lambda/4$, such mirrors can only be fabricated for the visible or near infrared ranges.

A Fabry-Perot resonator fed by a continuous generator, producing an exceedingly narrow line, can be used as a wavemeter. The length l can be varied, for example, by mounting one of the mirrors on a piezoelectric transducer driven by a slowly varying voltage. The output signal then has the shape indicated in Fig.4.13a. We can identify the high order mode if the cavity behaves as shown in Fig. 4.13b; fortunately, there are always a few modes with different amplitudes.

4.2.5 Weakly-coherent optical oscillators and a simplistic model of the atom

If we feed a cavity with radiation from a natural source, we have to consider the width of its HF spectrum. For $\Delta f/f = 10^{-6}$, the values of Δf in the

$$|T_m|^2 = \frac{4|t|^2}{e^{a\ell} - |\rho|^2 e^{-a\ell}}$$

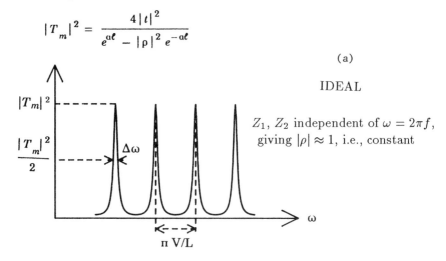

(a)

IDEAL

Z_1, Z_2 independent of $\omega = 2\pi f$, giving $|\rho| \approx 1$, i.e., constant

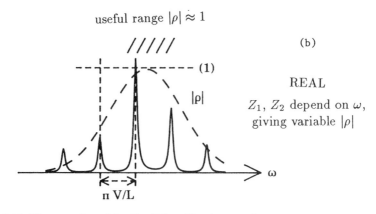

useful range $|\rho| \approx 1$

(b)

REAL

Z_1, Z_2 depend on ω, giving variable $|\rho|$

Fig. 4.13 Lines produced by the Fabry-Perot resonator.

visible range are approximately three times the separation between the cavity lines. A Fabry-Perot line is much narrower than the source line. If we can accurately identify the order n of a line, the scan produced by varying the length l gives the spectral line profile. The interferometer then allows us to obtain the high-frequency spectrum of the noncoherent natural source. Figure 4.14 illustrates the principle of this operation.

The recorded power spectrum can be described by

$$S(f) = S_0 \exp -[(f - f_0)/\Delta f]^2$$

or

$$S(f) = \frac{S_0}{1 + [(f - f_0)/\Delta f)]^2} \qquad (4.45)$$

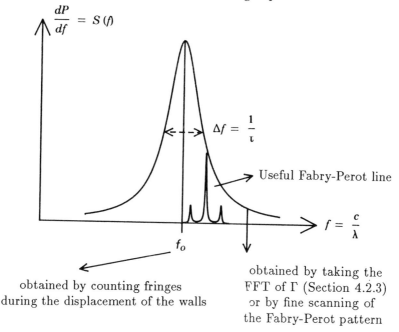

Fig. 4.14 Oscillator line with a short coherence time

These are similar to the spectrum of a weakly-damped resonator (Section 1.4). The resonator is excited by supplying it with energy W. It responds by emitting a damped wave with a time constant $\tau \approx 1/\Delta f$, and then returns to rest (it is de-excited). A very large number of resonators, independent of each other, will behave in the same way, generating wave trains with duration τ and random phases. There is no spatial coherence since each resonator emits randomly, and there is very weak temporal coherence since the same resonator can be randomly re-excited and retains the phase of its oscillation only during the time τ.

An atom can be described by this model in the first approximation. The impact of an electron (in a discharge) or the incidence of electromagnetic radiation will excite the atom which will thus receive energy W. It is de-excited by the emission of a train of electromagnetic waves of frequency f and duration τ. There is one, albeit essential, difference: the energy W absorbed by the atom is related to the radiation frequency f by $W = hf$, where $h = 6.6 \times 10^{-34}$ J.s is Planck's constant. The emitted energy is also W. It is the energy of the emitted photon. A schematic diagram, which will be gradually improved, is given in Fig. 4.15. The two states, ground and excited, are represented by horizontal lines and the energy is plotted along the vertical axis. Excitation (absorption of energy) and de-excitation (emission) are indicated by arrows.

This diagram is very encouraging for those looking for a frequency stan-

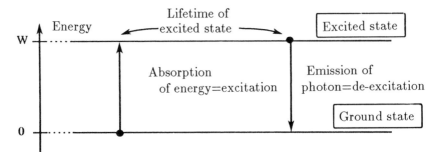

Fig. 4.15 Simplified model of an atom.

dard in the form of a resonator, as we have done since Chapter 1. The energy W, and hence f, are determined by the structure of the atom, which therefore seems to be a perfect natural standard. However, the emitted line is not infinitely narrow, as the diagram of Fig. 4.15 might lead us to believe. The excited state has a finite lifetime that broadens the line, giving it its natural width which is very small in comparison with the measured width.

Actually, the above diagram describes an isolated atom and the relative line width $\Delta f/f$ is the natural line width of the atom. In practice, we have to deal with a group of atoms, each of which interacts with the others. This is the prime cause of line broadening, whereby random perturbations modify f in different ways and not simultaneously for all the atoms. Changes in the frequency f can also be caused by external factors such as a magnetic field (Zeeman effect) or the electric field unavoidably present in a discharge (Stark effect). Finally, and most importantly, the atoms execute thermal motion in all directions. The wave emitted by a moving source has its frequency shifted by

$$\Delta f_D = \frac{vf}{c} \tag{4.46}$$

relative to the frequency emitted at rest (this is the Doppler effect) where V is the component of the thermal velocity in the direction of the observer. The mean thermal velocity is

$$\bar{v} = \sqrt{\frac{3kT}{m}} \tag{4.47}$$

where m is the mass of the atom, T is the absolute temperature and k is Boltzmann's constant $(1.4 \times 10^{-23}\,\mathrm{J/deg})$. The velocity v has random direction. The result is a Doppler broadening of the order of

$$\frac{\Delta f}{f} = \sqrt{\frac{3kT}{mc^2}} \tag{4.48}$$

A simple calculation ($T = 1000\,\text{K}$ and $m = 10^{-24}\,\text{kg}$) leads to a relative line width of the order of the observed value.

In practice, the relatively broad line examined with an interferometer is the resultant of a very large number of narrow lines with fluctuating frequencies. If we wish to build an atomic standard, we first have to minimise the Doppler effect. This can be done by using a beam of atoms with parallel velocities, and by observing at right angles to the beam. An important condition is that the atom must be isolated from its environment as much as possible. We will return to this in Chapter 6. It will nevertheless be necessary to interrogate the atom in order to establish its frequency. This is done by making it interact with a photon, which broadens the line and eventually sets a natural limit on the definition of the ideal lossless atomic resonator that maintains its frequency indefinitely with absolute precision.

4.3 MICROWAVE MEASUREMENTS USING COMPARISON OF OSCILLATORS

4.3.1 Active measurements

The principles of this method as applied to radio-frequency oscillators were developed in Chapter 2 and remain essentially valid in the microwave range. Features that are specific to microwaves relate to the characteristics of microwave oscillators and the corresponding detection techniques. The phase detector is linear only near quadrature and, whilst quadrature could be maintained during the measuring time for a quartz oscillator at, say, 5 MHz, this is out of the question for a microwave oscillator. Phase locking of the two sources is essential. Another feature follows from the importance of frequency fluctuations of microwave sources: the locking band must be relatively wide (a few kilohertz) in order to maintain the process.

There are two basic measuring structures. In the case of direct detection (Fig.4.16), the frequency of the standard microwave source O_2 is adjusted to the same value as the source under test O_1. Phase detection is carried out directly at the microwave frequency. In the case of intermediate-frequency detection (Fig. 4.17), the standard source is only used for translation of the signal to a lower frequency (usually, radio-frequency range). Another standard source of intermediate frequency (IF) must be available to perform demodulation. The radio-frequency oscillator or one of the microwave generators can then be locked.

The expression for the detected voltage is (*cf*. Section 2.3.1)

$$V = \frac{\mu G\left(p\right)}{1 + \mu K G\left(p\right) H\left(p\right)/p}\left(\Phi_1 - \Phi_r\right) \qquad (4.49)$$

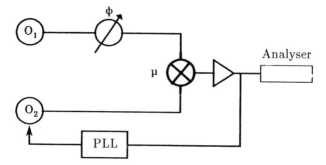

Fig. 4.16 Direct detection method.

where Φ_1 and Φ_r represent the phase fluctuations in the source under test and in the reference source respectively, μ is the gain of the phase detector (volts per hertz), $G(p)$ is the gain, $H(p)$ is the loop transfer function, K is the tuning sensitivity of the locked oscillator and $p = j2\pi F$ is the Laplace operation.

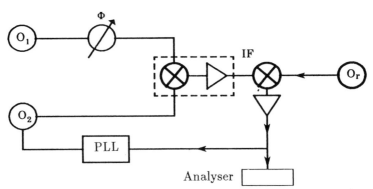

Fig. 4.17 Intermediate frequency method.

For low-gain operation, the output is proportional to the phase fluctuation:

$$V = \mu G\left(\Phi_1 - \Phi_r\right) \qquad \mu K G H\left(p\right) \ll p \qquad (4.50)$$

this condition can only be achieved for microwave oscillators at Fourier frequencies lower than a few kilohertz . To obtain the phase spectrum of the oscillator between, say, 10 Hz and 1 MHz, a pseudo-integrator has to be used. The dynamic range of this correcting filter will have to be very large, since the F^{-2} correction has to be introduced over at least two, if not three, decades (*cf.* Section 2.3.3).

The frequency spectrum can be examined with high gain, in which case

$$\mu K G A\left(p\right) \gg p \qquad V = \frac{\Phi_1 - \Phi_r}{K A\left(p\right)} \qquad (4.51)$$

but to obtain an analysis band of 1 MHz, a gain of at least 10^6 is necessary, and this presents a risk of instability in the locking loop. A pseudo-integrator is necessary in most cases.

In the active detection method and, for that matter, in passive frequency-to-phase conversion, the phase detector must be a diode mixer, since logic detectors cannot operate beyond 1 or 2 GHz. All mixers whose pass band for the IF output stretches up to the continuum can act as phase detectors. The input bandwidth can amount to several decades (say, $1 - 18$ GHz). The sensitivity μ for mixers handling signals of the order of a few dBm can be of the order of 100 mV/rad and is related to the mixer conversion losses which must be small. Since the output of the mixer decreases with increasing frequency, the same will apply to the active method. On the other hand, the ultimate sensitivity requirements become less critical, since the oscillator noise increases with frequency. A pre-amplifier, often integrated with the mixer, ensures a suitable power level at the output of the phase detector (a few dBm). This system can ensure phase locking of microwave oscillators, or RF oscillators, which will be VCOs in this case. The phase noise of the RF oscillator is additive to the residual noise of the measuring system. Its noise spectrum must be smaller than that of the microwave source. This condition is generally true even with a VCO if the analysis band width is not too large.

The ultimate sensitivity of the system is measured using a source of excitation adjusted to a suitable level, and by constructing a symmetric circuit that rejects source noise. In the direct detection method, a single source drives the two arms of the phase detector, whilst in the IF device, two symmetric circuits are used with two pre-amplifying microwave mixers driven by two identical oscillators which then supply the same intermediate frequency. An RF detection system delivers the noise signal to be analysed and another identical system supplies the correction voltage for the locking loop of the microwave oscillator (Fig.4.18).

Oscillator rejection depends on the symmetry of the system. When an IF pre-amplifier is employed, the phase-frequency relation is specific to each model and causes a transformation of the oscillator frequency noise into phase noise. To get to the residual noise of the stage, we have to be at maximum rejection, which is observed by modulating sinusoidally the frequency of one of the microwave oscillators operating in the locked mode. We can vary the beat frequency of the IF signal so as to obtain the minimum amplitude of the modulation line in the spectrum analyser. Oscillator noise rejection close to 30 dB has been achieved.

Figure 4.19 ilustrates residual noise measurement by the active method in the K band (at 24 GHz). The direct comparison method reveals a wide noise plateau ($S_\Phi = -155$ dB) extending to the upper limit of the range of analysis (100 kHz). Flicker noise appears at low frequencies and the cutoff frequency occurs at 2 kHz. The power level at the output of the microwave

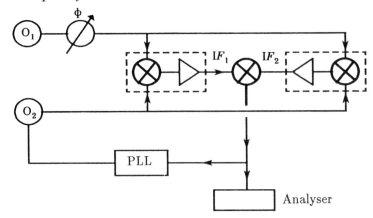

Fig. 4.18 Measurement of phase noise by the intermediate frequency method.

mixer is 7 dBm. In the intermediate frequency method, the phase noise level lies at -150 dB and flicker noise is the same as before. As the power of the signal under analysis decreases ($P \ll 1$ dBm), the ultimate sensitivity of the measuring system also decreases.

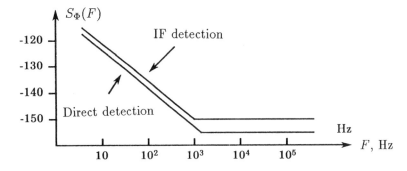

Fig. 4.19 Noise in the direct detection system and in the IF system.

The intermediate frequency system has the advantage of retaining high sensitivity up to input power of the order of 20 dBm (Fig. 4.20). This is attributed to the output pre-amplifier of the mixer, which provides a high level of phase detection. One could try to increase the strength of the signal under test by direct amplification, but the resulting gain is often only apparent, since the characteristic additive noise of the microwave amplifier tends to increase the measured noise level. Passive systems with frequency-to-phase conversion (Section 4.2.2) and active systems both have amplifiers at the phase detector outputs. Assuming that these devices are identical from the point of view of detection and amplification characteristics and of noise in the two circuits, and are characterised by a residual noise spectrum

$S_{\Delta V}(F)$, we find that the minimum phase noise spectrum is

$$\left(S_{\Delta\Phi}^{\min}\right) = \frac{S_{\Delta V}}{\mu^2} \tag{4.52}$$

for an active system and

$$\left(S_{\Delta F}^{\min}\right) = \frac{1}{F^2}\left(S_{\Delta\Phi}^{\min}\right) = \frac{1}{F^2}\left(\frac{f_0}{2Q}\right)^2 S_{\Delta V} \tag{4.53}$$

for a passive cavity system. The noise of the active system at the limit of its working band is the same as the lower limit of the noise in the passive system.

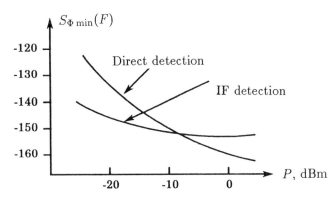

Fig. 4.20 Sensitivity of active circuits as a function of signal power.

We can now see (Fig. 4.21) the advantage of the active method for the analysis of phase noise near the carrier. It is best suited to the study of the phase noise of microwave oscillators with high spectral purity in the analysis range between a few hertz and several dozen or several hundred kilohertz.

The passive resonator system, with its lower sensitivity, but a wider bandwidth, is in general reserved for the testing of medium-quality oscillators.

We will conclude by mentioning a composite method that employs an active standard source, whose function is frequency translation, and a passive demodulation system as before. This arrangement can be very useful for testing oscillators in the millimetre range. Actually, it is not really possible to fabricate suitable high-Q cavities ($Q > 10\,000$) for resonant frequencies near 60–100 GHz. We can then advantageously use the heterodyne technique, by transferring the analysis to an intermediate band which can still be in the microwave range, in the L and X bands (say, 1–10 GHz).

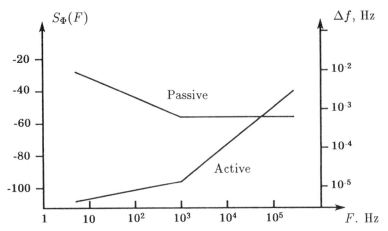

Fig. 4.21 Residual noise in active and passive circuits as a function of frequency.

4.3.2 Synthesis of microwave frequencies

The active method for testing microwave oscillators requires a standard frequency that is as spectrally pure and as stable as possible (better or at least equal to those of the source under test). There are no metrological oscillators with a preferred whole-number frequency, which can be the case at radio-frequencies. On the other hand, we may have to test an oscillator that represents the *state of the art* from the metrological point of view at any frequency, and it may not be possible to make a tunable standard source for the particular frequency range with at least the same spectral qualities as the oscillator. The solution is to use an intermediate frequency in the radio-frequency range. It is at this level that fine tuning will allow the demodulation of the oscillator signal, using the arrangement of Fig. 4.22, which can incorporate a radio-frequency synthesiser to perform the tuning. Frequency division can then be used to bring the spectrum of this interpolation signal below the spectrum of the oscillator under test (*cf.* Section 2.4).

We therefore have to produce a very stable standard microwave frequency that has high spectral purity and can be varied in steps of Δf in the range up to a few gigahertz. In the above arrangement, the maximum frequency step is $2F_i$. The phase noise level in the RF synthesisers operating up to 1 GHz is about -120 to -130 dB (0 dB = 1 rad^2/Hz). By dividing this synthesised signal by 10, we obtain a gain of 20 dB, sufficient to satisfy the specification of a very good microwave standard with phase noise level of about -150 dB. We thus obtain Δf (equal to $2F_i$) of $100 - 200$ MHz; the synthesised frequency f_s ($= 10F_i$) runs from 500 MHz to 1 GHz. In this Section we will consider the synthetic generation of discrete standard frequencies in the microwave range; in the next Section, we will examine

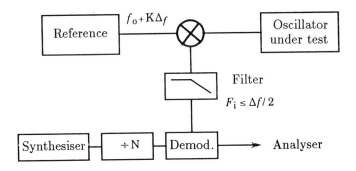

Fig. 4.22 Interpolation using intermediate radio-frequency.

the noise spectrum of these synthesised signals.

Synthesis by direct multiplication consists of using a multiplication chain that yields microwave frequencies that are multiples of the frequency f_0 of a standard oscillator. A working system will generally have several such stages. The nonlinear active elements are diodes (varactors, SRDs avalanche diodes and so on) driven at a very high level of excitation. This often requires a multiplication chain and pre-amplification at each stage, but a high multiplication factor must be avoided. A reasonable limit is $M_{max} = 5$. When the output signal from the multiplier, e.g., from the last multiplication stage, is used for locking a local oscillator, the power constraints are less critical and a high multiplication factor can be employed. *Comb* generators, incorporating the step recovery diodes (SRD), described in Chapter 2, can also be used in the lower part of the microwave range. Figure 4.23 shows the line intensifies at the output of such a multiplier, driven by a 1-GHz signal. Acceptable amplitudes ($> -20\,\mathrm{dBm}$) are obtained up to the K band. The noise level of signals synthesized by this multiplication process is the same as the noise of the reference signal times the multiplication factor (at least for perfect multiplication).

The microwave frequency f_0 delivered by a phase-locked oscillator can be divided by a factor N to produce a signal of the same frequency f_Q as a reference oscillator. The latter can be, say, a VHF quartz oscillator (50–200 MHz). The phase detector then produces the phase-locking signal for the microwave source. The frequency of the phase-locked oscillator is actually limited by a lack of commercial logic dividers beyond 2 GHz. A mixed system containing $\times M_0$ multiplication is shown in Fig. 4.24. The possible signal increment is at most equal to f_Q, and becomes $M_0 f_Q$ after multiplication. It can also be reduced by division of the oscillator frequency f_Q (by N_3). Synthesis by mixing employs multipliers, dividers and mixers with one or more phase-locked loops. Comparison with Fig.4.24 will show that, for example, mixing in the division chain can be used to multiply the

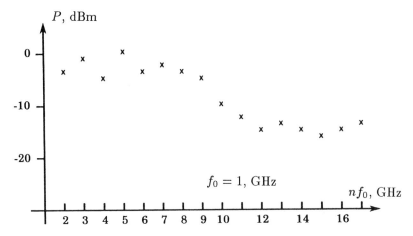

Fig. 4.23 Power level of lines produced by a comb frequency generator.

frequency $M f_Q$ of the VHF reference oscillator by M. The division chain then includes successive divisions by N_1 and N_2 (Fig. 4.25).

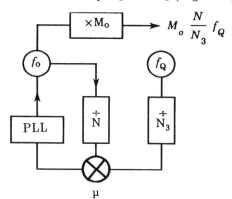

Fig. 4.24 Frequency synthesis by simple mixing.

The advantage of introducing the mixer becomes apparent when we consider the contribution of the frequency dividers to the phase noise of the microwave oscillator. Within the noise spectrum of the phase-locked oscillator, the noise of the dividers is multiplied by N^2, and this contribution is often predominant if the total division factor N is high. An incremental step of $1\,\mathrm{MHz}$ at $f_0 = 2\,400\,\mathrm{MHz}$ will need a divider with $N_3 = 50$ and $f_Q = 50\,\mathrm{MHz}$. The division factor becomes $N = f_0/\Delta f = 2\,400$, and the characteristic noise is increased by $67\,\mathrm{dB}$ in the circuit shown in Fig. 4.24. By using the circuit incorporating a mixer (Fig. 4.25), the same problem is solved by taking $N_1 = 10, N_2 = 10, M = 5$ and $N_3 = 50$. The increase in the divider phase noise is then equal to $20\log \mathrm{N_1 N_2}$, which amounts to $40\,\mathrm{dB}$. If we consider white phase noise with $S_\Phi = -155\,\mathrm{dB}$, produced for carrier frequencies between 10 and 100 kHz, we obtain a noise level of -95 dB for the microwave oscillator in the first case. For the second circuit, $S_\Phi = -120$ dB, which already is more acceptable for an oscillator operating in this frequency band.

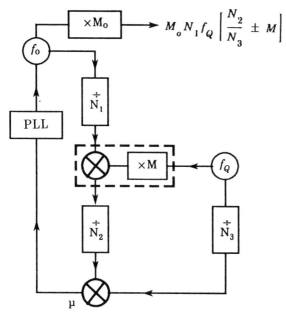

Fig. 4.25 Frequency synthesis with double mixing.

4.3.3 Phase spectrum of synthesised microwave signals

We now turn to the contribution to the phase noise spectrum of a microwave oscillator due to the different sources of noise in circuits containing mixers, dividers and multipliers. The structure is shown schematically in Fig.4.25. It is the phase-locked loop that was analysed in Chapter 2. The spectrum of the microwave phase-locked oscillator can be scanned in steps of $\Delta f = f_Q/N_3$ and is found to be

$$S_\Phi^a = \frac{S_\Phi^L}{|1 + G/pN_1N_2|^2} + \left(\frac{G}{p}\right)^2 \frac{S_{\Phi 1}/N_2^2 + S_{\Phi 2}}{|1 + G/pN_1N_2|^2} \qquad (4.54)$$

where S_Φ^L is the phase spectrum of the free microwave oscillator, $S_{\Phi 1}$ and $S_{\Phi 2}$ are the phase spectra of the VHF reference oscillator f_Q after multiplication by $M(S_{\Phi 1})$ or division by $N_3S_{\Phi 2}$, $G(p) = Kg\mu H(p)$ is the gain of the phase-locked loop μ (v/rad) is the characteristic of the phase detector, K (rad/volt) is the frequency control sensitivity of the microwave oscillator, and G and $H(p)$ are, respectively, the gain and the transfer function of the amplification loop. By introducing $G_T = G/pN_1N_2$ and the spectrum S_Φ of the oscillator generating f_Q, we obtain

$$S_\Phi^a = \frac{1}{|1 + G_T|^2} S_\Phi^L + \frac{N_1^2}{|1 + 1/G_T|^2} \left[M^2 + \left(\frac{N_2}{N_3} \right)^2 \right] S_\Phi \qquad (4.55)$$

This result was obtained by assuming a perfect process of division and multiplication. If we introduce the phase noise spectra of the dividers, S_{d1}, S_{d2}, S_{d3}, the additive noise of the multiplier S_M and represent the electronic noise by a source of noise V_n in the phase-locked loop, we find that the spectrum of the microwave oscillator becomes

$$S_\Phi^a = \frac{1}{|1 + G_T|^2} S_\Phi^L + \frac{N_1^2}{|1 + 1/G_T|^2} \left(S_{d1} + S_{M1} + M^2 S_\Phi \right)$$

$$+ \left| \frac{N_1 N_2}{1 + 1/G_T} \right|^2 \left(S_{d2} + S_{d3} + \frac{V_n^2}{g_v^2 \mu^2} + \frac{1}{N_3^2} S_\Phi \right) \qquad (4.56)$$

where g_v is the gain between the phase detector output and the noise source V_n.

The different spectra that appear in the above expression can be represented by models, and the relationship can then be processed by a computer. The phase noise of the frequency dividers plays a very important part in the spectrum of the phase-locked microwave oscillator, since it is multiplied by $(N_1 N_2)^2$. The spectrum of the ECL frequency dividers, which comes into play at high frequencies, has a phase noise plateau between $-150\,\mathrm{dB}$ and $-160\,\mathrm{dB}$. The spectrum of TTL logic dividers at low frequencies lies at $-165\,\mathrm{dB}$ (Fig.4.26). If the noise level of the phase-locked oscillator is to lie below $-120\,\mathrm{dB}$, the product $N_1 N_2$ must be less than 100. This can be achieved by a suitable choice of the multiplication factor M.

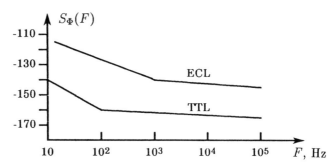

Fig. 4.26 Noise spectrum of frequency dividers.

Apart from their specific effect on the spectrum, frequency multipliers introduce an additive noise that is significant in the domain of flicker noise. Nevertheless, this adds no more than a few dB (between 1 and 5) to the theoretical spectrum; it can be more important when multiplying signals with a high noise level, but this is not really the case for the signals used in the metrology of frequencies.

The intrinsic noise of the electronic phase lock is essentially due to the pre-amplifier after the mixer. It is generally higher than the characteristic noise of the latter. In the expression for S_Φ^a, the electronic noise is multiplied, like divider noise, by the factor $N_1 N_2$. It is therefore essential to use a phase lock with very low noise.

As an example, Fig. 4.27 shows the spectrum of a signal synthesised at 2 522.5 MHz from a VHF reference oscillator (60 MHz) in steps of 500 kHz. It is clear that the main contribution is provided by divider and electronic noise, limited to less than -115 dB after 1 kHz. This relatively modest performance at 2.5 GHz is due to the small frequency step used in this example.

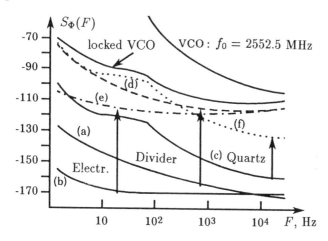

Fig. 4.27 Spectral density of phase fluctuations for a synthesised signal. Sources of noise and their respective contributions to the synthesised signal: divider (a, d), electronic (b, e) and quartz oscillator (c, f).

Better performance can be obtained with a comb generator and additive synthesis at microwave frequencies, avoiding the need for high multiplication or division factors. We again take the microwave standard frequency f_0 from a quartz or surface-wave VHF oscillator whose spectrum at 1 GHz, cleaned up near the carrier by locking to a metrological oscillator at 5 or 10 MHz, has the characteristic shown in Fig. 4.28a. From then on, we can use comb generators to cover the microwave range up to 18 GHz. The spacing between harmonics can be reduced by inserting a line spectrum with, say, 100 MHz spacing, obtained by division and re-injection into the circuit before multiplication (Fig. 4.28b). The power associated with each spectral line is obviously very small. Filtering is improved by using an oscillator locked to each of these lines; a YIG oscillator is perfectly suited to this purpose.

By introducing an RF interpolation signal into the phase-locked up loop of the YIG, it is possible to obtain a microwave reference frequency corresponding to the spectrum shown in Fig. 4.29. The noise of the interpolation signal is additive in the circuit shown in Fig. 4.30. It can be supplied by a VHF synthesiser after division by N (the resulting gain is $20 \log N$). The YIG oscillator at frequency $n f_0$ has practically the same relative phase noise spectrum as the microwave standard. This signal is therefore perfectly suited for translation to the radio-frequency range of the microwave oscillator under test. The analysis will be carried out in this domain by the classical methods described elsewhere.

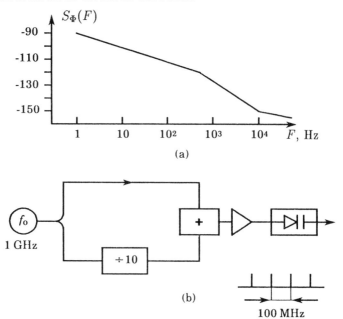

Fig. 4.28 Spectrum of 1-GHz quartz oscillator (*a*) and generation of a microwave comb spectrum with 100-MHz spacing (*b*).

4.4 COMPARISON OF OPTICAL FREQUENCIES

4.4.1 The laser oscillator

We saw in Section 4.2.5 that an atom can be compared to a resonant circuit.

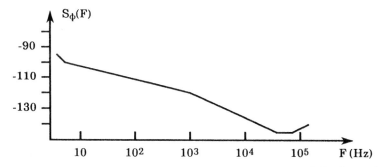

Fig. 4.29 Spectrum of YIG oscillator locked at 2 GHz to one of the lines of the microwave comb (*cf.* Fig. 4.28).

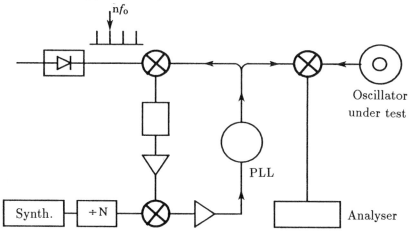

Fig. 4.30 Generation of a standard frequency for testing microwave oscillators.

It can emit an electromagnetic wave of frequency f (wavelength $\lambda = c/f$) in the coherence time τ. The wavelength can be compared with a standard length in an interferometer so long as the optical path difference over which interference takes place is less than $c\tau$. This allows the measurement of τ and, hence, of the line widths. The wavelength λ is obtained with the uncertainty

$$\frac{\Delta\lambda}{\lambda} = \frac{1}{f\tau} = \frac{\lambda}{c\tau} = \frac{\Delta f}{f} \tag{4.57}$$

Since c is fixed by convention, the relative uncertainty in λ is the same as in frequency. The usual order of magnitude of $\Delta\lambda/\lambda$ is 10^{-6}, so that τ is of the order of 10^{-9}s. It depends significantly on the environment (*cf.* Section 4.2.5.).

To measure the frequency directly, we have to have at our disposal feedback oscillators because the duration of the operation is longer than τ.

This assumes that we can supply energy to the atom in order to re-excite it whilst preserving its phase and thus maintain the oscillations. The phenomenon of stimulated emission, discovered by Einstein in 1917, provides the basis for this operation. An atom can be excited from the ground state by supplying it with well-defined energy W. It then returns spontaneously to the ground state by emitting a photon of energy $W = hf$. This becomes more frequent as the radiation frequency increases. On the other hand, when the excited atom interacts with a photon of frequency f, emitted by another atom, there is a high probability that it will be de-excited by emitting another photon and returning to the ground state. This is *stimulated emission* (Fig. 4.31). The final electromagnetic wave carries the energy $2W$ (the energy of the incident photon plus the energy of the emitted photon). The two photons must therefore be phase-coherent. The incident photon imposes its phase on the stimulated photon, the energies add and we have amplification.

The first spontaneous emission that takes place in a solid medium produces a photon with an arbitrary initial phase. All subsequent stimulated emissions, on the other hand, have *the same* phase. We thus have both temporal coherence, as in any oscillator, and also spatial coherence. The propagation of the phase-transporting photons forces the atoms to emit in phase. Unfortunately, spontaneous emission constantly takes place, whereby some atoms undergo spontaneous de-excitation, producing photons of arbitrary phase. The result is a perturbation of coherence, with the constant spontaneous emission playing the part of noise that causes phase fluctuations.

As we have seen, the interaction of photons with excited atoms produces stimulated emission and amplification. However, when a photon interacts with an atom in the *ground state*, it is absorbed as it excites the atom. This means that, to achieve global amplification in a solid medium with a very large number of atoms, the medium must contain more atoms in the excited state than in the ground state. This is not so in thermodynamic equilibrium in which the solid medium normally is. The medium has to be prepared by external action in order to alter its populations. This can be done by the selection of state (Stern, 1920) or by optical pumping (Kastler, 1950). These techniques will be described in Chapter 6. We now know how to produce states such as those indicated in Fig. 4.32.

Consider a medium consisting of atoms that emit at frequency f in a Fabry-Perot cavity resonating at the same frequency. If we can invert the population, we can produce an amplifying medium. When the absorption coefficient α of the medium (which must be less than 0 for amplification) reaches the value

$$\alpha_m = -\frac{1 - |\rho|^2}{2l} \qquad (4.58)$$

the Q factor of the system given by (4.44) becomes infinite and the oscillator locks and begins to generate. This is *L*ight *A*mplification by *S*timulated

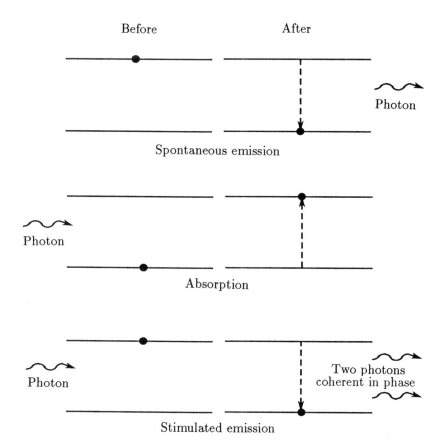

Before   After

Photon

Spontaneous emission

Photon

Absorption

Photon

Two photons
coherent in phase

Stimulated emission

Fig. 4.31 Three types of interaction between photon and atom.

*E*mission of *R*adiation, i.e. the laser. We have known how to build such systems since 1960.

The characteristic frequency of a group of atoms is not as well-defined as the above discussion might imply. Actually, we have to take into consideration the line width of the atom due to both its structure and, especially, perturbations by ambient atoms (Section 4.2.5.). The photons do not all have exactly the same frequency. They can nevertheless interact with one of the many atoms in a group if, for example, the velocity of the latter causes a Doppler effect within the frequency difference. The amplification coefficient α of the medium has a width $\Delta f \approx 1/\tau$ around f_0 and is negative if the population is inverted. To lock the oscillator, it is sufficient to have one of the lines of the Fabry-Perot cavity in the zone indicated in Fig. 4.33. These lines are spaced at $v/2l$ and are therefore very close together.

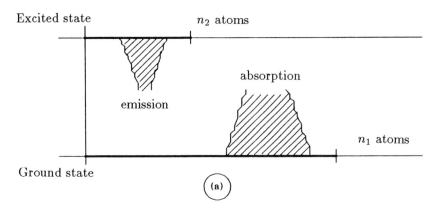

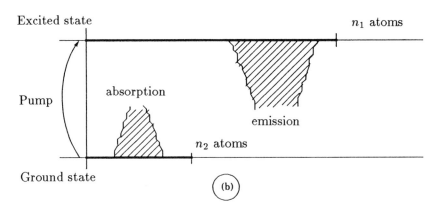

Fig. 4.32 Absorbing and amplifying media: (a) absorbing medium with $n_1 > n_2$, which is the special case of thermodynamic equilibrium; (b) amplifying medium with $n_2 < n_1$, which requires population inversion.

Several of them will satisfy the locking condition if

$$\frac{V}{2l} < \frac{1}{\tau} \tag{4.59}$$

where V is the phase velocity of the wave in the medium.

We can avoid this highly cumbersome phenomenon of multimode oscillations ($cf.$ Section 5.6) if the cavity walls are coated with a multilayer film (Section 2.4.4.). Instead of filling the cavity with media of impedance Z_1, we use an alternation of layers of impedance Z_1 and Z_1', and thickness $\lambda/4$ at the frequency f. The layer thicknesses are thus different if the phase velocity is not the same for Z_1 and Z_1'. The amplification coefficient α_m needed for locking becomes all the more important as $|\rho|$ becomes smaller, i.e., as we move away from the frequency corresponding to the $\pi/4$ layer thickness. A few dozen layers are required at this frequency to achieve $|\rho|$ of the order of 0.999, which is necessary for locking with a reasonable coefficient α_m. The multilayer coating acts as a *filtering* mirror, highly efficient at f and poorly reflecting at neighboring frequencies; its thickness amounts to a few dozen wavelengths. This is feasible in the visible ($\lambda = 0.5\,\mu\text{m}$) and the near infrared ($\lambda = 10\,\mu\text{m}$) ranges, but not in the microwave ranges in which a *mirror* of 50 quarter-wave layers would be necessary.

Once the oscillator locks, it enters a permanent régime. The Fabry-Perot line to which the oscillation is locked is theoretically a Dirac pulse at the frequency f. Spontaneous emission creates a noise background that broadens the the line. The discussion is similar to that of Section 3.4, and this is the way that oscillators operate in the optical spectrum. We will have the opportunity to return to these systems in Chapter 6.

4.4.2 The stabilised laser as a secondary frequency standard

The laser oscillation frequency f depends on the length of the cavity and the phase velocity V which in turn depends on the properties of the medium. It fluctuates under the influence of mechanical vibrations, temperature and gas pressure. The laser locks within a frequency band of the order of the natural line width (Fig. 4.33). If the oscillator line is extremely narrow, which is normal, its position will constantly vary. It is then impossible to envisage frequency measurements, which are necessarily quite long, without stabilising the characteristic frequency of the cavity during this interval of time in order to eliminate these fluctuations.

One of the mirrors defining the cavity is mounted on a piezoelectric transducer, driven by a slowly varying voltage V_0 which changes the length l. At any instant, we can thus achieve the compensation of frequency fluctuations due to perturbations, and therefore stabilise the frequency of the laser. This presupposes that the frequency variations can be easily detected by comparison with a frequency standard, and that a suitable electronic circuit is available to generate the control voltage.

Another Fabry-Perot cavity can be used as the comparison standard. The line is naturally narrow since $|\rho|$ has been adjusted to be as close as possible to unity at the particular frequency. The standard is placed in vacuum; it is not a laser, so that there is no need for an amplifying medium, i.e., $V = c$ and $\alpha = 0$. It is placed in a temperature-controlled enclosure and is mounted on supports designed to minimise vibrations. In the limit, the residual vibrations of the mirrors are due to thermal motion.

The above frequency standard is artificial and very difficult to build and to reproduce. It is much more convenient to use a natural reference provided by a characteristic point on an atomic resonance line. The frequency does not then fluctuate, but the line is broad and frequency conversion is not be easy to perform with precision (the maximum of a broad line and

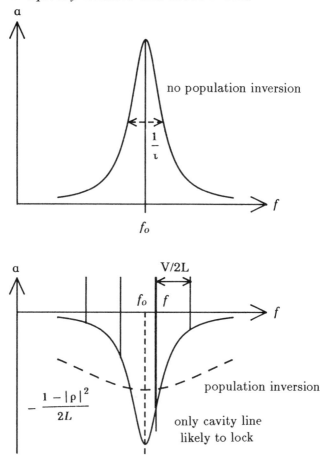

Fig. 4.33 Locking of the oscillations of atoms to a Fabry-Perot line in a cavity of length L.

the point of inflection are not easily found). Fortunately, nature helps in this operation by creating certain useful features on spectral lines, the most frequently employed being due to the saturation of absorption. The intensity of a laser beam is very high and, whilst absorption is described simply by linear laws at low intensities, its behaviour is much more complicated at high intensities. Indeed, the medium is saturated by a fraction of the incident intensity. All the atoms are then in the excited state, and the remainder of the light passes freely through the medium, since absorption is saturated. The additional line corresponding to the saturation effect is much narrower than the original spectral line and lies in a fixed position on the maximum. Saturated absorption is confined to atoms whose thermal velocities are perpendicular to the laser beam. The Doppler effect, the main cause of broadening, is thus eliminated.

There are other atomic reference points. They rely on the initiation of an atomic process that is as independent as possible of the ambient medium. The trick is to get as close as possible to the elementary concept of *all or nothing*: the energy levels are precisely defined and the photons are detected by the process of *all or nothing*. The slightest frequency fluctuation will prevent the interaction between the photon and the atom. Such techniques rely on the use of molecular beams and reduce the Doppler effect and the cascade effect in which several photons take part. They will be discussed in Chapter 6.

The laser has to be coupled to a standard frequency f_0. The use of a natural standard is only possible for lasers whose frequency is very close to the saturated absorption line. Stabilised lasers provide secondary standards whose frequencies are distributed randomly, depending on a coincidence between the lines of a substance A, easy to lock to a laser because we know how to achieve population inversion, and the lines of a substance B with a strong saturated absorption line; fortunately these coincidences are frequent. To achieve frequency stabilisation, the high-intensity laser light is very slowly frequency modulated by applying a voltage

$$V = V_0 + \varepsilon \sin 2\pi Ft \qquad (4.60)$$

to a transducer (F is of the order of 1 kHz). This light is then sent to the frequency standard (Fig. 4.34).

The output signal $v(t)$ is amplitude modulated after frequency-to-amplitude conversion by the standard. Synchronised detection is then used to extract from this amplitude variation the term of frequency F, which is used to establish the mean control voltage V_0. For example, if V_0 is such that the laser frequency f is equal to f_0, the amplitude-modulated signal from the standard contains only the frequency 2F which is eliminated by the synchronous detector (Fig. 4.34). The control circuit must therefore generate a voltage V_0 regulated by the output of the synchronous detector, which will isolate the signal of frequency F, if it exists; V_0 remains constant as long as this control signal is zero. The appearance of an output signal causes a variation of V_0, leading back to the previous state. The very low frequency fluctuations of the control voltage V_0 are easy to analyse and can be directly related to the frequency fluctuations of the laser. Figure 4.35 shows a few examples.

CO_2 lasers can oscillate in 300 lines (if we use the range of C and O isotopes) between 26 and 32 THz. They can be stabilised on each line by saturated absorption in CO_2. The performance attained is nevertheless limited (relative stability of 10^{-12} over 100 sec; reproducibility 10^{-10}). The OsO_4 molecule offers numerous coincidences near 29 THz with the CO_2 laser lines, which results in much better performance (10^{-14} over 100

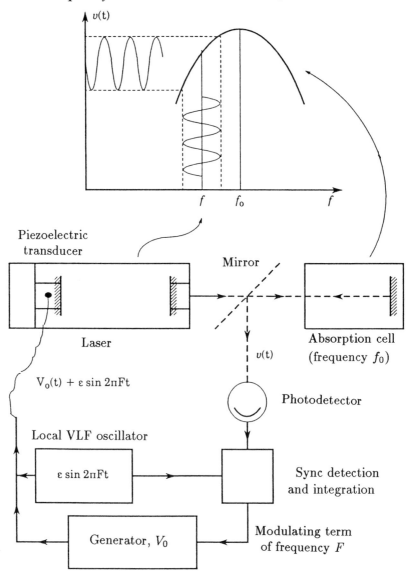

Fig. 4.34 Principle of laser frequency stabilisation.

sec). The HeNe laser operating at $3.39\,\mu$m was the first stabilised laser. It was used in determinations of the speed of light. The reproducibility of compact lasers appears to be limited to 10^{-9}. Higher performance devices, using the E line of methane, offer a reproducibility close to 10^{-12}. HeNe(CH_4) lasers have achieved short-term stability of the order of 10^{-15} for one-second integration. HeNe lasers operating at 633 nm and 612 nm,

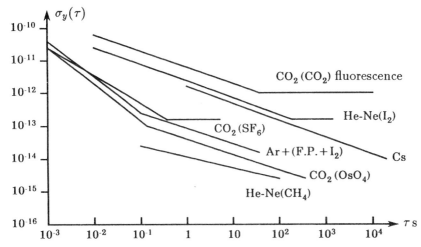

Fig. 4.35 Relative frequency fluctuations of different stabilised lasers (the stability of a cesium clock is also shown for comparison).

and the argon laser stabilised by saturated absorption in iodine, are used in distance metrology and in high-resolution spectroscopy. Devices employing an absorption cell outside the laser cavity have a reproducibility close to 10^{-12}; with an internal cell, a reproducibility of 10^{-10} can be obtained. Any laser emitting between 500 nm and 650 nm can be locked to an iodine transition; this property could be exploited to produce a series of reference frequencies in the visible range.

4.4.3 Comparison of stabilised laser frequencies

Whenever possible, stabilised lasers are based on one of the processes just described. A heterodyne chain that mixes oscillations in a nonlinear circuit, and produces composite frequencies, can then be constructed and filters can isolate low frequencies that can then be processed by methods suited to the microwave range. The optical frequencies can thus be related to the standard caesium clock.

The main difficulty lies in making nonlinear mixers that receive signals with adjacent frequencies f_1 and f_2 and generate a signal of frequency $|f_1 - f_2|$. Such devices are also found to generate harmonics: driven by a signal of frequency f, they produce frequencies mf. The signal to noise power ratio of these harmonics must be high enough for measurement, and this limits the order of multiplication that can be used. The performance of currently available nonlinear devices can be used to divide the spectrum into a number of regions. Figure 4.36 illustrates the relative importance of harmonics obtained with a Schottky diode at 70 and 100 GHz.

Consider a photodiode receiving two laser beams with frequencies f_1 and f_2. Its nonlinear properties and its small bandwidth will ensure an

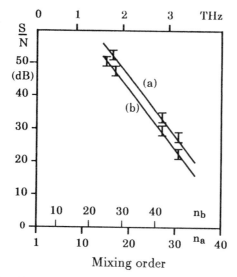

Fig. 4.36 Signal to noise ratio in the 100–kHz band as a function of the mixing order obtained with a small Schottky diode: (*a*) incident frequency 100 GHz, (*b*) incident frequency 70 GHz.

output current at the difference frequency. The system functions if f_1 and f_2 lie between 400 and 600 THz and if $|f_1 - f_2|$ is less than 100 GHz. A nonlinear crystal, driven by the frequency f, generates usable harmonics as long as f and $2f$ (limiting ourselves to the second harmonic, which is the only practical case) are in the pass band. For $LiNbO_3$, this occurs between 100 and 500 THz. Such devices can be used to construct a chain linking three stabilised lasers in optical and near infrared ranges (Fig. 4.37).

In this example, $f_1 = 130$ THz is generated by a dye laser emitting near 2.3 mm. It is multiplied by 2 and is mixed in a photodiode with the frequency f_2 of a HeNe laser, which is of the order of 260 THz. The low frequency output signal from the photodiode allows the accurate measurement of $|2f_1 - f_2| = F_1$. The frequency f_2 is also multiplied by 2 and is mixed in another photodiode with a signal of frequency f_3 of the order of 520 THz from another dye laser. The quantity $|2f_2 - f_3| = F_2$ can then be measured accurately. In this way, we can connect lines lying between 100 THz and the visible range, provided a sufficient number of stabilised lasers is available. This is where the difficulty lies because low multiples of a frequency (2 or 3) have to coincide with another frequency within 100 GHz and the composite signal has to be processed by microwave techniques at F_1 and F_2.

Metal-insulator-metal diodes (Fig. 4.38) can be used to mix frequencies or generate harmonics between about 1 and 100 THz. They consist of a tungsten point resting on a thin oxide layer deposited on a metal substrate

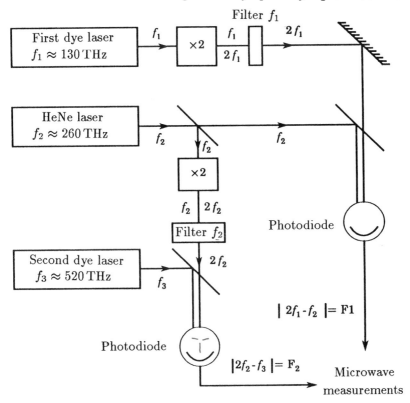

Fig. 4.37 Measuring chain linking optical and microwave ranges.

(nickel, cobalt, niobium...). The formic acid laser emits near $f_1 = 716\,\text{GHz}$ (in the far infrared) and an alkali metal laser emits near $f_2 = 4.25\,\text{THz}$. An optical system focuses the beams on the diode junction. The composite frequency $|6f_1 - f_2|$ must be small. It is mixed with f_3 emitted by a stabilised klystron until the generated frequency $|6f_1 - f_2 - f_3|$ lies in the radio-frequency range in which counting methods can be used (f_3 must be of the order of 45 GHz). An arrangment of the type shown in Fig. 4.38 will therefore allow the linking of frequency sources within a band between a few hundred gigahertz and a few dozen terahertz. Stabilised microwave generators provide additional standards that facilitate such measurements, by increasing the number of reference points provided by the (rare) stabilised lasers.

The low-frequency output of the MIM diode can also be used to stabilise the frequency of a laser by locking it to an harmonic or subharmonic of a stabilised laser, following the method described previously. Saturated absorption of an OsO_4 line can be used to stabilise a CO_2 laser ($f_1 \approx 30\,\text{THz}$). The methyl alcohol laser operates near $f_2 \approx 425\,\text{THz}$ and the

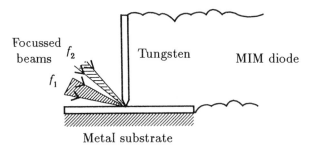

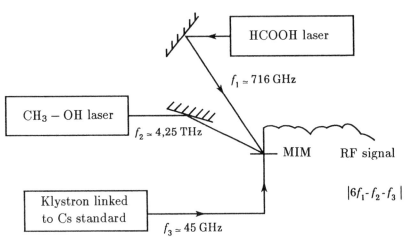

Fig. 4.38 Principle of frequency measurement between 1 and 10 THz.

formic acid laser near $f_3 \approx 716\,\text{GHz}$. The chain illustrated in Fig. 4.39 generates a signal of approximately 8 GHz with the aid of the MIM diode which produces the combination $f_1 - 7f_2 + f_3$.

A frequency-stablized klystron can generate a low-frequency signal with the help of a heterodyne. This can then be used to stabilise one of the two lasers, usually the one with the higher frequency. This will stabilise a formic acid laser on one of the harmonics of a stabilised klystron, using the 9th harmonic (from a Schottky diode) of a 7957-GHz klystron tied to the caesium clock. This chain is illustrated in Fig. 4.39. It allows the measurement of the CO_2 laser frequency stabilised by a saturated-absorption line of OsO_4 ($f = 29\,\text{THz}$).

4.4.4 Applications of frequency measurement in the optical range

The measuring chains described in the last Section can be used to measure the frequency of a stabilised laser. As technology advances, the frequency spectrum acquires an increasing number of reference of points, and the

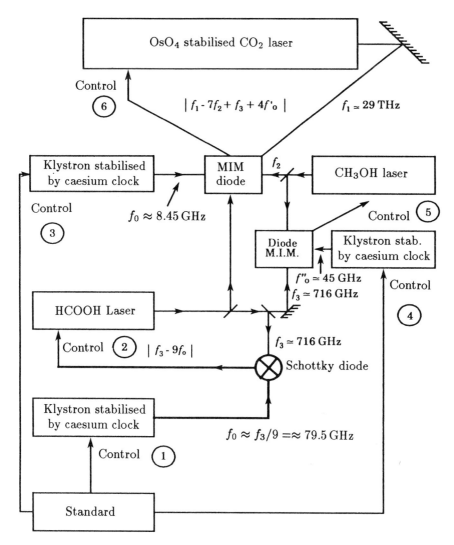

Fig. 4.39 Measurement of the frequency of a CO_2 laser stabilised by OsO_4 ($\approx$ 29 THz).

design of such chains becomes easier. However, amplification is not yet possible in the optical spectrum because lasers do not produce sufficient power. We are therefore led via the heterodyne to the analysis of radio-frequency signals to which we can apply spectral and time-domain analysis. We can then reverse the process and return from the spectrum of the LF signal to the spectrum of the signals from which it was generated. This is not easy because the mixing of spectra in a nonlinear circuit is a highly complex process. Much work remains to be done before we can measure the laser line shape.

The development of optical analogue telecommunication will require detailed knowledge of carrier frequencies, and this will constitute an application of the above methods of measurement. However, it is in the domain of distance metrology that laser frequency measurements are particularly interesting. The laser line is much narrower than that of the atom from which it originates. The coherence time is much longer. Moreover, we can make an interferometer with a large path difference and use it to measure long distances (100 m with an uncertainty of 0.01 mm). We can also measure the Doppler effect during the displacement of one of the two interferometer mirrors, and use integration to deduce the distance travelled. But we must be aware of the fact that the phase velocity in the atmosphere is not c but c/n, so that it is also necessary to measure the refractive index n.

The precision of an interferometer is inversely proportional to the path difference between interfering beams. The above technique will therefore allow very high resolution spectroscopy as compared with non-coherent sources. Indeed, we can explore in detail the shape of emission lines. More particularly, we can carry out this work on the krypton emission line which is remarkably narrow. This line was previously the primary standard of length, but has now become a secondary standard since c was fixed by convention. It is nevertheless a useful tool in the calibration of interferometers working in the visible range ($\lambda_{Kr} = 0.5651289 \pm 4 \times 10^{-7} \mu$m).

5

Classical frequency standards

5.1 REVIEW OF RESULTS

We measure a quantity by comparing it with another quantity of the same nature, chosen as a standard. In previous chapters we described the methods available for frequency comparison in different spectral ranges, and also the applications of such measurements. In several instances (Sections 1.4 and 1.5), we gave simple information about oscillators that seemed stable enough to be used as standards. We shall now collect together these pieces of information and then specify the nature of the problems that have to be solved before we can build a system supplying stable frequency. We will raise this oscillator to the status of a *standard* by describing its oscillation in terms of a law as rigorous as possible, which will define the nature of the standard. All other oscillators will then be described by more approximate laws established by the methods described previously.

To establish whether an oscillator can be chosen as a standard, we construct several examples of it, identical in principle, and we constantly compare them by the methods described earlier, assuming that the statistical properties of their fluctuations are the same for all of them (*cf.* Section 1.1.3). This yields the spectrum and variance of the oscillator. We will choose as standard the oscillator whose variance has the longest plateau, smallest ordinate and best reproducibility.

An oscillator is a high-grade resonator whose losses are compensated by an electronic circuit (Sections 1.4 and 1.5). It is subject to fluctuations and drifts due to external factors. If the models that we have described are reliable, we now understand the mechanism of the different frequency variations, and can try to correct them. The most important component of the oscillator is therefore the resonator: it determines the output frequency. Some resonators can be described by models taken from classical physics. They are mainly LC circuits, piezoelectric resonators and electromagnetic cavities. Others are atoms that must be described by quantum physics. This chapter is devoted to classical resonators; Chapter 6 will deal with atomic resonators.

We shall consider resonator models that are valid for piezoelectric transducers and cavities, and will note the negative effects of nonlinearities and of mode coupling. We will then examine the influence of the feedback system on frequency in the simple but useful case of a resonator with two variables of state. This will allow us to describe the operation of quartz oscillators and the klystron. These are the oscillators whose qualities give them the status of secondary standards. However, resonators made artificially are difficult to reproduce. Atomic clocks, which use natural resonators, are the only devices that can reach the status of a primary standard.

We will continue our discussion of classical oscillators by analysing the effect of an external driver, or of coupling between two modes in multimode operation. We then have to use more than two variables to describe the system. In general, there are several persistent oscillation frequencies. Noise is always present in the circuit and complicates the situation by broadening the spectral lines (Sections 1.5 and 3.4.2). The result is that an oscillation represented by closely spaced but distinct lines can appear as a random phenomenon with a continuous spectrum. In fact, weak noise is restricted to producing random coupling between two adjacent modes and when there is a lot of it, the circuit alternates randomly between one mode and another, and the situation is very difficult to describe.

5.2 PROBLEMS POSED BY THE RESONATOR

5.2.1 The lossless resonator with two variables of state: nonlinear effects

The ideal oscillator is a loss-free resonator that is perfectly isolated. It has a fixed frequency. The simple pendulum is an interesting example with which we begin our discussion; the capacitive circuit is also a useful model (*cf*. Section 1.4). We shall now show that these simple and often effective concepts must be considered in some depth to account for the frequently observed phenomena.

The ideal loss-free resonator is described by a variable v satisfying the equation

$$\ddot{v} + \omega_0^2 v = 0 \tag{5.1}$$

if we can make a linear approximation to the laws of physics. The second derivative is a complex mathematical tool poorly suited to numerical calculations. We prefer to increase the number of variables (and hence of equations), so that all the equations are of the first order. We therefore set

$$q = v \qquad p = \dot{v} \tag{5.2}$$

The state of the resonator is therefore described by two *variables of state* (q, p) that obey the equations of *state*

$$\dot{q} = p, \quad \dot{p} = -\omega_0^2 q \tag{5.3}$$

The behaviour of the system is well represented on the p, q plane by a phase trajectory that is a function of time. In the above simple case, the combination

$$\dot{p}\dot{q} - \dot{q}\dot{p} = p\dot{p} + \omega_0^2 q\dot{q} = \frac{1}{2}\frac{d}{dt}\left(\omega_0^2 q^2 + p^2\right) = 0 \qquad (5.4)$$

has the first integral

$$p^2 + \omega_0^2 q^2 = h = \omega_0^2 q_0^2 + p_0^2 \qquad (5.5)$$

The constant is proportional to the total energy of the system. The phase trajectories are ellipses. For each value of the integration constant h, i.e., for the initial values q_0, p_0, there is a closed phase trajectory, i.e., a cycle (Fig. 5.1) with a period of $2\pi/\omega_0$.

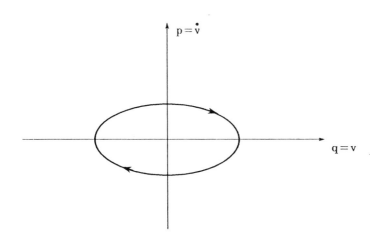

$p = \overset{\bullet}{v}$

$q = v$

Fig. 5.1 Phase trajectory of an undamped linear oscillating system.

A more thorough analysis of the resonator shows that the above description is too simplistic. The laws of physics cannot in general be described correctly by linear relationships.

The simple pendulum discussed in Section 1.4 is an example. Its potential energy is (Fig. 1.6)

$$W_p = mgl\left(1 - \cos\vartheta\right) \qquad (5.6)$$

and its kinetic energy is given by

$$W_c = \frac{1}{2}ml^2\dot{\vartheta}^2 \qquad (5.7)$$

where ϑ is the deflection. Conservation of energy then leads to

$$mgl\,(1 - \cos\vartheta) + \frac{1}{2}ml^2\dot{\vartheta}^2 = h \tag{5.8}$$

or

$$\ddot{\vartheta} + \frac{g}{l}\sin\vartheta = 0 \tag{5.9}$$

The nonlinear equations of state are

$$\dot{q} = p$$

$$\tag{5.10}$$

$$\dot{p} = -\omega_0^2 \sin q$$

where $q = \vartheta, p = \dot{\vartheta}$ and $\omega_0^2 = g/l$.
    The first integral is

$$\frac{1}{2}p^2 - \omega_0^2 \cos q = h \tag{5.11}$$

The complexity of this example derives from the fact that the potential energy is not a quadratic function of the variable q, but a more complicated expression. The resonator is therefore be described by equations of state of the form

$$\dot{q} = p$$

$$\tag{5.12}$$

$$\dot{p} = f(q)$$

A combination that can be integrated still exists. We have

$$p^2 + F\,(q) = h \tag{5.13}$$

$F(q)$ being a primitive of $f(q)$. In elasticity theory, linearity is only a first approximation (Hooke's law) and nonlinear terms have to be introduced to describe a real mechanical resonator. In electromagnetic theory, the electrical and magnetic properties of matter are described by linear laws, but only to a first approximation. Ferrites, ceramics and so on are really nonlinear phenomena. We therefore have to adopt equations such as (5.12), but we will often be able to expand $f(q)$ around the position of rest $q = 0$ at which the potential energy is a minimum and is conventionally set equal to zero. A realistic approximation that is valid for small amplitudes is given by

$$\dot{q} = p$$

$$\tag{5.14}$$

$$\dot{p} = -\omega_0^2 q + aq^2 + bq^3 \ldots$$

For example, the pendulum is then described by

$$\dot{q} = p$$

$$\dot{p} = -\omega_0^2 q + \frac{\omega_0^2}{6} q^3$$

(5.15)

The fact that the first integral exists shows that there is a cycle, i.e., a periodic régime. However, these cycles are no longer ellipses; the periodic régime is no longer described by a sinusoid, but by a periodic function that can be expanded into a Fourier series. The period and the Fourier coefficients can be calculated for the general case (5.12) by phase mapping methods. We will return to this later.

In practice, we always try to approach the linear régime, the system is devised so that parameters $a, b, ...$ are as small as possible and we work with small amplitudes. We can then proceed by successive approximation. For example, let us consider the pendulum once again. In the first approximation, if we neglect the nonlinear effect (very small amplitudes), we obtain

$$q = \vartheta_0 \sin \omega t \qquad \omega = \omega_0 = \sqrt{g/l}$$

(5.16)

If we retain the term containing q^3 in (5.15), sinusoidal oscillations with pulsatance ω will will be accompanied by the harmonic with pulsatance 3ω. We therefore try as a second approximation, a solution of the form

$$q = \vartheta_0 \sin \omega t + \vartheta_3 \sin (3\omega t + \Phi)$$

(5.17)

where $\vartheta_3 \ll \vartheta_0$. Substituting this in (5.15), we obtain

$$\Phi = 0 \qquad \vartheta_3 = -\vartheta_0^3/192$$

(5.18)

and

$$\omega^2 = \omega_0^2 \left(1 - \frac{\vartheta_0^2}{8}\right)$$

(5.19)

The second approximation thus becomes

$$\vartheta = \vartheta_0 \sin \omega t - \frac{\vartheta_0^3}{192} \sin 3\omega t$$

(5.20)

This result was obtained by neglecting the higher harmonics with pulsatance 5ω or greater. The third approximation consists of three terms with pulsatances $\omega, 3\omega$ and 5ω. The result is

$$\omega^2 = \omega_0^2 \left(1 - \frac{\vartheta_0^2}{8} + \frac{\vartheta_0^4}{1536}...\right)$$

$$(5.21)$$

$$\vartheta = \vartheta_0 \sin \omega t - \frac{\vartheta_0^3}{192}\left(1 + \frac{21}{192}\vartheta_0^2\right) sin 3\omega t + \frac{\vartheta_0^5}{36864} sin 5\omega t$$

This is a better approximation, but we have had to neglect 7ω.... We thus obtain the successive approximations in the form of limited expansions in powers of the amplitude. This most important result is very annoying to the resonator designer because it shows that the frequency is a function of amplitude. This is so for all nonlinear resonators. Constant frequency implies constant amplitude. We must also remember that a nonlinear resonator will generate some inconvenient harmonics, as we shall see later.

The method of successive approximations is clumsy, but effective, so long as the amplitudes are small. A more general approach can be illustrated by the considering pendulum. The first integral is

$$\frac{1}{2}p^2 - \omega_0^2 \cos q = h \qquad (5.22)$$

The shape of the corresponding phase curves is shown in Fig. 5.2.

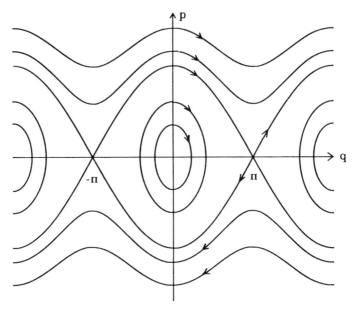

Fig. 5.2 Phase trajectories of an undamped nonlinear oscillator described by the equation is $\ddot{\vartheta} + \omega_0^2 \sin \vartheta = 0, q = \vartheta, p = \dot{\vartheta}$.

There are two regimes. The first oscillates between q_0 and $-q_0$ and occurs for $|h| < \omega_0^2$. We then have

$$\dot{q} = p = \sqrt{2h + 2\omega_0^2 \cos q} = \omega_0 \sqrt{2}\sqrt{\cos q - \cos q_0} \qquad (5.23)$$

where $h = -\omega_0^2 \cos q_0$ The period, which depends on q_0, is given by

$$T = \frac{2\sqrt{2}}{\omega_0} \int_0^{q_0} \frac{dq}{\sqrt{\cos q - \cos q_0}} \tag{5.24}$$

This is the well-known elliptic integral (tabulated since the 19th century). Proceeding in this way, we eventually obtain the function $q(t)$ and its representation by a Fourier series. When $|h| > \omega_0^2$, the pendulum rotates endlessly.

The phase plane therefore contains two types of trajectory (Fig. 5.2). For $|h| < \omega_0^2$, we have closed curves (cycles) around points of stable equilibrium ($q = 2\pi n, p = 0$), which are ellipses near these points. For $|h| > \omega_0^2$, the trajectories are open and represent rotation, which is another type of periodic motion in the variable p. If p_0 is the value of this variable of state (related to the velocity) for $q = \pi$, the period of rotation is given by

$$T_R = 4 \int_0^{\pi/2} \frac{dq}{\sqrt{p_0^2 + 2\omega_0^2 \sin^2(q/2)}} \tag{5.25}$$

An intermediate regime is described by a closed orbit as it runs from $q = -\pi$ to $q = +\pi$. However, it is covered in an infinite time T that corresponds to $q_0 = \pi$. It is an homoclinic orbit, connecting in an infinite time points of unstable equilibrium [$q = (2n + 1)\pi, p = 0$]. This trajectory separates the two régimes of oscillation and rotation and is called for this reason *separatrix*.

A system apparently as simple as the pendulum, whose isochronic small-amplitude oscillations were made use of in early frequency measurements, is in reality a highly complex device. Designs of resonators must not be confined to the linear approximation, and every effort must be made to understand the effects of nonlinear terms. Apart from exceptional cases, the frequency is always a function of the structure of the resonator and of the amplitude of the oscillations. Equations such as (5.10) are encountered, for example, in the description of oscillators using Josephson diodes, and in phase feedback loops. This topic is therefore not as academic as it may appear at first sight.

5.2.2 Frequency analysis of the nonlinear damped resonator with two variables of state

Resonators are always damped. A term proportional to velocity can describe this phenomenon in the first approximation. If we use $q = \vartheta, p = \dot\vartheta$ as the variables of state, the equation

$$\ddot\vartheta + 2\alpha\dot\vartheta + \omega_0^2\vartheta = 0 \tag{5.26}$$

becomes

$$\dot{q} = \dot{p} \qquad p = -\omega_0^2 q - 2\alpha p \qquad (5.27)$$

The phase curve is now a spiral which, for weak damping, winds itself around the origin – a point of stable equilibrium (Fig. 5.3). We will call the point $(0,0)$ a *point attractor* or focus. The surface of attraction is like an infinite basin, with all the trajectories reaching the focus for all initial conditions (q_0, p_0). The nonlinear resonators introduced in the previous Section are damped. For example, a pendulum executing large-amplitude oscillations can be described by

$$\dot{q} = p \qquad \dot{p} = -\omega_0^2 \sin q - 2\alpha p \qquad (5.28)$$

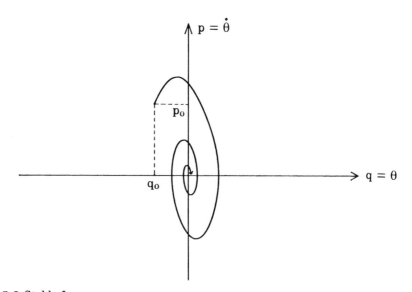

Fig. 5.3 Stable focus.

This is a more complicated case, but it can be tackled by a graphical method.

In general, a system with two variables of state can be described by

$$\dot{q} = F(p, q) \qquad \dot{p} = G(p, q) \qquad (5.29)$$

A first integral giving the equation of a cycle is then no longer available, and a periodic regime does not exist. To integrate, we specify the initial conditions (p_0, q_0) at $t = t_0$ and the integration step Δt. The coordinates of the adjacent point are then given by

$$p_1 = p_0 + G(p_0, q_0)\,\Delta t, \quad q_1 = q_0 + F(p_0, q_0)\,\Delta t, \quad t = t_0 + \Delta t \quad (5.30)$$

and we repeat the operation. A computerised calculation gives the state of the system at times $t_0 + n\Delta t$ and, hence, the parametric phase trajectory. If the curve returns to the initial point at times $t_0 + n\Delta t$, we have a resonator with period $n\Delta t$. This general method suffers from two flaws: (1) rounding errors are inherent in all recursive algorithms, since the computer has a limited number of digits to display the data and has to round off the numbers; step n uses the results of step $n-1$, so that there is a risk of accummulating errors and (2) sampling takes place in steps of Δt and the periodic regime is identified by returns to the initial state; all that the computer can do is to establish this to the nearest digit, and there may well be situations where the phase trajectory passes very close to the starting point and then definitely diverges.

This method is general and difficult to implement. However, it is the only one that can be applied to complicated systems. We will return to it later.

Figure 5.4 shows examples of phase trajectories obtained in this way for the damped pendulum. They wind around points of stable equilibrium ($q = 2\pi n, p = 0$) and represent damped oscillations about these points. However, if the initial conditions are very different, there are first several complete circuits, an attraction to the neighbourhood of a point of stable equilibrium and, finally, damped oscillation about this point. This is a complicated phenomenon that is very sensitive to initial conditions and can occur after two or three turns. The bifurcation from one regime to another depends on a small variation of initial conditions.

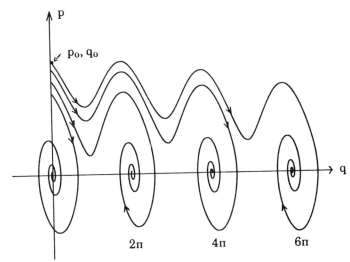

Fig. 5.4 Phase trajectories of a nonlinear damped oscillator (damped pendulum).

Some noise introduces an uncertainty into the initial conditions, these phenomena cannot be predicted with certainty. For the simple pendulum, the phenomenon is not serious because all the equilibrium values of q, i.e., $(2\pi n)$, correspond to the *same* position. Thus, although we know where we will get to, we cannot predict with certainty how many turns this will take. A very small *variation of the initial conditions*, a slight noise perturbation, can make us miss a turn and thus introduce an important change in the descriptions of the motion. We will meet these concepts again later, whenever nonlinear terms have to be used to describe a system.

Analysis of the response of a linear resonator to a sinusoidal signal can be used to determine the resonator parameters (Section 1.4). The method can be extended to the nonlinear case. We shall confine our attention to the simple case where a single third-degree term is sufficient to describe the nonlinearity. We thus have

$$\dot{q} = p$$

(5.31)

$$\dot{p} = -\omega_0^2 q + bq^3 - 2\alpha p + A \sin \omega t$$

This system corresponds to the D"uffing equation. Such equations contain a time-dependent term and are no longer autonomous. Graphical integration can still be used, but it gives more complicated results since time is involved. The phase diagram is three-dimensional (q, p, t). The initial point is p_0, q_0, t_0 and the system is periodic if it returns to this point at times $t_0 + mT$ (Fig. 5.5).

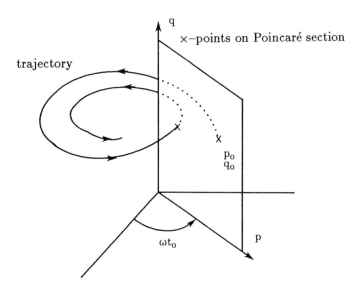

Fig. 5.5 Poincaré section.

We shall therefore be interested in points corresponding to $t_0, t_0 + T, t_0 + 2T, t_0 + 3T, \ldots t_0 + nT$; this group of points depends on t_0 and is called a Poincaré section of the system. A periodic solution will therefore have a Poincaré section containing a finite number n of points, but this number can be very large, i.e., the period nT can be long. The frequency will then be a submultiple of the driving frequency $1/T$. We will return to these delicate matters later. Here, we are merely concerned with demonstrating the complexity of the phenomenon and its vocabulary.

It is tempting to look for a periodic régime with the same frequency as the driving signal, although it is not certain that it exists; a transient state is a possibility and there may even be other periodic states. Let us ignore these difficulties for the moment, and try to describe the system by the simple expression

$$q = Q \sin (\omega t - \Phi) \tag{5.32}$$

If we substitute this in (5.31), we obtain

$$\left[(\omega_0^2 - \omega^2)^2 - \frac{3}{4}\beta Q^2 \right]^2 + 4\alpha^2 \omega^2 = \frac{A^2}{Q^2} \tag{5.33}$$

where we have neglected terms involving $\sin 3\omega t$ and due to the nonlinearity. Figure 5.6 shows Q as a function of the amplitude A of the driving signal. When A is very small, the amplitude Q of the forced oscillation is small; the resonance curve is at first classical, but then bends slightly to the right or the left, depending on the sign of β. When A is large, the forced oscillation has a large amplitude, the nonlinearity then play a more important role and the curve is more complicated. A frequency sweep that imposes a particular direction on the variation of ω causes amplitude jumps, e.g., from M to N at ω_A, as the frequency is increased. The jump occurs between N' and M' at ω_B as the frequency is reduced. The locus of maxima of the resonance curves is a parabola known as the anisochronic curve. These results can be used to detect the nonlinearities of a resonator and to determine its parameters from the resonance curves (and hence to perfect the model used).

Although these results are incomplete, they do illustrate the difficulty of the problem. Suppose we drive the system at frequency ω_1 and wait for it to reach its steady state; we then slowly increase the frequency to ω_A. This gives Q_M^2. But what will happen if we suddenly apply a sinusoidal signal of pulsatance ω_A? The abrupt application of a sinusoidal pulse subjects the resonator to a broad spectrum whose components are damped out gradually, but which ones disappear first? The low frequency or the high frequency ones? In other words, once the different frequency transients disappear, will the steady state involve Q_M^2 or Q_N^2. There are in fact two possible steady states; which one will be established will depend on the history of the applied signal (which is not a sinusoid). If we wish to go further, we have to take into account the harmonics, and start with, say, a term in 3ω. The calculations then become impossible, but there is worse to come. Let us suppose that a signal of pulsatance $\omega/3$ is present in the system at a given time. The nonlinearities will generate harmonics, including the third harmonic with pulsatance ω. The latter will coincide with the driving pulsatance. This will produce resonance, i.e., a large-amplitude forced oscillation. As the different harmonics of a signal have amplitudes that are mutually related, the resonance that increases the amplitude of the third harmonic will also increase amplitude of the fundamental with pulsatance $\omega/3$. The unwanted harmonics using will not be eliminated by a low-pass filter alone, because they are submultiples. They can be so numerous that a quasi-continuous spectrum, i.e., chaos, is produced; we will return to this later. The nonlinearities of damped resonators are difficult to eliminate and produce complex phenomena. We have thus gone a long way from the simplistic description of Section 1.4.

5.2.3 The multimode resonator and mode coupling

Most practical resonators exhibit propagation phenomena. In Section 4.2 we described the propagation of electromagnetic waves in the Fabry-Perot cavity. In Chapter 1 we saw that the gravitational force, which controls the oscillations of a pendulum, could be replaced in practice by an elastic force due to displacement waves. Piezoelectric quartz provides a remarkable example of propagation in an elastic resonator (cf. Section 5.4). In these systems, resonance is due to standing waves established within a distance L in the system. There is therefore an infinite number of resonant frequencies because a whole number of half-wave-lengths must fit into the length L. The resonator is a three-dimensional object whose length L can be chosen in three different ways, each with an associated integer. We must therefore expect a triple infinity of characteristic frequencies, specified by three indices. The calculation is simple in the case of an empty parallelepiped with metal walls and side lengths A, B, C. The frequencies are given by

$$f_{m,n,p} = \frac{c}{2}\sqrt{\frac{m^2}{A^2} + \frac{n^2}{B^2} + \frac{p^2}{C^2}} \qquad (5.34)$$

where m, n, p are integers. This ensemble of characteristic frequencies constitutes a line spectrum. The lines have no width if there is no damping, which is evidently unrealistic. Damping broadens the lines, as we saw earlier. If the lines of the multimode resonator are very close together, broadening due to damping will transform the line spectrum into a totally unusable continuum (Fig. 5.7). In practice, the resonator must be designed so that the greatest possible number of unwanted modes is eliminated. This is done in the Fabry-Perot by using plane parallel mirrors (Section 4.2.4). Resonances are then possible only in the direction perpendicular to the

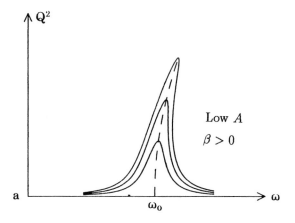

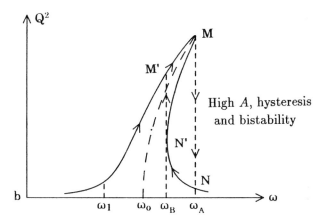

Fig. 5.6 Nonlinear resonance: (a) excitation by moderate A, (b) excitation by large A (hysteresis and bistability).

mirrors ($n = p = 0$). Additional measures (multilayered mirrors) allow the generation of a relatively simple spectrum with lines of different amplitude and width. However, apart from exceptional cases, the propagation resonator is characterised by the spectrum shown Fig. 5.7, with several lines in the useful frequency band in which we attempt loss compensation in order to produce an oscillator. We must therefore determine if the resonator will lock to a single frequency or to several frequencies, which is not desirable.

The LC circuit is a simple practical model of single-mode lossless resonance. By adding a conductance G we take the losses into account and have an excellent equivalent circuit diagram with two variables of state.

The same operation can be applied to a multimode resonator, except that we then need several LCG circuits to account for the different resonant frequencies. Figure 5.8 shows an example of a circuit representing a two-mode resonator in the form of two coupled circuits resonating at ω_1 and ω_2, respectively. Four variables of state are now needed to set up the equations. In the general case of N linear variables of state, the equations of state become

$$(\dot{q}) = (A)(q) \tag{5.35}$$

where the elements of the matrix A will be denoted by $A_{n,m}$.

Let us now examine whether this system has sinusoidal solutions of the form $e^{j\omega t}$ in the undamped case. The corresponding frequencies are the roots of the determinant

$$|A_{n,m} - j\omega \delta_{mn}| = 0 \tag{5.36}$$

where $\delta_{m,n}$ is the Kronecker symbol ($\delta_{mn} = 1$ for $m = n$ and $\delta_{mn} = 0$ for $m \neq n$).

In general, there will be several characteristic frequencies. In the case of Fig. 5.8 with $G_1 = G_2 = 0$, we obtain

$$\left(\frac{\omega^2}{\omega_1^2}\right)^2 (1 + K_1 + K_2) - \frac{\omega^2}{\omega_1^2}\left[1 + K_2 + (1 + K_1)\left(\frac{\omega_2}{\omega_1}\right)^2\right] + \left(\frac{\omega_2}{\omega_1}\right)^2 = 0 \tag{5.37}$$

which is illustrated in Fig. 5.8 for an interesting case.

There are two resonant frequencies whose difference depends on the coupling (via C_0) between the two circuits. The degree of coupling has a crucial affect on frequency. Initially, we supply energy W to the system by charging the capacitors. The system responds in a complicated manner since its oscillations are described by two sinusoids with frequencies that are generally not related by a simple ratio. The notion of frequency, established in time-domain analysis, is difficult to apply by observing a quasi-periodic signal. Counting the oscillations is no longer a wholly meaningful procedure.

The phase trajectories must now be drawn in four-dimensional space. The use of such a complicated representation is not obvious at first sight, but it becomes simpler in the case of lossless resonators ($G_1 = G_2 = 0$) for which energy is conserved. The first integral still exists and takes the form of a relationship between four variables of state. For given energy W_0, it is sufficient to fix three variables, for example, q_1, q_2 and i_1, and the fourth, i_2, can be determined immediately. Further simplification is possible because the amplitudes of the two sinusoidal oscillations with the two possible frequencies are proportional to one another. The phase trajectory is a very complicated curve on the q_1, i_1 plane (for example). However, in three-dimensional space (q_1, q_2, i_2) it is inscribed on a torus whose large

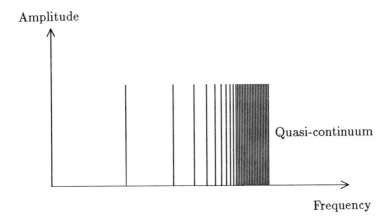

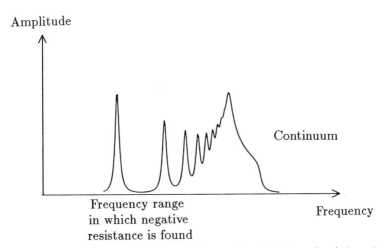

Fig. 5.7 Line spectrum of a resonator with (bottom) and without (top) damping.

and small circumferences correspond to the first and second characteristic frequencies, respectively. These concepts are very useful in the advance towards a theoretical description, and can be extended to conservative systems with N variables of state.

Actually, we are primarily interested in the damped resonator. We will therefore seek solutions such as e^{pt}, which amounts to writing

$$|A_{n,m} - p\delta_{m,n}| = 0 \qquad (5.38)$$

and searching for the complex roots. The position and width of resonance peaks and the parameters of the model can then be found by comparing the above curves with measurements. When it is given energy W_0, the circuit response is described by a sum of damped sinusoids with different

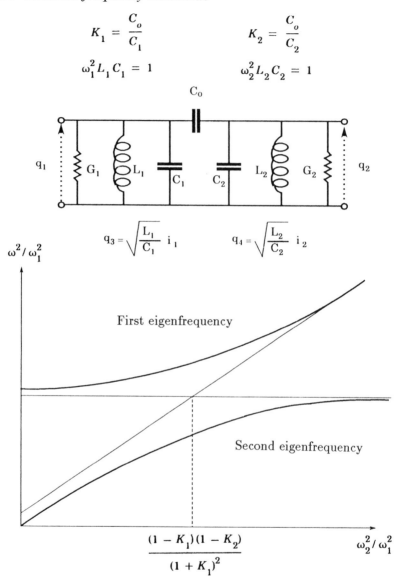

$$K_1 = \frac{C_o}{C_1} \qquad\qquad K_2 = \frac{C_o}{C_2}$$

$$\omega_1^2 L_1 C_1 = 1 \qquad\qquad \omega_2^2 L_2 C_2 = 1$$

$$q_3 = \sqrt{\frac{L_1}{C_1}}\, i_1 \qquad\qquad q_4 = \sqrt{\frac{L_2}{C_2}}\, i_2$$

Fig. 5.8 Model of a two-mode resonator.

damping constants. Observations of such a signal do not lead to anything in practice because it is no longer possible to count periods, as was done for the linear damped resonator with two variables of state.

In general, to cover the whole range of frequencies, we have to use multi-mode resonators. To integrate such devices in a system which, by providing feedback, results in an oscillator, the devices have to be modelled by comparing them with a classical circuit. This often requires more than two variables of state and leads to a time-domain description that is very complicated even for free oscillation. The simplest model is made by coupling a number of oscillatory circuits, one for each mode to be represented. However, circuit coupling produces some coupling between the modes, and it is always difficult to isolate a single mode. Finally, we recall that nonlinear effects are also present in multimode resonators, including elastic resonators; the resonance curves are not symmetric bell-shaped peaks to which a simple definition of frequency can be applied. It is always difficult to extract a well-defined frequency from an analysis of a resonator.

5.3 AN OSCILLATOR WITH TWO VARIABLES OF STATE

5.3.1 Negative resistance

All resonators exhibit losses whereby stored energy decreases. It is therefore necessary to couple the resonator to an active circuit that compensates such losses. Indeed, this is the way to produce an oscillator. In the electrical model of an oscillator, losses are represented by a positive classical conductance. On the other hand, a negative conductance (or negative resistance) can be a dipole that supplies energy; it therefore contains a source of energy that forces the current up the potential gradient, in contrast to nature which drives the current down the potential gradient. Figure 5.9 illustrates these definitions.

Electronics allows us to build such devices. A beam of electrons gains energy by traversing a constant potential difference. The electrons can be slowed down by a suitably directed electric field, and the device that creates this field absorbs the energy of the electrons. If it is a resonator, it will be able to compensate its losses in this way, but the electron beam is continuous whilst the field created by the resonator is alternately decelerating and accelerating in succesive half-periods, i.e., the resonator loses and gains energy in these half-periods, respectively. This means that the decelerating effect must be emphasised by modulating the beam density at the frequency of the resonator. A circuit with negative resistance that compensates losses will therefore contain a static generator that accelerates the electron and a feedback loop that performs amplitude modulation of the beam so that more electrons are slowed down and fewer electrons accelerated in the resonator.

Figure 5.10 shows a simple circuit consisting of a negative-conductance dipole made up of an operational amplifier and a feedback loop. We have

$$i = -G\frac{1 - 1/A - 1/\rho GA}{1 + \rho G/A + 1/A}v = -G_1 v \qquad (5.39)$$

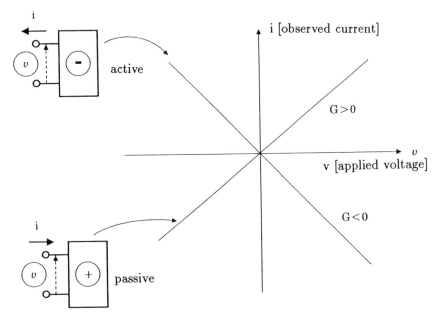

Fig. 5.9 Positive and negative conductance.

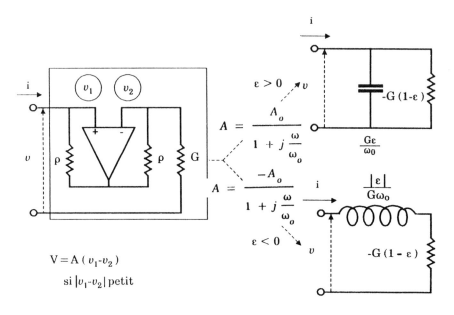

$$V = A\,(\upsilon_1 - \upsilon_2)$$

si $|\upsilon_1 - \upsilon_2|$ petit

Fig. 5.10 Elementary diagram of a dipole with negative conductance (or negative resistance).

If the gain A is high, we have in effect $G_1 = G$, i.e., a negative-impedance converter. This simplistic model ceases to work when A is finite and depends on frequency. A relation such as

$$A = \frac{A_0}{1 + j(\omega/\omega_0)} \tag{5.40}$$

represents in the first approximation the variation of gain with frequency. It corresponds to the differential equation

$$\frac{dv}{dt} + \omega_0 v = \omega_0 A_0 (v_1 - v_2) \tag{5.41}$$

In a sinusoidal regime, and if A_0 is large enough, it can be shown that

$$i = \left[-G(1 - \varepsilon) + j\frac{G\varepsilon}{\omega_0}\omega \right] v \tag{5.42}$$

$$\varepsilon = \frac{2}{A_0} + \frac{1}{\rho G A_0} + \frac{\rho G}{A_0} \tag{5.43}$$

which shows that the dipole is described by a negative conductance associated with inductive or capacitive parasitic elements, depending on the way in which the amplifier is connected (Fig. 5.10).

By connecting this negative conductance to the terminals of the resonator of Fig. 1.8 we compensate the loss conductance, but we also slightly modify the capacitance, i.e., the resonance frequency. If we look for a solution such as e^{pt}, we find that p is complex with positive real part if $G_1 > G_0$, which is the condition for oscillation. The first stage of oscillator design is thus to couple an active component to the resonator and then describe the circuit by a linear model. The final stage consists of looking for the locking condition, i.e., for a pair of roots of the form $k^2 + j\omega$. Let us suppose for the moment that there is only one pair, the other values of p being real and negative, or complex with negative real parts. In Section 5.6 we shall consider the difficulties that arise from the existence of several possible locking states. This cannot be the case for systems with two variables of state, since the critical values of p are given by a second degree equation.

The amplitudes of the oscillations will thus increase, the saturation of the amplifier will come into play and an expression of the form

$$A = \frac{A_0}{k} \tanh k (v_1 - v_2) \tag{5.44}$$

will represent quite correctly the gain of a real amplifier at low frequencies. An analysis of the circuit leads to a relation of the form shown in Fig. 5.11.

For weak signals, the previous model still holds, but with a negative conductance G. The opposite situation prevails for large signals: the amplifier saturates and the circuit acquires a positive conductance. The relationship between current i and voltage v is now a cubic of the form

$$i = -Gv + \beta v^3 \tag{5.45}$$

which provides an acceptable description at low frequencies. There is also the corresponding expression for voltage as a function of current:

$$v = -Ri + \alpha i^3 \tag{5.46}$$

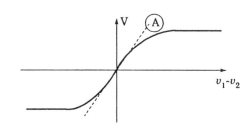

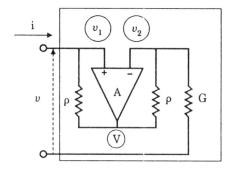

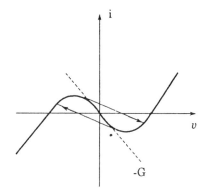

Fig. 5.11 Nonlinear characteristic of an amplifier with negative resistance.

Of course, saturation phenomena are intimately linked to frequency dispersion. Saturation is a nonlinear effect which means that A as a function of frequency cannot be described by a simple harmonic represented by a complex number. We have to go back to the differential equations and treat the problem as a whole. The parasitic capacitance (or self-inductance) that is also nonlinear is described by an expression of the form

$$i = C_0 \left(1 - \delta v + r v^2 + ...\right) \frac{dv}{dt} \tag{5.47}$$

A full analysis of the circuit shows that it is governed by a nonlinear differential relationship between i and v:

$$i = F\left(v, \frac{dv}{dt}..., \int v \, dt, ...\right) \tag{5.48}$$

Our second task is to set up this equation. For example, in the case of Fig. 5.11, a suitable model is

$$i = -Gv + \beta v^3 + C_0 \left(1 + \gamma v^2\right) \frac{dv}{dt} \tag{5.49}$$

In practice, the phenomena are complicated by the amplifier time constant (or delay) $1/\omega_0$. For large signals, the characteristic of the circuit exhibits hysteresis as shown in Fig. 5.11.

5.3.2 A simple method: linearisation of the 'first harmonic'

Our aim is to make an oscillator, i.e., to couple a negative conductance to a very high Q resonator, and obtain a periodic regime. This regime will be quasi-sinusoidal, because of the selectivity of the resonator. We can then represent the signal by a Fourier series with low harmonic content, and use the nonlinear differential relationship describing the circuit to relate the coefficients. This description will only be usable in the steady periodic state or in a slightly perturbed state. A slowly-varying state of this kind is called adiabatic. The underlying assumption is that the relative variation of the amplitude over one period is negligible, i.e.,

$$\frac{1}{V} \frac{dV}{dt} \frac{2\pi}{\omega} \ll 1 \tag{5.50}$$

The first step is to put

$$v = V \sin \omega t \tag{5.51}$$

and then substitute this in (5.48), retaining only the fundamental term of angular frequency ω in the response i. We thus obtain

$$i = A \cos \omega t + B \sin \omega t \tag{5.52}$$

$$A = \frac{\omega}{\pi} \int_0^{2\pi/\omega} F\left(V \sin\omega t, \omega V \cos\omega t, ..., -\frac{V}{\omega}\cos\omega t, ...\right) \cos\omega t \, dt$$

$$(5.53)$$

$$B = \frac{\omega}{\pi} \int_0^{2\pi/\omega} F\left(V \sin\omega t, \omega V \cos\omega t, ..., \frac{V}{\omega}\cos\omega t, ...\right) \sin\omega t \, dt$$

In the case of the example described by (5.49), we have

$$i = \left(-G + \frac{3}{4}\beta V^2\right) V \sin\omega t - C_0 \left(1 + \frac{3}{4}\gamma V^2\right) \omega V \cos\omega t \qquad (5.54)$$

where we have retained only terms oscillating at the angular frequency ω.

Since we have linearised the system, we can use the complex number notation and describe the circuit by an admittance $Y(j\omega, V)$, such that

$$i = |Y(j\omega, V)| \, V \sin(\omega t + \Phi) \qquad (5.55)$$

which obviously corresponds to

$$|Y| = \frac{1}{V}\sqrt{A^2 + B^2} \qquad \tan\Phi = \frac{B}{A} \qquad (5.56)$$

The above method can be extended to the case of a slowly-varying slow transient and anything let through by the small bandwidth of the resonator. We start with a slowly-varying quasi-sinusoidal state of the form $e^{(a+j\omega)t} = e^{pt}$, remembering that we have to be careful before exploiting all the resources of the symbolic formalism when the system is nonlinear. A full analysis of transients is only possible by going back to the nonlinear differential equations.

We can improve the steady-state solution by expanding v into a Fourier series:

$$v = \sum_{n=1}^{\infty} V_n \sin(n\omega t + \Phi_n) \qquad (5.57)$$

The calculation is generally impractical except in the fortunately frequent case where the amplitude of the harmonics decreases very rapidly with their order. We can then confine our attention to the first few terms of the series. If we take a nonlinear pure conductance, such that

$$i = -Gv + \beta v^3 \qquad (5.58)$$

and if we put

$$v = V[\cos\omega t + \varepsilon\cos(3\omega t + \Phi)] \qquad (5.59)$$

we obtain, retaining only the first power of ε,

$$i = \left[-G + \frac{3}{4}\beta V^2 + \frac{3}{4}\beta V^2 \varepsilon \cos \Phi\right] V \cos \omega t - \left(\frac{3}{4}\beta V^2 \varepsilon\right) V \sin \omega t$$

$$+ \left[-GV\varepsilon \cos \Phi + \frac{1}{4}\beta V^3 + \frac{3}{2}\beta V^3 \varepsilon \cos \Phi\right] \cos 3\omega t +$$

$$\left(GV\varepsilon \sin \Phi + \frac{3}{2}\beta V^3 \varepsilon\right) \sin 3\omega t \qquad (5.60)$$

where we have neglected harmonics of frequency 5ω and higher. As can be seen, there are no even harmonics. When the nonlinear analysis of a circuit is too complicated to lead to a relationship of the form $i = F(v)$, the above relationships can be found directly by experiment, using a sinusoidal v with adjustable ω and V. The current i can then be measured, using a narrow filter centered on ω, and the amplitude and phase of the sinusoidal response signal can also be determined. This yields $|Y|$ and Φ. The correct representation of the linearised admittance is found by performing measurements for different ω and V. The circuit is thus investigated in terms of linear equations, which allow the use of complex notation (the admittance matrix for a multipole, for example). The coefficients are then functions of amplitude and frequency. This type of analysis involves the linearised circuit and is based on classical methods. Everything therefore appears to be simple if a single amplitude is involved in the linearisation. This will be so in the case of oscillators with two variables of state which we will study in the next Section. We will then discover the differences that are encountered when we cannot neglect all the harmonics (Section 5.3.4) and, later, when several amplitudes are involved in the linearisation process (Section 5.6).

5.3.3 Oscillator with two variables of state

The simplest oscillator is that modelled by an LCG_0 resonator with a negative conductance (Fig. 5.12) which can be described by the simple equation

$$i = -Gv + \beta v^3 \qquad (5.61)$$

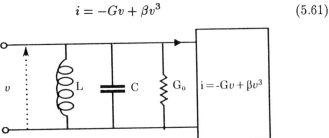

Fig. 5.12 LC resonator with a negative resistance.

A parasitic capacitance, assumed linear ($\gamma = 0$), is included in C. If we linearise the nonlinear component, we obtain

$$i = \left(-G + \frac{3}{4}\beta V^2\right) v \tag{5.62}$$

If we analyse the circuit by the classical method, we find for the sinusoidal regime

$$jC\omega + \frac{1}{jL\omega} + G_0 - G + \frac{3}{4}\beta V^2 = 0 \tag{5.63}$$

and hence

$$\omega = \frac{1}{\sqrt{LC}} = \omega_0 \qquad V = \sqrt{\frac{4g}{3\beta}} = V_0 \qquad (g = G - G_0) \tag{5.64}$$

The steady-state frequency and amplitude are thus fixed, and G must be positive ($G > G_0$). For small signals, the negative conductance takes over, the circuit becomes unstable and the oscillator begins to operate.

A resonator with a negative nonlinear conductance in the linearised regime has a total conductance given by

$$g_1 = G_0 - G + \frac{3}{4}\beta V^2 = \frac{3}{4}\beta \left(V_0^2 - V^2\right) \tag{5.65}$$

When $V = V_0$, we have $g_1 = 0$. Compensation takes place and the oscillator reaches its steady state. When a perturbation changes V from V_0 to $V_0 + \Delta V$, the total admittance which was zero becomes $(3/2)\beta V_0 \Delta V$. It is positive (i.e., it damps) if ΔV is positive and its effect is to reduce V, which returns to V_0. When ΔV is negative, the admittance is negative and the amplitude increases; the steady state is stable. We can approximate the beginning of the transient state after the perturbation by

$$v = V_i e^{\alpha t} e^{j\omega t} = V_i e^{pt} \qquad (V_i = V_0 + \Delta V) \tag{5.66}$$

If α is small, we have

$$\alpha = -\frac{3}{4}\frac{\beta V_0}{C}\Delta V \qquad \omega = \omega_0 \tag{5.67}$$

where ΔV and α have opposite signs. This is the condition for stability.

The above calculation gives only the initial behaviour, just after the perturbation. For a more complete calculation, we describe the transient state by

$$v = V(t) \sin \omega_0 t \tag{5.68}$$

where the amplitude $V(t)$ varies slowly as compared with $\sin \omega_0 t$. We can approximate the signal by a sinusoidal variation for a few periods. This enables us to determine the accumulated energy W (proportional to V^2) and then the variation ΔW per period. As in Section 1.4, we find that

$$\frac{\Delta W}{2\pi W} = -\frac{1}{2\pi W}\frac{dW}{dt}\frac{2\pi}{\omega_0} = -\frac{2}{\omega_0}\frac{1}{V}\frac{dV}{dt} \qquad (5.69)$$

The linearised circuit oscillates with

$$Q = \frac{1}{g_1}\sqrt{\frac{C}{L}} = \frac{2\pi W}{\Delta W} \qquad (5.70)$$

so that, substituting for g_1, we obtain

$$\frac{1}{V}\frac{dV}{dt} = -\frac{\omega_0}{2Q} = \frac{3\beta}{8C}\left(V_0^2 - V^2\right) \qquad (5.71)$$

and, after integration,

$$V = \frac{V_0}{\sqrt{1 + e^{-t/\tau}}} \qquad K = \left(\frac{V_0}{V_i}\right)^2 - 1 \qquad \tau = C/g \qquad (5.72)$$

This formula is illustrated in Fig. 5.13 and shows that V tends to the steady-state amplitude, whatever the initial conditions; τ gives the order of magnitude of the duration of the transient. If τ must small, so that the transient is rapidly damped out, we must choose high g, i.e., $G \gg G_0$; the effect of the negative conductance is then be important. The oscillator is said to be 'hard' and is rich in harmonics, as we shall see (Section 5.3.4).

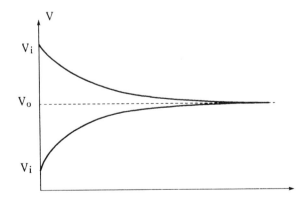

Fig. 5.13 Transitions towards the steady state V_0 of an oscillator when $V_i > V_0$ and $V_i < V_0$.

There are two types of trajectory in the phase space $(\dot{v}, v)$. They are illustrated in Fig. 5.14. When locking is impossible $(G < G_0)$, the trajectories are spirals that wind themselves around the point of stable equilibrium, i.e., the origin. The basin of attraction is infinite. All the trajectories beginning at a point located in a small domain (a small circle, for example) arrive at the point 0 which is a point *well*. The dimensions of this domain decrease until it becomes a point. It is an area (two dimensions) that shrinks to a point (which can be considered as an infinitely small area). If locking takes place $(G > G_0)$, the situation is different. The phase trajectories are spirals that wind themselves around the cycle that corresponds to the steady state régime with amplitude V_0. All the points that were initially in a small two-dimensional domain end up on a segment of the cycle, i.e., a one-dimensional domain.

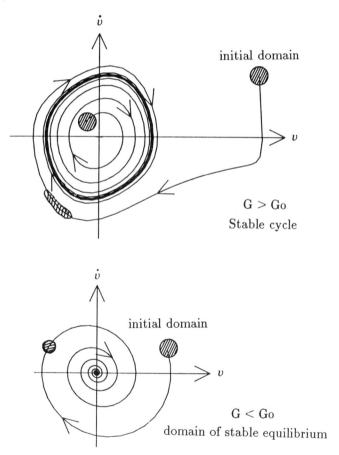

Fig. 5.14 Phase plane trajectories.

The simple oscillator examined above has two states. The first corresponds to a point attractor (the origin) and an infinite surface of attraction. The phase domain shrinks in all its dimensions. After the threshold $(G = G_0)$ at which bifurcation takes place, we obtain a cyclic attractor state. The basin of attraction is still infinite, but the phase domain spreads over the cycle, i.e., it shrinks in only one dimension. These concepts, first introduced by Lyapunov, will be useful later.

We can generalise the above analysis to any system with two variables of state:

$$\dot{q} = F(p, q)$$

$$(5.73)$$

$$\dot{p} = G(p, q)$$

To determine whether this system can be an oscillator, we first have to find a position of unstable equilibrium. We start with

$$\dot{q} = \dot{p} = 0$$

$$(5.74)$$

$$F(p, q) = 0 \qquad G(p, q) = 0$$

Next, we assume that a solution exists and, for convenience, take $p = 0$ and $q = 0$ (or transform the coordinates). In order to test for instability at this point, we suppose that p and q have small amplitudes, so that we can linearise the system:

$$\dot{q} = \left(\frac{\partial F}{\partial q}\right)_0 q + \left(\frac{\partial F}{\partial p}\right)_0 p = A(0)q + B(0)p$$

$$(5.75)$$

$$\dot{p} = \left(\frac{\partial F}{\partial q}\right)_0 q + \left(\frac{\partial F}{\partial p}\right)_0 p = C(0)q + D(0)p$$

An exponential solution with a growing amplitude is possible and represents the starting of an oscillator with an increasing amplitude if

$$A(0) + D(0) < 0 \qquad A(0)D(0) - B(0)C(0) = \omega_0^2 > 0 \qquad (5.76)$$

This is the condition for locking. Everything is therefore simple as long as there is one and only one equilibrium position that can be made unstable. The next step is to find the periodic steady-state solution by means of the first-harmonic approximation. Substituting

$$q = a\cos\omega t + b\sin\omega t \qquad p = c\cos\omega t + d\sin\omega t \qquad (5.77)$$

in the linearised equations as in the last Section, and taking F and G to be periodic functions with pulsatance ω replaced by the fundamental, we obtain four linear equations for the four variables a, b, c, d. The solution exists only for a value of ω that is readily found (by equating to zero the determinant of the coefficients). Finally, we have to establish whether the steady state is stable. This is done by replacing a by $a + \Delta a$ (and similarly for the other amplitudes) and examining the perturbations (Δa, ...); the steady state is stable if they decrease.

5.3.4 Two more methods for the oscillator with two variables of state

(*a*) *The harmonic balance.* The nonlinear terms generate harmonics, so that a more realistic model is a signal containing the fundamental and a few harmonics. Let us try

$$v = V\left[\cos \omega t + \varepsilon \sin\left(3\omega t + \Phi\right)\right] \tag{5.78}$$

where we have introduced 3ω and not 2ω because of the odd-power characteristic of the expression for the total conductance g_1 introduced in the previous Section. Substituting this expression for v in the characteristic equation, we obtain

$$i = -gv + \beta v^3 \tag{5.79}$$

and if ε is small compared to unity, and v is not very different from V_0, we find that the current in the total nonlinear conductance is

$$i = 2g\Delta V \cos \omega t + gV_0 \varepsilon \sin\left(\omega t + \Phi\right) \tag{5.80}$$

$$+ gV_0 \varepsilon \sin\left(3\omega t + \Phi\right) + \left(g\frac{V_0}{3} + g\Delta V\right)\cos 3\omega t$$

where

$$V_0 = \sqrt{\frac{4g}{3\beta}} \qquad \text{and} \qquad V = V_0 + \Delta V \tag{5.81}$$

However, v is also the voltage applied to the LC circuit (Fig. 5.15). The linear calculation is easy and leads to an expression for the current flowing in the circuit. Assuming that ω is close to $1/\sqrt{LC}$, we thus obtain

$$i_1 = -2V_0 C\Delta\omega \sin \omega t + \frac{8}{3}C\omega_0 \varepsilon V_0 \cos\left(3\omega t + \Phi\right) \tag{5.82}$$

where

$$\omega = \frac{1}{\sqrt{LC}} + \Delta\omega = \omega_0 + \Delta\omega \tag{5.83}$$

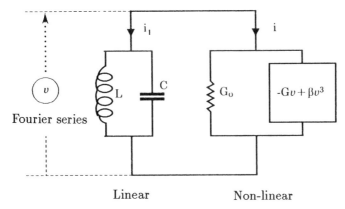

Linear   Non-linear

Fig. 5.15 Principle of the method of harmonic balance.

Comparing the above two expressions for the current (the method of harmonic balance), we find that

$$2g\Delta V + gV_0\varepsilon \sin \Phi = 0$$

$$gV_0\varepsilon \cos \Phi = 2CV_0\Delta\omega \tag{5.84}$$

$$gV_0\varepsilon \cos \Phi = \frac{8}{3}C\omega_0\varepsilon V_0 \sin \Phi$$

$$gV_0\varepsilon \sin \Phi + \left(\frac{gV_0}{3} + g\Delta V\right) = -\frac{8}{3}C\omega_0\varepsilon V_0 \cos \Phi$$

These four equations express the fact that $i = -i_1$ and give

$$\varepsilon = -\frac{g}{8C\omega_0} \qquad \Delta\omega = -\frac{g^2}{16C^2\omega_0}$$

$$\tag{5.85}$$

$$\Phi = +\frac{3g}{8C\omega_0} \qquad \Delta V = -\frac{V_0 3g^2}{128C^2\omega_0^2}$$

The steady-state signal from the generator is therefore given by

$$v = \sqrt{\frac{3g}{3\beta}}\left[\left(1 - \frac{3g^2}{128C^2\omega_0^2}\right)\cos\omega t - \frac{g}{8C\omega_0}\sin\left(3\omega t + \frac{3g}{8C\omega_0}\right)\right] \tag{5.86}$$

$$\omega = \omega_0\left(1 - \frac{g^2}{16C^2\omega_0^2}\right)$$

These results are still complicated, despite the considerable approximations that have been made. Nevertheless, they show that $g = G - G_0$

plays an essential part in the generation of the harmonics. A purely sinusoidal oscillation corresponds to a negligible g, i.e., quasi-perfect loss compensation in the case of negative resistance. However, the amplitude is then infinitesimal unless the nonlinear terms are also negligible. We are thus approaching the ideal case that, unfortunately, cannot be achieved. In practice, g appears in the first order in the amplitude of the harmonics, and in the second order in the variations in amplitude and pulsatance. Finally, we note that the perturbation decay time τ increases with g in the steady state. A quasi-sinusoidal oscillator, poor in harmonics, can be described as *soft*. It is very sensitive to noise.

We can now continue the frequency analysis of an oscillator by this method. We take a Fourier series in which the number of terms is compatible with the computational power available. This fixes the order of the highest harmonic that we will retain. We shall deal separately with the nonlinear part in which we will retain only harmonics compatible with the initial conditions. We will then compare the harmonics. In general, there are more equations than unknowns and we have to use a *least squares* procedure. The calculation is long and does not yield information about the transient state.

(b) Time-domain analysis; the piece-wise linear model. In the case of the circuit of Fig. 5.15, we can write Ohm's law in the form of the differential equation

$$C\dot{v} + \frac{1}{L} \int v dt + G_0 v - Gv + \beta v^3 = 0 \tag{5.87}$$

so that

$$\ddot{v} - \gamma \left(1 - \varepsilon^2 v^2\right) \dot{v} + \omega_0^2 v = 0 \tag{5.88}$$

where

$$\omega_0^2 = \frac{1}{LC} \quad \gamma = \frac{G - G_0}{C} = \frac{g}{c} \quad \varepsilon = \sqrt{\frac{3\beta}{g}} = \frac{2}{V_0} \tag{5.89}$$

This is the well-known equation of Van der Pol. There have been many attempts to integrate it. Graphical methods can be used to examine the transient state, and the time-domain is interesting. A quasi-sinusoidal solution with pulsatance ω_0 and amplitude V_0 if g is small. This agrees with the previous results.

We now turn to the time-domain analysis of the oscillator with two variables of state in the simple case that will be useful later; it will also throw further light on the only mechanical system with effectively negative conductance, namely, the escapement mechanism introduced in Chapter 1 (Fig. 1.10).

The electrical analogue of this system is shown in Fig.5.16 and its conductance can be positive or negative. For non-zero values of v, the circuit is damped and has the positive classical conductance g_0. During the short time when the variable changes from $-\eta$ to $+\eta$. The system is active for the short time during which the variable changes from $-\eta$ to $+\eta$ and has a negative conductance. Figure 5.16 illustrates the phenomenon for constant conductances. The characteristic is then a set of straight linear segments, and piece-wise linear analysis becomes possible.

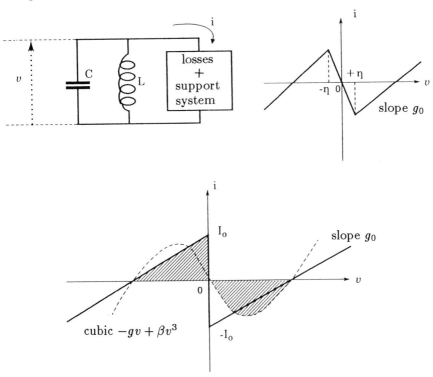

Fig. 5.16 Electrical analogue of the escapement mechanism.

Consider the simple limiting case $\eta \to 0$ (Fig. 5.16). To compare time-domain analysis with frequency analysis, for which we used we used a cubic, we put

$$g_0 = \frac{g}{2} \qquad I_0 = \frac{g\sqrt{g}}{2\sqrt{\beta}} \qquad (5.90)$$

The circuit equations now become

$$\ddot{v} + 2\alpha\dot{v} + \omega_0^2 v = 0 \qquad (v \neq 0) \qquad (5.91)$$

$$\ddot{v} + \left[2\alpha - \frac{2I_0}{C}\delta(v)\right]\dot{v} + \omega_0^2 v = 0 \quad (v = 0)$$

where

$$2\alpha = \frac{g_0}{C} \qquad \omega_0^2 = \frac{1}{LC} \qquad (5.92)$$

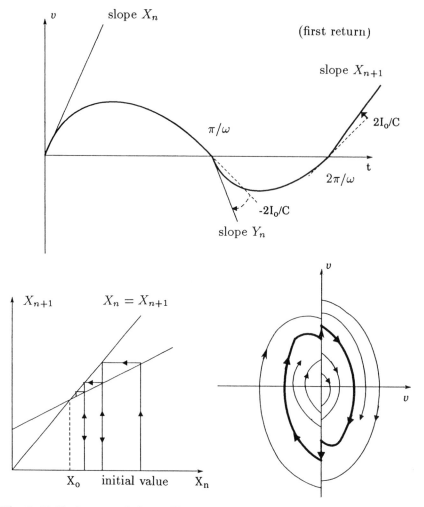

Fig. 5.17 Trajectory of the oscillator with a discontinuous conductance (Fig. 5.16).

We now take a trajectory defined by the initial conditions $v = 0$, $\dot{v} = X_n > 0, t = 0$ (Fig. 5.17). The first equation has the solution

$$v = \frac{X_n}{\omega} e^{-\alpha t} \sin \omega t \tag{5.93}$$

which is valid up to $t = \pi/\omega$. The curve eventually turns over and crosses the time axis at which point v goes to zero and the second equation becomes valid. Integration then gives (at $v = 0$)

$$\Delta \dot{v} = \frac{2I_0}{C} \frac{\dot{v}}{|\dot{v}|} \tag{5.94}$$

This *shock* produces an increase in the absolute value of $\dot{v}$ which, being negative, becomes even more negative (Fig. 5.17). The initial conditions for the next integration step therefore become

$$v = 0 \quad t = \frac{\pi}{\omega} \quad \dot{v} = -X_n e^{-\alpha\pi/\omega} - \frac{2I_0}{C} = Y_n \tag{5.95}$$

The first equation, once again valid, allows us to continue the calculation until the next crossing of $v = 0$, and so on.

The successive values $X_n, X_{n+1}\ldots$ can be linked by the recurrence relation

$$X_{n+1} = X_n - \left(e^{-\alpha\pi/\omega} + 1\right)\left[X_n\left(1 - e^{-\alpha\pi/\omega}\right) - \frac{2I_0}{C}\right] \tag{5.96}$$

and the steady state is given by

$$X_{n+1} = X_n \tag{5.97}$$

so that the amplitude becomes

$$X_n = X_0 = \frac{2I_0}{C\left(1 - e^{-\alpha\pi/\omega}\right)} \tag{5.98}$$

The frequency is obviously $\omega/2\pi$. Figure 5.17 illustrates the solution. When α is small, the signal is quasi-sinusoidal and the amplitude is given by the approximate expression

$$V_0 = \frac{2I_0}{C\alpha\pi} = \frac{4}{\pi}\frac{g}{\beta} \tag{5.99}$$

This agrees to within an order of magnitude with the the frequency analysis. On the other hand, when α is large, the spectrum contains a large number of harmonics. We then find that the amplitude of the third harmonic agrees with the result obtained by frequency analysis.

If the initial value of X_n differs from X_0, the solution tends to the previous state, as shown by the ratio of between X_{n+1} to X_n. The duration of the transient increases with α. Indeed, $(X_{n+1} - X_n)/X_n$ represents the reduction in the amplitude per period. When α is small and the amplitude close to that of the steady state, we have

$$\frac{1}{V}\frac{dV}{dt} \approx \frac{X_{n+1} - X_n}{X_n T} \approx \alpha\left(\frac{X_0 - X_n}{X_0}\right) \tag{5.100}$$

which is in good agreement with frequency analysis.

5.4 THE QUARTZ OSCILLATOR: AN EXCELLENT SECONDARY STANDARD

5.4.1 The quartz crystal

Piezoelectric resonators, of which the quartz resonator is a special case, are plates or rods cut from high-grade crystals along a given crystal direction. The geometrical shape is crucial because the frequency depends on one or several dimensions of the solid which acts as a resonant cavity for acoustic waves. The piezoelectric crystal is excited by an alternating electric field produced between electrodes deposited on (or near) the solid. The finite crystal vibrates in certain modes characterised by well-defined frequencies and Q; the electrodes are used to excite and the detect the resonances.

The aim is to produce a resonator with frequency as constant and as stable as possible (for the so-called time-frequency applications), but since as frequency depends on external factors such as temperature, pressure, force, acceleration and so on, the resonator can also be regarded as a receiver.

The resonantor can be represented by an equivalent electrical circuit. The most commonly used circuit is shown in Fig. 5.18. It is valid in the neighbourhood of a resonant frequency. The diagram clearly shows that there is a series resonance frequency f_s and a parallel resonance frequency f_p:

$$f_s = \frac{1}{2\pi\sqrt{L_1 C_1}} \qquad f_a \approx f_s \left(1 + \frac{C_1}{2C_0}\right) \tag{5.101}$$

where C_1 and C_2 are proportional to the electrode areas. For example, the fifth overtone 5-MHz resonator with an AT cut has the following parameters:

$$C_0 = 4\,\mathrm{pF} \quad \frac{C_0}{C_1} = 3600 \quad L_1 = 9\,\mathrm{H} \quad R_1 = 95\,\Omega \quad Q = \frac{L_1 \omega}{R_1} = 3 \times 10^6$$

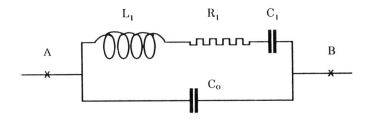

Fig. 5.18 Equivalent circuit diagram of the quartz resonator.

In first-order approximation, the reactance of the crystal is that of the static capacitance C_0, except near resonances where it can be deduced from the equivalent diagram.

The resonant frequency depends on the nature of the mechanical vibrations and on the dimensions of the crystal. By using bending modes, it is possible to operate near the kilohertz range; with high thickness-shear modes it is possible to reach several hundred megahertz. The frequency range is therefore very large.

The resonant frequency is a function of temperature, which is by far the most important characteristic of a resonator. Applications rely on minimum temperature dependence (control of frequency) or a dependence that is as linear as possible (thermometry).

The relationship between the resonant frequency and temperature T is usually described by an expansion around a temperature T_0:

$$\frac{f_T - f_{T_0}}{f_{T_0}} = a_0 \left(T - T_0\right) + b_0 \left(T - T_0\right)^2 + c_0 \left(T - T_0\right)^3 \qquad (5.102)$$

where a_0, b_0, c_0 depend on the direction of the cut and the vibration mode . Bechmann showed that the above expression is valid for a quartz resonator between $-200°C$ and $200°C$; the constants a_0, b_0, c_0 depend on the angles ϑ and φ at which the crystal is cut (Fig. 5.19). The curves representing this variation are therefore very useful to choose the cutting directions used in practice.

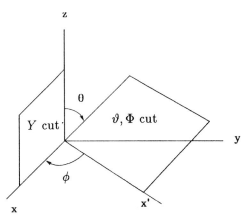

Fig. 5.19 Y cut and (ϑ, φ) cut (x, y and z are the axes of symmetry of the quartz crystal).

When $a_0 = b_0 = 0$ for a given vibration (shearing) mode, we obtain a useable cut. The frequency versus temperature function is then a cubic. The AT cut is obtained by simple rotation through an angle ϑ around the

x axis of the crystal from a Y cut (perpendicular to the y axis). The so-called double rotation cuts, obtained from the Y cut by rotation through an angle φ around the z axis, followed by rotation through an angle ϑ around the x' axis, are often used. When they are combined with low temperature sensitivity, low sensitivity to constraints and so on, they are generally superior to the AT cuts.

The Q factor of a resonator is given by

$$Q_i = \frac{\bar{C}}{\omega \eta_s} \tag{5.103}$$

where $\bar{C}$ is an elastic coefficient that depends on the crystal cut and the vibration mode, ω is the pulsatance of the vibration and n_s is the viscosity of the material. $\bar{C}$ is of the order of $10^{11} \text{N}/\text{m}^2$ and η is of the order of a few $10^{-3} \text{Ns}/\text{m}^2$. This corresponds to Q of the order of 3×10^6 at 5 MHz.

Actually, the above Q_i, which assumes a perfect resonator, is an intrinsic upper limit, since the true Q_r is such that:

$$\frac{1}{Q_r} = \frac{1}{Q_i} + \frac{1}{Q_1} + \frac{1}{Q_2} + \frac{1}{Q_3} + ... + \frac{1}{Q_n} \tag{5.104}$$

where each Q_n corresponds to a phenomenon that increases damping (mounting, loading by electrodes and so on).

The increasing demand for quartz resonators over the last ten years has led to considerable advances, comparable to those achieved in 1948 by Sykes who introduced deposited electrodes and, especially, by Warner in 1952 (energy trapping).

Recent advances have been due to better theoretical understanding of the resonator and to greatly improved technology. Actually, it has become necessary to reconsider completely the fundamental idea of the volume resonator such as that shown in Fig. 5.20.

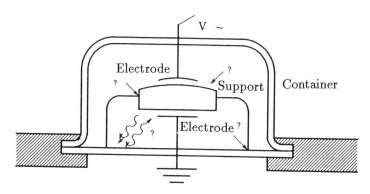

Fig. 5.20 The quartz crystal and its environment.

The better theoretical understanding of the resonator has been backed up by more accurate measurements. The old theoretical model of the infinite flat plate has been replaced by three-dimensional models, some of which are non-linear. Similarly, a much better understanding of mode-coupling phenomena has been achieved. Finally, a new thermal model of the resonator was recently proposed. It gives a much better representation of true conditions and allows the development of the so-called fast heating resonators.

A better treatment of limiting conditions and their technological implementation is considered crucial. Figure 5.20 shows that the crystal is a solid bounded by surfaces and connected to the external world by supports and electrodes. New technologies (ultrasonic machining, chemical and ion-beam machining, radiation treatment and so on) have therefore been essential in the fabrication of samples and supports, and also in the development of new cleaning and decontaminating techniques for crystals.

There have been improvements in the quality and commercial availability of crystals but, despite recent efforts, it is certainly in this area that the greatest efforts will be necessary to ensure that manufacturers can readily obtain a material of suitable quality. If there is little problem with very high purity crystals and medium quality resonators, an effort will certainly be needed to produce high-purity crystals with low dislocation density for advanced technological applications. These crystals are clearly a developing market.

In view of all this, new high-performance resonators have been developed, including the so-called flat pack ceramic resonators, BVA resonators, composite resonators and miniaturised resonators for the watch making industry. We note the current trend towards miniaturisation. Indeed, the clock and watch industries, particularly the latter, require resonator volumes of less than a few cubic millimeters. These resonators are being produced with a unit price of a fraction of a pound (250 to 300 millions are manufactured worldwide per annum). There is very considerable interest in different applications (low power consumption oscillators, rapid heating and so on). Furthermore, we note current developments in telecommunications whereby frequencies of the order of 100 MHz, a few hundred megahertz or even of the order of a gigahertz, are being approached.

The quartz crystal is the basis of the best resonators that can be described by classical physics. State of the art technology offers us frequencies with high long-term stability and Q factors as high as 10^6, but it is difficult to venture beyond a few hundred megahertz with the above volume-wave devices. Surface-wave technology is now available, using propagation velocities of the order of a few thousand metres per second. These waves can be trapped between two arrays of metal grids or grooves etched on the quartz wafer, thus providing an elastic resonator similar to the Fabry-Perot model. High modes of vibration can be used to reach frequencies beyond the gigahertz.

5.4.2 Electronics for quartz oscillators

It would be tedious to consider here all the possible types of oscillator. There are many types and very detailed descriptions can be found in the literature. We will briefly summarise the most common versions. Logic circuits can be used although we will mention only simple gate oscillators (there are also multiple-gate oscillators that are generally less stable). The basic diagram is given in Fig. 5.21. These oscillators are simple to make and although less stable than the oscillators considered below, they are satisfactory for many applications (in particular those with CMOS gates and frequencies of a few megahertz).

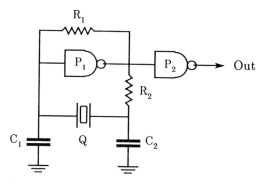

Fig. 5.21 Gated oscillator.

Active analogue components can be employed as in Section 5.2. The basic circuit is illustrated in Fig. 5.22 (without the supplies), the earth being connected to the emitter (Pierce), to the collector (Colpitts) or base (Clapp).

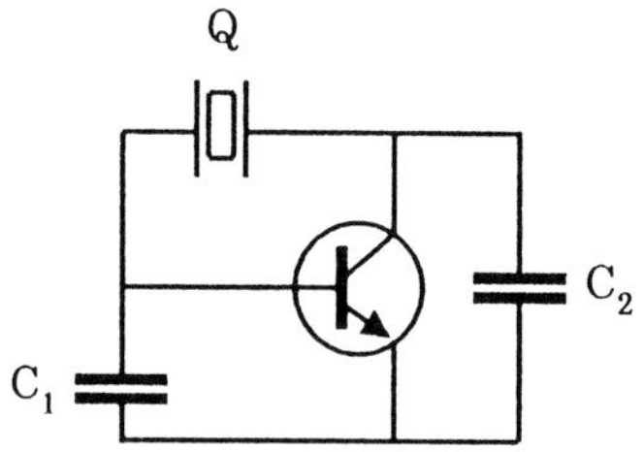

Fig. 5.22 Classical oscillator circuit.

The Pierce oscillator is used at higher frequencies with impedance inversion (up to 75 MHz) or the so-called *grounded-base* circuit (up to 150 MHz). The oscillators with even higher frequencies rely on special techniques that depend largely on the components and technologies employed.

One of the most important advances made during the last decade has been the introduction of measuring techniques that allow the quartz resonator and the associated electronics to be tested separately. It is now possible to identify the contributions of the electronics and of the resonator to resonator aging and stability. Putting it crudely, we can say that aging is mostly associated with the resonator. Short and medium term stability of a quartz oscillator is determined by amplifier noise and the $1/f$ noise of the resonator.

Stability can be improved by using the resonator at a higher power level, so long as the other performance indicators do not degrade (and this depends mainly on the resonator and, in particular, its departure from isochronism.)

5.4.3 Performance of quartz oscillators

The discovery of piezoelectricity by Pierre and Jaques Curie in 1880 was followed by a forty-year wait for the first quartz oscillator made by W.G. Cady. The oscillator incorporated three vacuum tubes and a quartz resonator in a feedback circuit. The Pierce oscillator was invented in 1921 and the first radio station controlled by reference to a quartz oscillator came on air in 1926 (this was the New York transmitter). Before the World War II the quartz oscillator was the subject of increasing interest because of its applications in military communications, but manufacturing remained at very low levels. Important advances had to wait for World War II during which around 150 million oscillators were manufactured after systematic industrial, scientific and technological development. Further rationalisation and development took place after 1945, especially in the USA (the beginning of the Frequency Control Symposium and of important scientific and technical organisations under the aegis of IEEE), in France, in England and in Japan. Recent advances in the understanding of resonators and oscillators have lead to the development of new oscillators whose performance has improved by more than two orders of magnitude in the last twenty or thirty years (Figs. 5.23 and 5.24).

Millions of oscillators are manufactured every year from the very high diffusion oscillators to the very small number of oscillators with performance close to the rubidium atomic clocks. Virtually all the time-frequency devices contain a quartz oscillator. They are widely used in civil and military communications, in radio navigation and radar, in different systems for satellites, planes, boats, and missiles, in frequency, position and velocity feedback systems, in metrological instruments (including chronometers), in television, minicomputers and so on. The frequencies run from a few kilohertz to a few hundred megahertz, and stabilities range from a few parts in 10^5 to a few parts in 10^{13}. The quartz oscillator market is by far the most important market for radio-frequency oscillators. The need for improved

Stability (short term)

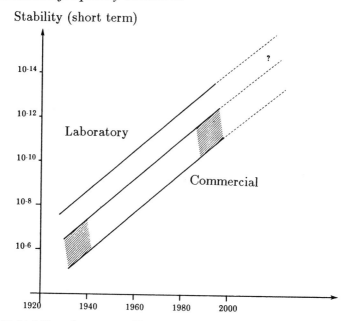

Fig. 5.23 Stability of quartz oscillators.

per day

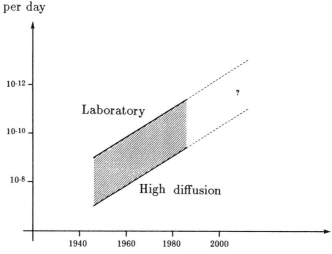

Fig. 5.24 Aging of quartz oscillators.

performance is often matched by the advantages of mass production. This has substantially accelerated research in the field with the result that important progress has been made under constant pressure from customers.

There is obviously a big difference between a quartz oscillator in a watch and the quartz oscillator controlling the short-term stability of a hydrogen maser whose performance is better by a factor of 1,000 or 10,000. Similarly, there is a huge difference between the simple time piece and temperature-compensated or temperature-controlled precision oscillators. There are also oscillators for specific applications (watches, television and so on). The table below lists the average performance of devices ranging from the simple time piece to the high-precision oscillator. We ought to add that the oscillators usually employed in watches are high-grade devices that are contained (and this includes the power supply) in less than $1 \, cm^3$!

Current advances are focussed on improved performance, but also on improved response to ambient changes. For example, sensitivity to acceleration and vibration has been reduced by at least an order of magnitude during the last decade. It is possible to attain a sensitivity of less than $10^{-10}/g$. The influence of external temperature can be reduced to less than 10^{-12} per °C and of the external pressure to less than 10^{-10} to 10^{-11} per bar. The last twenty years have brought considerable advances in the performance of quartz oscillators as much at the scientific as at the technical and industrial levels. In view of recent and important innovations in ideas and techniques in this field, further progress may be expected. Quartz oscillators are small, reliable and relatively cheap. Their short-term stability remains unmatched, and definite improvements have been achieved in respect of aging and sensitivity to the environment. The subject is ripe for further important developments

5.5 THE ELECTROMAGNETIC CAVITY OSCILLATORS

5.5.1 The electromagnetic resonator

Piezoelectric resonators are difficult to construct above $\sim 1 \, GHz$. In contrast, electromagnetic cavity resonators are then readily fabricated. These cavities have metal walls and linear dimensions of the order of half the wavelength in the fundamental mode. They are well suited to the frequency range $\sim 1 - 100 \, GHz$, but it is still necessary to have high Q and a stable enough characteristic frequency before we can consider making a practical oscillator.

The energy stored in an electromagnetic cavity oscillates between the electrical and magnetic forms, but their mean value remain constant in time. The magnetic energy density $B^2/2\mu_0$ is not the same at every point in the cavity. We can define a mean of B^2 in the whole volume (we will denote it by $\overline{B_v^2}$) and express the energy in terms of it:

$$W = \frac{1}{2\mu_0}\overline{B_v^2}V \qquad (5.105)$$

where V is the volume of the cavity. Energy is lost by currents induced in the walls, and it is the magnetic induction B tangential to the metal wall

Comparison of different families of quartz oscillators

Characteristic	Simple time reference	Mass produced commercial models	Accurate commercial models	Laboratory prototypes
Volume	1 to 15 cm^3	10 to 20 cm^3	$\approx 1\,000$ cm^3	$\approx 1\,000$ cm^3
Aging	1 to 50 $\times 10^{-6}$ cm^3 per annum	10^{-7} per day 10^{-6} per annum	3 to 8×10^{-11} per day	3×10^{-10} - 10^{-9} per annum modelled
Short term stability $\tau = 10^3$ s $\tau = 1$ s $\tau = 100$ s		10^{-9}	$< 10^{-9}$ $< 10^{-12}$ $< 10^{-12}$	10^{-11} $< 5 \times 10^{-13}$ $< 2 \times 10^{-13}$
Phase noise 1 Hz 10 Hz 10^4 Hz			- 118 dB - 152 dB - 140 dB	- 122 dB - 152 dB - 164 dB
Accel. sensitivity	2×10^{-9}/g	1 to 2×10^{-9}/g	3 to 10×10^{-10}/g	3×10^{-11}/g max. SC 7×10^{-11}/g max. AT
Final frequency obtained to	non-thermostated	2×10^{-8} after 1 h	10^{-9} after 1 h	10^{-9} after 40 min
Reproducibility	10^{-6} to 10^{-5}	10^{-7}	10^{-9} to 10^{-10}	10^{-10} to 10^{-11}

that gives rise to these currents. We can show that the energy lost by Joule heating of the wall per unit area per period $1/f$ is

$$\Delta W_1 = \sqrt{\frac{\pi}{\mu_0 f \sigma}} \frac{B^2}{2\mu_0} \tag{5.106}$$

where σ is the conductivity of the wall material. The total energy lost per period is

$$\Delta W = \frac{1}{2\mu_0} \sqrt{\frac{\pi}{\mu_0 f \sigma}} \overline{B_s^2} S \tag{5.107}$$

where $\overline{B_s^2}$ is the mean of B^2 over the whole area S of the cavity. This gives

$$Q = \frac{2\pi W}{\Delta W} = \frac{2}{\delta} \left(\frac{V}{S}\right) \frac{\overline{B_v^2}}{\overline{B_s^2}} \tag{5.108}$$

and the skin depth is

$$\delta = \frac{1}{\sqrt{\pi \mu_0 \sigma f}} \tag{5.109}$$

The volume-to-area ratio V/S of the enclosure is of the order of the linear dimensions of the cavity, i.e., $\lambda/2$. Since the mean squares of B are not very different, we obtain

$$Q \approx \frac{\lambda}{\delta} = \frac{5 \times 10^9}{\sqrt{f}} \tag{5.110}$$

When $\sigma = 6 \times 10^7$ mho/m (copper), we can expect $Q = 10^4 - 10^5$, i.e., potentially excellent oscillators for the frequency range of interest, provided the mean resonator frequency is stable.

To increase Q, we can concentrate the electromagnetic energy at the centre of the cavity in order to make B^2 as small as possible along the walls. A small piece of material with high permittivity ε or a high magnetic permeability μ is placed in the cavity and traps a large fraction of the energy in the form of stationary waves. The fields are weak in the neighbourhood of the metal walls that are there to prevent the system from radiating. It is obviously advisable to choose the materials so that they do not introduce additional losses. One standard method employs a ferrite ball (yttrium iron garnet or YIG) whose high permeability can be adjusted by means of a constant external magnetic induction. The resonant frequency can be tuned in this way.

Unfortunately, the walls of such cavity resonators become very hot because of Joule heating, their dimensions are temperature dependent and their shape is sensitive to vibrations, so that they cannot be used as standards. The oscillator must therefore be controlled by a clock, which requires the adjustment of frequency. The frequency of the YIG resonator can be adjusted by external magnetic induction. We shall see in the next Section that the device in which oscillations are maintained by an electron beam, i.e., the reflex klystron, also has this essential characteristic property.

Before we proceed from the resonator to the oscillator, we have to introduce an active circuit, which presents two problems. The resonator and the active system have to be coupled so that the frequency is essentially governed by the resonator, which is not easy at microwave frequencies at which the slightest discontinuity (connection) introduces parasitic elements. Furthermore the active element must be able to operate at these high frequencies. The most ingenious arrangement is a cavity containing an electron beam that maintains the oscillations. The resonator must have a strongly capacitive zone in which an alternating electric field predominates and a strongly inductive zone in which the magnetic energy is concentrated. Figure 5.25 illustrates this type of cavity, known as a *rhumbatron*. In the notation of the figure, the fundamental mode frequency is

$$f \approx \frac{c}{2\pi} \frac{\sqrt{dh}}{2R} \qquad (5.111)$$

which gives 3 GHz for dimensions of 1 cm Moreover

$$Q \approx \frac{R}{\delta} \qquad (5.112)$$

which gives values of the order of 10^4–10^5. These cavities incorporate grids in the capacitive part, which are crossed by the electron beam and in the exchange the energy takes place. This is the principle of the klystron.

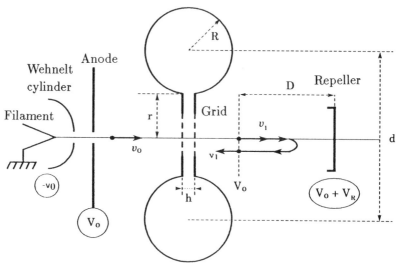

Fig. 5.25 Diagram of a klystron.

5.5.2 The principle of the reflex klystron

An electron gun produces a continuous beam of electrons (Fig. 5.25). The electrons arrive at the first grid with velocity

$$v_0 = \sqrt{\frac{2eV_0}{m}} \tag{5.113}$$

where V_0 is the accelerating voltage.

The electrons cross the electric field between the two grids. If the transit time which is of the order of h/v_0 is small compared to the period $2\pi/\omega$ (this requires a high accelerating voltage), the crossing is quasi-instantaneous. The velocity at exit from the cavity is then

$$v_1 = v_0 \left(1 + \frac{V_1}{2V_0} \sin \omega t_0 \right) \qquad (V_1 \ll V_0) \tag{5.114}$$

The electrons are slowed down by a negative electrode known as the *repeller* whose potential $V_0 + V_R$ is high enough to reflect are all electrons back to the grid system. The transit time taken by an electron between the second grid $(t = t_0)$ and back again (t_1) is simple to determine and is given by

$$t_1 = t_0 + \frac{4D}{v_0} \frac{V_0}{V_R} \left(1 + \frac{V_1}{2V_0} \sin \omega t_0 \right) \qquad (V_R > V_0) \tag{5.115}$$

During a short interval dt_0 the beam transports charge $I_0 dt_0$ where I_0 is the beam current. After reflection, this charge is contained within the time interval dt_1, obtained by differentiating the above equation. The returning current I at the time t_1 is

$$I(t_1) = I_0 \frac{dt_0}{dt_1} = \frac{I_0}{1 + 2\beta \cos \omega t_0} \tag{5.116}$$

where

$$\beta = \frac{\omega D V_1}{v_0 V_R} \tag{5.117}$$

and t_0 is related to t_1 – the only useful variable – by (5.115).

$I(t_1)$ is a periodic signal that we can expand into a Fourier series and retain only the fundamental term. If the cavity has a high enough Q, we can write

$$I(t_1) = A \cos \omega t_1 + B \sin \omega t_1 \tag{5.118}$$

where

$$A = \frac{\omega}{\pi} \int_0^{2\pi/\omega} I(t_1) \cos \omega t_1 dt_1 \qquad B = \frac{\omega}{\pi} \int_0^{2\pi/\omega} I(t_1) \sin \omega t_1 dt_1 \tag{5.119}$$

Changing the integration variable and replacing t_1 by t_0, we can express the current in terms of the Bessel function of order 1:

$$I = 2I_0 J_1(2\beta) \cos \omega (t_1 - \vartheta) \tag{5.120}$$

where

$$\vartheta = \frac{4DV_0}{v_0 V_R} \tag{5.121}$$

so that the voltage between the grids at the time t_1 is $V_1 \sin \omega t_1$, whilst the component of the current with the same frequency is given by the above expression. This is exactly as if there was an admittance Y (ratio of current to voltage) between the two grids. Hence

$$Y = \omega \vartheta \frac{I_0}{V_0} \frac{J_1(2\beta)}{2\beta} e^{-j(\omega\vartheta - \pi/2)} \tag{5.122}$$

The admittance Y is in parallel with the cavity admittance which is given by

$$Y_0 = G_0 \left(1 + 2jQ\frac{\omega - \omega_0}{\omega_0}\right) \tag{5.123}$$

in the neighbourhood of the resonant frequency. Sinusoidal oscillations will be maintained if the total admittance is zero, which leads to the two equations

$$ctg\omega\vartheta = 2Q\frac{\omega - \omega_0}{\omega_0} \qquad \frac{J_1(2\beta)}{2\beta} = -\frac{G_0 V_0}{\omega\vartheta I_0 \sin \omega\vartheta} \tag{5.124}$$

The first equation determines the locking frequency which is in practice very close to ω_0 and depends on ϑ and, hence, on V_R. The klystron is therefore an oscillator whose frequency can be tuned by means of an external voltage. The second equation determines the value of β and, hence, of the voltage V_1, the amplitude of the resulting oscillations. It has a solution if I_0 is greater than the locking threshold. Several modes of operation are possible; we work with the fundamental by adopting the smallest possible value of I_0.

5.5.3 Secondary standards at microwave frequencies

Reflex klystrons can cover a range extending from a few gigahertz to 220 GHz with power outputs between a few milliwatts and several watts. The electron-beam tuning plateau lies between 10 MHz and 100 MHz; it allows easy locking. By distorting the elastic cavity wall, we can achieve mechanical tuning of the frequency between a few hundred megahertz and a few gigahertz. These simple, robust and relatively low-cost tubes are commonly used as oscillators.

The two-cavity klystron relies on a principle similar to that discussed in the last Section. The beam crossing the first cavity is velocity modulated and covers a distance D to reach the second cavity. This produces a current modulation. The second cavity is therefore crossed by a modulated beam which can transfer its energy if the electron bunches have a suitable phase relative to the electric field. An oscillator is produced by coupling the two cavities while preserving the phase relation (in the reflex klystron this coupling is achieved by merging the two cavities). These two-cavity klystrons can operate as amplifiers or oscillators and are clearly superior to reflex klystrons as far as noise is concerned. The FM noise spectrum in the X band (10 - 12 GHz) at 1 kHz carrier can be lower than 110 dB/Hz. When it is long-term locked to a quartz oscillator harmonic, the two-cavity klystron is the best secondary standard at microwave frequencies.

To obtain other standards, we can couple an active device to a cavity in accordance with the classical, lumped-parameter oscillator circuit. There are many devices that can operate in the microwave frequency band . Some are extensions of medium-frequency devices, whilst transistors with very small dimensions can still operate near 20 GHz. Other systems rely on physical processes similar to that underlying the klystron (e.g., transit-time modulation of beam current). Transit-time diodes (IMPATT and Gunn) can be used to reach 100 GHz.

In contrast, the principle of the klystron can be exploited in more depth by incorporating the electron beam in the cavity. The basic idea is to use a slow-wave guide with the electron beam propagating along its axis. Fine tuning then concentrates the electrons in zones of decelerating electric field, so that the beam energy is transferred to the electric field which becomes stronger. This is the principle of the travelling-wave tube. Everything happens as if the guide had a negative damping coefficient. To produce an oscillator, it is sufficient to have a reaction, for example, by closing the guide on itself. This is the principle of the magnetron. We can also place an amplifying guide between two mirrors, which is the principle of the laser, rarely used at microwave frequencies,(*cf*. Chapter 4). Finally – and this is the best solution – we can use the reverse propagation of particular modes to ensure the necessary reaction. This is the principle of the carcinotron, which continues to operate up to 300 GHz. These oscillators do not all have the properties that would clearly qualify them as secondary standards.

5.6 SOME COMPLEX PHENOMENA IN OSCILLATORS

5.6.1 Single-mode oscillation in a multimode resonator

Resonators with acoustic (quartz) or electromagnetic cavities are multimode devices. The electronics that produces the negative conductance can operate over a spectrum broad enough to contain several characteristic frequencies, and oscillations occur at different frequencies. In reality, the different oscillations are coupled and some are suppressed. Singlemode operation is usable as long as it is stable. We shall now examine the behaviour of the circuit of Fig. 5.26 which involves two coupled modes. We will determine whether operation in a single sinusoidal mode is possible. We put

$$v_1 = V_1 \sin \omega t \qquad v_2 = V_2 \sin (\omega t + \varphi) \qquad (5.125)$$

which will allow linearisation. Clearly, this method can only be used if the signals are sinusoidal; a linear combination of sinusoidal signals with different frequencies will not allow this simplification.

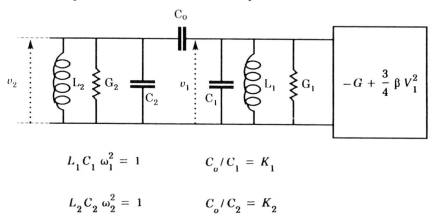

$$L_1 C_1 \omega_1^2 = 1 \qquad\qquad C_o / C_1 = K_1$$

$$L_2 C_2 \omega_2^2 = 1 \qquad\qquad C_o / C_2 = K_2$$

Fig. 5.26 Oscillator containing a two-mode resonator.

The behaviour of the negative-conductance circuit of Fig. 5.26 is readily established after linearisation. It is described by

$$i = \left(-G + \frac{3}{4}\beta V_1^2\right) v_1 \qquad (5.126)$$

We seek a solution such as (5.124), and obtain two real equations:

$$[(1 + K_1)\omega^2 - \omega_1^2]\, [(1 + K_2)\omega^2 - \omega_2^2] - K_1 K_2 \omega^4 -$$

$$-\frac{G_2 \omega^2}{C_1 C^2}\left[G_1 - G + \frac{3}{4}\beta V_1^2\right] = 0$$

$$[(1 + K_1)\omega^2 - \omega_1^2]\left[G_1 - G + \frac{3}{4}\beta V_1^2\right] + \frac{G_1 G_2}{C_1}\,[(1 + K_2)\omega^2 - \omega_2^2] = 0$$

$$(5.127)$$

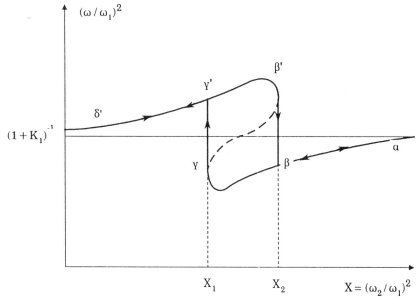

Fig. 5.27 Frequency of an oscillator with two strongly-coupled modes.

The oscillation frequency is obtained by eliminating V_1 between these equations. It is shown in Fig. 5.27 as a function of $X = (\omega_2/\omega_1)^2$.

The circuit therefore exhibits simple behaviour for values of X outside the interval $[X_1, X_2]$, i.e., for frequencies ω_1 and ω_2 that are very different. A single frequency is possible, in which case the oscillator is a single-mode device. We can now perform a perturbative calculation analogous to that of Section 5.3.3 and show that this solution is stable. The interpretation is simple. If the characteristic frequencies of the two coupled resonators are very different, single locking takes place. The oscillation frequency is close to

$$\omega \approx \frac{\omega_1}{\sqrt{1 + K_1}} \qquad (5.128)$$

The oscillator selects from the two possible frequencies the frequency closest to ω_1, i.e., the frequency of the resonator that maintains the negative conductance. The other frequency does not allow locking to take place, which can be verified by by small-signal analysis of the circuit. The corresponding poles have negative real parts. Since the locking frequency depends in a complicated manner on several parameters that are difficult to control, such an oscillator will qualify as a standard standard only when a detailed study, such as that outlined in the next Section, confirms the above preliminary analysis.

The problems are much more complicated when $(\omega_1/\omega_2)^2$ lies between X_1 and X_2. There are then three possible frequencies for the single-mode regime. They correspond to three different amplitudes. Stability arguments then show that the intermediate-frequency regime is unstable and the two others are stable as long as they do not exist simultaneously. We must therefore distinguish two possible cases.

Let us first consider the circuit working with a very high value of X. Only one frequency is then possible and we have single-mode operation. Coupling can be adjusted to alter the separation between the characteristic frequencies, e.g., by acting on the self-inductance L_2. The single-mode regime continues up to the value of X_1 that corresponds to the path α, β, γ inf Fig. 5.27. At γ we observe a frequency jump, and an amplitude jump occurs at γ, which takes us to the branch $\gamma'\delta'$, and the single-mode regime continues. Clearly, the intermediate branch corresponds to unstable operation. If we follow the curve in the reverse direction, i.e., for X increasing, the frequency and amplitude jumps occur at β', and we have hysteresis. The single-mode oscillator, locked to a frequency close to ω_1, tends to perist in this state. If it is forced to move away from it, e.g., by adjustment of its parameters, it jumps to another regime that brings it closer to ω_1.

Difficulties appear when the oscillator is started with X lying between X_1 and X_2. Two stable regimes are then possible, but neither is yet established. The linear small-signal theory shows that both can start, since all the poles have positive real parts and that either can be stable in single-mode steady state. More advanced theory shows that the simultaneous initiation of both modes produces a coupling between them, with one mode suppressing the other and becoming predominant. However, the theory does not allow the two cases to be easily distinguished because the initial conditions are crucial here. Figure 5.28 illustrates the situation. The ball rolling with friction on the inclined section can stop in either of the two valleys. A very small change in initial conditions can cause a large difference in the final state because everything depends on the last crossing of the position of unstable equilibrium, i.e., the intermediate peak. The ball can go over with near-zero velocity or, on the contrary, it can reverse its direction at the very last moment and run back. This extreme sensitivity to the initial conditions will be encountered again in the next Section.

5.6.2 Time-domain analysis of oscillators with more than two variables of state: bifurcation and the risk of chaos

The above method reveals only the sinusoidal steady state, if it exists. It cannot tell us anything about transient phenomena or quasi-periodic multimode régimes. The only way to advance our discussion is by time-domain analysis with simple models such as the circuit shown in Fig. 5.19, which can be treated by the piecewise-linear approximation

The diodes produce sharp points on the piecewise-linear characteristic. This allows the simulation of the curvature of the characteristic, which frequently plays an important part in practice.

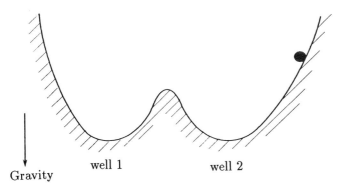

Fig. 5.28 A system with two positions of stable equilibrum (the ball is thrown in and its motion is damped by gravity and friction).

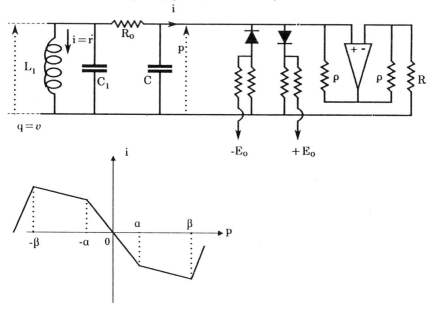

Fig. 5.29 Simple oscillator with three variables of state.

The circuit of Fig. 5.29 can be described by three variables of state, namely, p, q, r that satisfy the following equations:

$$\dot{p} = \frac{1}{CR_0}p - \frac{1}{CR_0}q + \frac{1}{L}i(p)$$

$$\dot{q} = \frac{1}{C_1 R_0}p - q - \frac{1}{C_1}r \qquad (5.129)$$

$$\dot{r} = \frac{q}{L}$$

where i is a function of p, represented by straight sections. These equations are readily integrated for $|p| < \alpha$, then for $\alpha < |p| < \beta - \alpha$ and, finally, for $|p| > \beta$. The phase trajectory is obtained by linking curve segments described by exponentials.

The p, q, r phase space (Figure 5.30) has three dimensions. Step-by-step integration is simple and efficient in two dimensions, but cannot readily be used in three dimensions. However, the geometrical representation is still useful, as we shall see. Actually, the oscillator of Fig. 5.29 is described by three variables of state and is very simple; most of classical multimode oscillators require at least four variables of state, in which case phase space is too complicated to use, and there is no simple geometrical representation to help with our discussion.

Let us start with the point A, which corresponds to $p = \alpha$, and two arbitrary initial values q_0 and r_0. After one turn, we reach B, which again corresponds to $p = \alpha$ and to values q_1 and r_1 that depend, of course, on q_0 and r_0. A periodic state is established if after the first turn

$$q_1 (q_0, r_0) = q_0 \qquad r_1 = (q_0, r_0) = r_0 \qquad (5.130)$$

The solution of these equations provides us with the initial conditions $q_0 = Q, r_0 = R$ that generate a periodic regime (and, hence, the frequency f, too). This nonsinusoidal regime has numerous harmonics with frequencies that are simple multiples of f.

The next step is to determine whether this regime is stable. This can be done by a perturbative calculation with initial conditions $Q + \Delta Q, R + \Delta R$, as was done earlier. However, the calculations are only feasible when powerful computers are available. We start with arbitrary initial values and compute the successive returns illustrated in Fig. 5.30. In this way we can establish whether the successive points $A, B, C, \ldots$ tend towards the point Q, R that corresponds to the steady state. This state is stable and Q, R is a point attractor on a basin of attraction.

The oscillator cycle is a skew curve that can be described by its projections on the p, r and q, r planes. It contains a loop.

The above equations may not have a stable solution, in which case we have to look for periodic solutions that occur on second return. We therefore calculate q_2 and r_2 where

$$q_2 (q_0, r_0) = q_0 \qquad r_2 = (q_0, r_0) = r_0 \qquad (5.131)$$

Let Q_2, R_2 be a solution of these equations. If we start at this point, the first return takes us to Q_2', R_2' and the second, obviously, to Q_2, R_2. It is clear that if we start from Q_2', R_2', we reach Q_2, R_2 on first return and again Q_2', R_2' on the second. The periodic steady state reached on second return can be obtained for two initial values. There are therefore two point

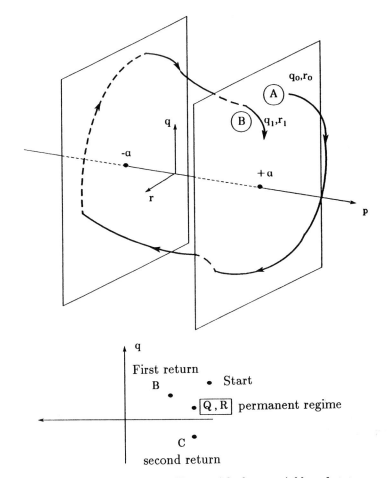

Fig. 5.30 Phase space of an oscillator with three variables of state.

attractors or none. Stability now has to be investigated by examining the transient state. The two-loop cycle and the frequency must be calculated. Finally, the curved surfaces of attraction associated with each attractor are determined, although the boundaries are often difficult to establish.

When the state with a single point attractor becomes unstable, it is generally succeeded by a state with two stable point attractors. This transition phenomenon is known as bifurcation. The one-loop cycle with frequency f_0 gives way to a cycle with two loops, and the frequency becomes $f_0/2$. However, the regime is not sinusoidal, i.e., harmonics are present and include the frequency f_0 (Fig. 5.31). The number of lines in the spectrum increases, other bifurcations can occur, the situation becomes very complicated and we can imagine the appearance of stable state states corresponding to an

n-point attractor. The curved surfaces of attraction are very difficult to determine, and the q, r phase plane is filled with a very large number of attractors, both stable and unstable, depending on the values of the circuit parameters.

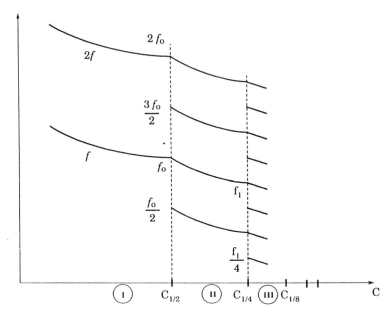

Fig. 5.31 Bifurcation.

Bifurcation is readily observed, e.g., by varying the capacitance C in the circuit of Fig. 5.29. It is a troublesome phenomenon because it causes frequency demultiplication. It can be avoided by a suitable choice of parameters. Unfortunately, bifurcations can become more and more numerous, and get closer and closer, with the spectrum acquiring an increasing number of closely spaced lines. When the frequency spacing is smaller than the resolving power of the spectrum analyser, the spectrum becomes effectively continuous. When the bifurcations become so frequent that the required parameter change is smaller than the precision with which this parameter can be set, we no longer know which regime we are in. This is the state of apparent chaos, resulting from an amplification of the uncertainties. The oscillator then behaves as a source of coloured noise.

More advanced analysis reveals the strange phenomena that constitute true chaos. The number of bifurcations continues to increase and tends to infinity for *a particular value of the tuning parameter*. The spectrum is then rigourously continuous (in the apparent and readily observable chaos, the spectrum is continuous only because of the overlap of very closely-spaced spectral lines). The point attractors, now infinite in number, produce a complicated figure on the q, r plane, known as a *strange attractor*. An infinite set of points constitute a curve (in one dimension), a surface (in two dimensions) or any other *fractal* structure with *fractional* dimension. Whilst apparent chaos is a *blur* due to the mixing of a very large number of modes under the influence of background noise, real chaos is a strange phenomenon. The oscillator appears to function as a noise generator, but this is not at all the definitive disorder observed when random noise due to the superposition of a very large number of elementary phenomena is examined. Indeed, we need only change the tuning parameter very slightly to return to a very simple oscillation.

There is no theory at present that can predict the onset of true chaos from the form of the basic equations. Chaos can set in as soon as the nonlinear equations involve more than two variables of state. It seems that the probability of the onset of chaos increases with the number of variables of state. However, even in the absence of chaos (even apparent chaos), we often encounter bifurcation, which becomes a disagreeable phenomenon in a circuit in which a particular frequency has to be maintained. The parameters must therefore be carefully chosen to avoid bifurcation.

5.6.3 Locking to an external signal

An oscillator cannot be completely protected from its environment. It is physically supported by it, it is supplied by an external source of energy and it has to provide us with a signal, which implies coupling to a measuring instrument. The oscillator will therefore be constantly influenced by its environment. As a first approximation, this can be modelled by an external generator.

We begin with the simple case of a sinusoidal driving signal of pulsatance ω_1 (Fig. 5.32), and perform a time-domain analysis by the piecewise-linear method used in the simplified form in Section 5.3.4. The system is described by

$$\ddot{v} + 2\alpha\dot{v} + \omega_0^2 v = \frac{I_1\omega_1}{C}\cos\omega_1 t \qquad (v \neq 0)$$

$$\ddot{v} + \left[2\alpha - \frac{2I_0}{C}\delta(v)\right]\dot{v} + \omega_0^2 v = \frac{I_1\omega_1}{C}\cos\omega_1 t \qquad (v = 0) \qquad (5.132)$$

At time t_i, we start with $v = 0$ and derivative $\dot{v} = X_n$, which allows us to take the first step in the calculation. The solution of the first equation is

$$v = Ae^{-\alpha t}\sin(\omega t + \Phi) + \frac{I_1\omega_1}{C\sqrt{(\omega_0^2 - \omega_1^2)^2 + 4\alpha^2\omega_1^2}}\sin(\omega_1 t + \Phi) \quad (5.133)$$

where

$$\omega = \sqrt{\omega_0^2 - \alpha^2} \qquad \tan\Phi = \frac{\omega_0^2 - \omega_1^2}{2\alpha\omega_1} \qquad (5.134)$$

$I_1 \sin \omega_1 t$

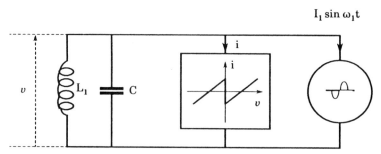

Fig. 5.32 Simple oscillator locked to an external reference frequency.

The initial conditions t_i and X_n allow us to determine the integration constants A and Φ and then the time of first crossing of the point $v = 0$, as well as the slope $\dot{v}$ at this point (Fig. 5.33). The crossing of $v = 0$ introduces a break in the slope of the curve, as we saw in Section 5.3, which leads to the initial conditions for the second branch of the curve. The calculation thus proceeds step by step.

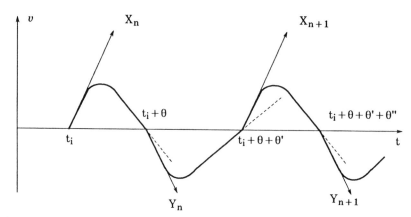

Fig. 5.33 Transient behaviour of a locked oscillator.

We now look for particular solutions, e.g., a periodic state of pulsatance ω_1. The branch that has been calculated then represents half a period. We must have $\vartheta = \pi/\omega_1$ and, simultaneously, Y_n must equal $-X_n$, so that the next step is identical to the previous one with the sign reversed. These two conditions allow the determination of the starting time t_0 and the initial value of the derivative X_0 that give this periodic state the same frequency as the driving signal, i.e., the oscillator becomes locked to the external signal. It can be shown that the equations have solutions provided

$$\frac{I_1}{I_0} \geq \frac{\sqrt{(\omega_1^2 - \omega_0^2)^2 + 4\alpha^2\omega_1^2}}{\omega\omega_1} \frac{|\sin \pi\omega/\omega_1|}{\cosh \pi\alpha/\omega_1 + \cos \pi\alpha/\omega_1} \qquad (5.135)$$

where, to calculate Φ, we use the fact that $\cos(\omega_1 t_i + \Phi)$ lies between -1 and +1.

Figure 5.34 shows the locking zone near ω. If the driving frequency is equal to that of the isolated oscillator, locking has obviously been achieved. If the frequencies are different, the amplitude has to exceed a threshold. The oscillator then 'forgets' its frequency and locks to the frequency of the external generator. As long as ω_1 is close to ω, we have

$$I_1 \geq \frac{4I_0}{\pi\alpha} |\omega_1 - \omega| \tag{5.136}$$

where α is small, i.e., we are dealing with a high Q resonator. For given driving amplitude, the size of the locking zone increases with α, i.e., with decreasing resonator Q. A good oscillator is difficult to lock except for the neighbourhood of its characteristic frequency.

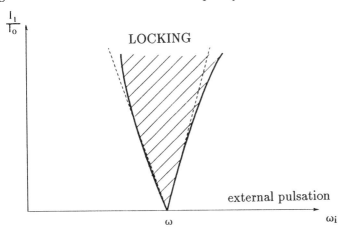

Fig. 5.34 The locking zone.

The next step is to determine whether the locked regime is stable. We proceed as in Section 5.6.2, starting with arbitrary initial conditions and consider the evolution of the system in terms of the coordinates X_n and sampling time (Fig. 5.35). It is interesting to fold this diagram a number of times equal to a multiple of $2\pi/\omega_1$ to bring out the point attractor corresponding to locking. The figure illustrates this process: the point with coordinates X_0, t_0 represents the locking attractor.

Locking does not occur for small-amplitude driving signals, which do, nevertheless, disturb the operation of the oscillator. The oscillator is a nonlinear circuit that receives a signal of pulsatance ω_1 which *beats* with the oscillator frequency. Full analysis, illustrated in Fig. 5.36, shows that there are parasitic lines and that the oscillator frequency tends to approach

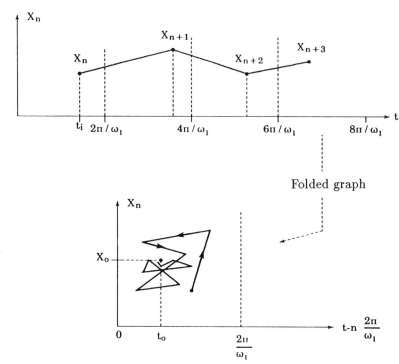

Fig. 5.35 Attractor of a locked oscillator.

the driving frequency. All periodic external factors are harmful to the maintainance of a well-defined frequency.

Continuing the calculation, we can attempt to compute step by step a regime of frequency $\omega_1/4\pi$, i.e., half that of the driver. This is done in four steps (Fig. 5.33), each step providing the initial conditions for the next one. As we cross the point $v = 0$ for the second time in the same direction (second return), the time is $t_i + \vartheta + \vartheta' + \vartheta'' + \vartheta'''$ and the slope is X_{n+2}, so that the half-frequency state is defined by the following conditions:

$$\vartheta + \vartheta' + \vartheta'' + \vartheta''' = \frac{4\pi}{\omega_1} \qquad X_{n+2} = X_n \qquad (5.137)$$

The initial conditions obtained by solving these equations determine the position of the attractor. We saw in a similar case in Section 5.6.2 that there are two possible solutions. There are two attractor points. The half-frequency regime is observed only if it is stable, which can be established with the help of a computer, as in the case of the simple locking. The basins of attraction, which are very complicated, can also be obtained. We note that these regimes of frequency demultiplication demand that the amplitude of the driving signal exceeds a threshold that can be very small if the driving frequency is close to half the oscillator frequency.

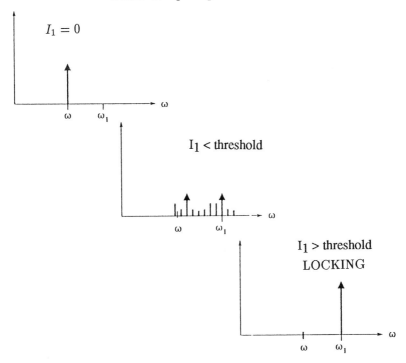

Fig. 5.36 Spectrum of locked oscillator.

Any periodic external signal is harmful to the operation of an oscillator. Even a very small amplitude can generate parasitic lines, cause a frequency drift and produce frequency demultiplication. Just as in the case of Section 5.6.2, bifurcation corresponding to demultiplications by four or by eight can occur. Demultiplications by three and other odd numbers is also possible. Apparent chaos appears as soon as the numerous frequencies due to beats are sufficiently close together.

5.6.4 Effect of background noise on an oscillator

An oscillator incorporates numerous sources of noise. Active elements such as transistors introduce Schottky noise, passive elements produce thermal noise and the environment makes its own contributions. This is one of the most important problems in the physics of oscillators. The interfering signal has a very small amplitude and a broad spectrum. Obviously, it will not give rise to a locking phenomenon, but we have seen that a very small amplitude sinusoidal driving signal can generate multiple lines and frequency variations. We may therefore expect that a driving signal with a continuous spectrum will generate a continuous spectrum. Other difficulties arise from the fact that noise entering the circuit within the spectral interval $\omega, \omega + \Delta\omega$ can be due to low frequencies translated by the circuit nonlinearities (*cf.* Section 3.4). Finally, in the case of a linear circuit, we know how to replace all the sources of noise by a single equivalent source at the input. However, the situation is not then as simple as in the case of a nonlinear circuit.

We can nevertheless attempt to use linearisation to obtain a few results. Let us assume that all the sources of noise have been collected together in a single generator placed at the input of the circuit and characterised by its power spectral density. We will adopt the simple model of an oscillator illustrated in Fig. 5.37.

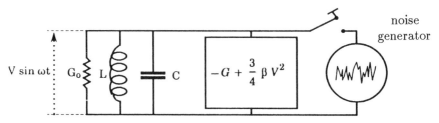

Fig. 5.37 Oscillator incorporating a generator of noise.

A resonator with

$$Q = \frac{\omega_0}{\Delta\omega_0} = \frac{1}{G_0}\sqrt{\frac{C}{L}} \tag{5.138}$$

is associated with a negative conductance. The noise can be represented symbolically by a single generator of random current. Noise is removed by opening the switch. Linearisation is then simple, as we saw earlier. The linearised conductance becomes $-G + 3/4\beta V^2$ and Kirchhoff's law leads to

$$\left[jC\omega + \frac{1}{jL\omega} + G_0 - G + \frac{3}{4}\beta V^2 \right] v = 0 \tag{5.139}$$

which gives the amplitude and the oscillation frequency of the steady-state sinusoidal oscillation without noise:

$$\omega = \frac{1}{\sqrt{LC}} \qquad V = \sqrt{\frac{4g}{3\beta}} \qquad (g = G - G_0) \tag{5.140}$$

The spectral line has infinitesimal width and noise cannot enter the oscillator under these operating conditions. The power output is given by

$$P = \frac{1}{2}G_0 V^2 = \frac{2gG_0}{3\beta} \tag{5.141}$$

where we have assumed that the combined losses, symbolically represented by G_0, are entirely due to the oscillator output signal. In fact, we should also take into account losses due to the Joule effect.

Noise is introduced by closing the switch shown in Fig. 5.37. This leads to frequency and amplitude fluctuations, so that the signal becomes quasi-sinusoidal. We saw in Chapter 1 that the power spectrum is then no longer a Dirac line, but a bell-shaped curve of very small width. From then on, noise can enter the circuit. Roughly speaking, we can say that the first noise pulses cause a fluctuation of the signal which ceases to be perfectly sinusoidal. Noise continues to enter and, hence, to broaden the spectrum by making the signal random. If there is some reaction mechanism that limits this phenomenon, a steady state is established with a narrow but nonzero bandwidth. The input signal will be noisy and the output signal quasi-sinusoidal.

In the absence of noise, the output is sinusoidal and the linearised total conductance is zero. In the quasi-sinusoidal regime, this conductance is very small; let us call it g_0. The circuit is then a very selective amplifier with bandwidth

$$\Delta\omega = \frac{g_0}{C} \tag{5.142}$$

This circuit is described by the differential equation

$$C\ddot{v} + g_0\dot{v} + \frac{1}{L}v = \frac{di}{dt} \tag{5.143}$$

where $i(t)$ is the random current from the generator of equivalent noise. We will describe the latter by its power spectral density $I^2(\omega)$.

The noise output of an amplifier can be calculated by a classical procedure. We identify the input signal with noise entering the selective amplifier. The output power is

$$P = \frac{1}{2}G_0 \int_0^\infty \frac{\omega^2 I^2(\omega)\, d\omega}{g_0^2 + \left(\omega C - \frac{1}{\omega L}\right)^2} = \frac{\pi \omega_0^2 I^2(\omega_0) G_0}{4C^2 \Delta\omega} \tag{5.144}$$

This calculation assumes that $\omega^2 I^2(\omega)$ is comparable with $\omega_0^2 I^2(\omega_0)$, which is valid because the bandwidth of the circuit is small. If we omit the negative conductance, the noise entering the resonator is

$$\Delta P_0 = \frac{\pi \omega_0^2 I^2(\omega_0)}{4C} \tag{5.145}$$

We can then simply express the oscillator bandwidth $\Delta\omega$ in terms of its the coherence time, as set out in Chapter 1. This gives

$$\Delta\omega = \frac{A(\omega_0)}{P} \frac{\omega_0^2}{Q^2} \tag{5.146}$$

and, by introducing the spectral density of the equivalent noise power at the input, we obtain

$$A(\omega_0) = \frac{\Delta P_0}{\Delta\omega_0} \tag{5.147}$$

This simple calculation gives the oscillator bandwidth and an expression for the power spectrum. The oscillator will obviously improve as the noise level is reduced. The power output rises and the resonator Q increases. Since the relation is quadratic, a gain by a factor of 10 in Q improves the oscillator bandwidth by a factor of 100.

This theory reveals several aspects of the problem. When it is easy to measure Q, P and ω_0, it is more difficult to measure the oscillator bandwidth (as we saw in Chapter 3). On the other hand, only such measurements give the spectral density of the equivalent input noise generator, thus indicating whether the equivalent generator is a valid concept.

6
Atomic frequency standards

6.1 INTRODUCTION

The resonators used in the oscillators described so far were macroscopic. Their characteristic frequency, and hence that of the oscillator, depends on their dimensions, on the properties of the material from which they are made and on the acoustic or electromagnetic mode employed. It is therefore impossible to construct an oscillator whose output frequency can be made equal to a predetermined value to a very high degree of precision, e.g., to a relative accuracy of 10^{-10}. The frequency must therefore be calibrated against a reference frequency and eventually adjusted by means of a fine tuning device often provided with the resonator. Moreover, the oscillators considered so far exhibit aging which causes a slow drift of their frequency, even in unperturbed environment. More or less frequent calibration may be essential.

A reference frequency as stable and as accurate as possible must therefore be available for measurements on oscillators with poorer stability and accuracy. A hierarchy is thus built up with a source at the top whose nature and properties are such that it qualifies as an absolute frequency standard. This ultimate reference is linked to the definition of the second, i.e., one of the basic units of the Systeme International d'Unités (SI).

The astronomical definition of the second was abandoned in 1967 for reasons discussed in the next Chapter, and the 13th General Conference on Weights and Measures adopted the following definition: 'the second is the duration of 9 192 631 770 oscillation periods of the radiation corresponding to the transition between the two hyperfine-structure levels of the ground state of the caesium–133 atom'. The device which allows this definition is the caesium beam clock. We note that 'clock', 'time standard', and 'frequency standard' are used interchangeably to describe the same device. Strictly speaking, they distinguish between different applications of devices such as time measuring, time keeping or reference frequency instruments, respectively. Some of these devices have a volume of the order of 30–40 dm^3 and are produced commercially at a rate of 500–1000 units per annum.

Other devices of much greater volume are produced and used in specialised laboratories that aim to attain a better definition of the second.

They are the caesium-beam primary standards. In 1990, only five countries processed such primary standards. They were Canada, the United States of America, Japan, Germany and the USSR.

Other atoms, such as hydrogen and rubidium, are also used to make practical atomic clocks.

The fundamental principles that underlie the operation of caesium, hydrogen and rubidium clocks will be summarised below. We will describe their structure, the physics of their operation and typical performance. We will briefly mention their most frequent applications and will indicate the motivation and prospects of current research which is concerned as much with atomic frequency standards as with the use of ions suspended in vacuum by means of suitable fields.

We shall present descriptions of atomic clocks together with a detailed analysis of their physical principles and their operation as reference frequency sources. The developments of frequency standards and their novel applications are described in the proceedings of the following annual conferences: Annual Symposium on Frequency Control (USA), Precise Time and Time Interval (PTTI), Applications and Planning Meeting (USA) and the European Time and Frequency Forum.

6.2 SOME PHYSICAL PROPERTIES EXPLOITED IN ATOMIC STANDARDS OF FREQUENCY AND TIME

6.2.1 Universality and invariance of atomic properties

There are two current postulates. First, atomic properties are universal, i.e., they do not vary from place to place, apart from relativistic effects associated with the motion of atoms and the gravitational potential which they experience. Second, atomic properties are independent of time, i.e., fundamental constants do not vary with time, so that interatomic couplings that determine the differences between energy levels can be considered to be constant. There is no experimental evidence against these postulates, so that the atomic frequency standard is assumed to produce a uniform time scale.

It is clear that the atoms must be disturbed as little as possible by interactions with their neighbours. This is the reason why the number density of atoms in atomic clocks is always kept very low to limit the number of interatomic collisions. This is achieved by using an *atomic beam* or in a very low pressure cell. Other sources of disturbance are also minimised, as we shall see later.

6.2.2 Fine and hyperfine structure of atomic spectra

Some atomic physics will now be recalled to introduce ground and excited states of the alkali metal atoms, on the one hand, and hyperfine structure of the ground state, on the other.

(a) Fine structure.

Alkali metal atoms have a single electron in their outermost shell. The inner shells are full. The Hg^+ ion has a similar electronic structure, and the hydrogen atom has only one electron.

Every electron has an intrinsic property known as *spin*. The spin of the electron cloud of an alkali metal atom is equal to that of the single electron. The corresponding spin quantum number S is therefore equal to 1/2. An angular momentum $\hbar S$ is associated with the electron spin ($\hbar$ is Planck's constant divided by 2π). The electron also has orbital angular momentum with quantum number L. The total angular momentum $\hbar \vec{J}$ is the sum of the spin and orbital angular momenta:

$$\hbar \vec{J} = \hbar \vec{S} + \hbar \vec{L} \qquad (6.1)$$

Angular momentum addition rules require that the quantum number J must take the values $L + \frac{1}{2}$ and $|L - \frac{1}{2}|$. In what follows, we will confine our attention to $L = 0$ and $L = 1$.

For $L = 0$, the energy of the single electron is a minimum and the atom is said to be in the ground state. Classical spectroscopic notation describes this as the $^2S_{\frac{1}{2}}$ state. The exponent in front of the letter is called the multiplicity. Its value is 2 for all the alkali metals. The letter S must not be confused with the spin quantum number. It is a symbolic representation of the value of L. The subscript indicates the value of J. When a number is given before this designation, it is the number of the outermost atomic shell.

An atom can be raised to an excited state by supplying energy to it. The first excited state corresponds to $L = 1$. States with $J = 1/2$ and $J = 3/2$, correspond to sublevels with slightly different energies. They are the fine-structure levels $^2P_{1/2}$ and $^2P_{3/2}$ that are used in optical pumping which we will discussed later. Transitions between the $^2S_{1/2}$ and $^2P_{1/2}$ levels, on the one hand, and $^2S_{1/2}$ and $^2P_{3/2}$, on the other, produce the closely spaced D_1 and D_2 lines in the emission and absorption spectra of the alkali metals. The wavelengths of the fine-structure doublets of hydrogen, rubidium, caesium and the Hg^+ ion are listed in Table 6.1.

(b) Hyperfine structure

The proton and the neutron also have spin $1/2$. The spin quantum number of a nucleus I is obtained by combining the spins of its nucleons. It is zero when the nucleon number, i.e., the mass number, is even, and to a multiple of $1/2$ when the mass number is odd (*see* Table 6.1).

Table 6.1. Properties of atoms used in atomic frequency and time standards. The hyperfine transition frequency of caesium–133 is fixed by definition. The uncertainties in the measured values of the hyperfine transition frequency of the other atoms (ions) are indicated

Atom (ion)	At. wt.	I	F	λD_1 (nm)	λD_2 (nm)	f (Hz)
H	1	1/2	0;1	121,6	121,6	1 420 405 751 770 ±0,003
Rb	85	5/2	2;3	794,8	780,0	3 035 732 440 ±3
Rb	87	3/2	1;2	794,8	780,0	6 834 682 612,8 ±0,5
Cs	133	7/2	3;4	894,3	852,1	9 192 631 770
Hg +	199	1/2	0;1	194,2	165,0	40 507 347 996,9 ±0,3

The total angular momentum of an atom, $\hbar\vec{F}$, is the sum of the electronic and nuclear angular momenta. It is therefore defined by an expression similar to (6.1). The corresponding quantum number is F. In the ground state, it can have the values $|I - 1/2|$ and $I + 1/2$ (*see* Table 6.1).

The projection of the angular momentum $\hbar\vec{F}$ on the quantisation axis takes the discrete values $\hbar m_F$, where m_F is the magnetic quantum number associated with F. Its $2F+1$ possible values are $-F, -F+1, ...0, ...F-1, F$. When a magnetic field is present, it defines the quantisation axis, and each of the two hyperfine-structure levels splits into $2F + 1$ sublevels. Magnetic moments are associated with the spins of the nucleus and the electron. The nuclear magnetic moment is

$$\vec{\mu}_I = g_I\mu_B\vec{I} \tag{6.2}$$

where μ_B is the Bohr magneton and g_I the nuclear Lande factor. The magnetic moment of a single electron is

$$\vec{\mu}_J = -g_J\mu_B\vec{S} \tag{6.3}$$

where g_J is the Landé factor of an electron bound to an atom. The value of g_J is very close to 2; that of g_I is three orders of magnitude smaller.

The mutual magnetic energy of these magnetic moments determines the energy difference between the hyperfine structure levels in the ground state in the absence of the magnetic field. This magnetic coupling is weak, so that the transition frequency between the hyperfine structure levels in the ground state lies in the microwave range (Table 6.1).

We shall see that the transitions between hyperfine states can be observed directly and are low enough for the probability of spontaneous emission to be extremely low. In other words, an electron occupying the upper hyperfine structure level in the ground state has an extremely long lifetime (of the order of one year). Transitions between the hyperfine structure levels of the ground state can therefore be produced only by absorption or by stimulated emission.

The hyperfine structure levels can be split by a static magnetic induction B_0. This is known as the *Zeeman effect*. The sublevel energy can be calculated exactly in the case of alkali metal atoms in the ground state. It is given by the Breit–Rabi formula

$$E\left(F, m_F\right) = -\frac{hf_0}{2\left(2I + 1\right)} - g_I\mu_B B_0 m_F \pm \frac{hf_0}{2}\left(1 + \frac{4m_F}{2I + 1}x + x^2\right)^{1/2}$$

(6.4)

where

$$x = \left(g_J + g_I\right)\mu_B B_0/hf_0$$

(6.5)

f_0 is the hyperfine transition frequency in zero field. The positive sign applies to $F = I + 1/2$ and the negative sign to $F = I - 1/2$. This formula gives the frequency of a given hyperfine transition as a function of the magnetic induction. Figures 6.1 and 6.2 show graphs of $E(F, m_F)$ for hydrogen and caesium atoms, respectively.

Equation (6.1) and Fig. 6.1 and 6.2 clearly demonstrate some remarkable properties. The level energies corresponding to $m_F = 0$ do not vary appreciably when the magnetic induction is low. The transition frequency between these levels depends on magnetic induction only in the second order. This so-called *clock transition* is exploited in atomic clocks. The basic expression is

$$f - f_0 = K_0 B_0^2$$

(6.6)

where $K_0 = 2773 \times 10^8\,\mathrm{Hz\,T^{-2}}$ for hydrogen and $427.45 \times 10^8\,\mathrm{Hz\,T^{-2}}$ for caesium. Typically, $B_0 = 10^{-7}\,\mathrm{T}$ for the hydrogen clock and $B_0 = 6 \times 10^{-6}\,\mathrm{T}$ for the caesium clock. The corresponding values of $f - f_0$ are $2.77 \times 10^{-3}\,\mathrm{Hz}$ for hydrogen and $1.5388\,\mathrm{Hz}$ for caesium.

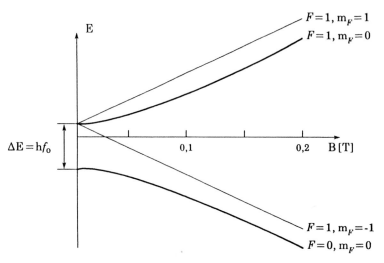

Fig. 6.1 Hydrogen levels

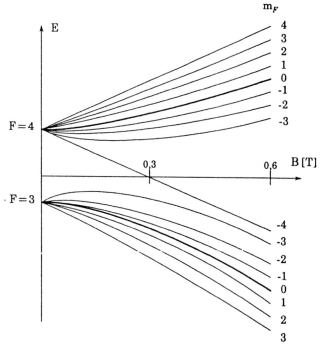

Fig. 6.2 Caesium levels.

For values of the magnetic induction whose order of magnitude is defined by the above data, the level energy for $m_F \neq 0$ varies linearly with B, so that, for each value of F, the transition frequency f_z corresponding to adjacent levels is proportional to B_0:

$$f_z = K_z B_0 \tag{6.7}$$

For hydrogen, $K_z = 1.39908 \times 10^{10} \text{Hz T}^{-1}$ whilst for caesium K_z is close to $3.5 \times 10^9 \text{Hz T}^{-1}$. In a weak field, f_z lies in the low-frequency domain. Moreover, transitions with $\Delta F = \pm 1$, $\Delta m_F = 0$ can be observed when the nuclear spin is greater than $1/2$. The frequency f_z' of these transitions is given by

$$f_z' - f_0 = m_F K_z' B_0 \tag{6.8}$$

and lies in the microwave range. In the case of caesium, $K_z' = 700.84 \times 10^7 \text{Hz T}^{-1}$. Measurements of certain transition frequencies therefore yield the magnetic induction and, hence, the shift of the clock frequency (6.6).

In strong magnetic fields, the energy levels split into two groups. The physical reason for this is that the gain in electron energy is more important than the energy of the electron-nucleus interaction. The magnetic behaviour of the atom therefore becomes similar to that of its outermost electron. The energy is greater when the electron spin is antiparallel to the magnetic field B_0, and lower when the spin is parallel to B_0. We note that sublevels with $m_F = 0$ belong to different groups. It is therefore possible to sort the atoms in accordance with the occupancy of the levels between which the clock transition takes place.

Another important property of the above hyperfine structure levels is that their energy depends on the square of the applied electric field. This effect can be ignored for the electric fields encountered in practice.

We note that excited states also have a hyperfine structure, but the level splitting is smaller than in the ground state. We will not include this structure in the qualitative description of optical pumping that will follow later.

6.2.3 Selection rules

The condition $E_1 - E_2 = h f_0$ expresses the conservation of energy in the interaction between an atom and electromagnetic radiation, but it is not a sufficient condition. Selection rules have to be obeyed as well. They limit the number of possible transitions and in part determine the structure of the atomic clock. For transitions between hyperfine structure levels, the selection rules are as follows:

(1) when the microwave magnetic field is parallel to the static magnetic induction B_0, the only transitions that are allowed are those satisfying the conditions $\Delta F = \pm 1$ and $\Delta m_F = 0$; in particular, this rule applies to clock transitions between levels with magnetic quantum number $m_F = 0$

(2) When the alternating magnetic field is perpendicular to the static magnetic field, the only transitions that are allowed are those satisfying

either $\Delta F = 0$, $\Delta m_F = \pm 1$ or $\Delta F = \pm 1$, $\Delta m_F = \pm 1$; these transitions belong to the low-frequency domain when they are produced between sublevels with the same value of F, and to the microwave domain otherwise.

6.2.4 Selection of state

(a) Preparation of atomic system

Let us show first that the observation, under good conditions, of transitions between hyperfine structure levels in the ground state necessitates a preparation of the atomic system.

Einstein showed that, when electromagnetic radiation interacts with an atomic system, the probabilities of absorption and of stimulated emission are equal. This means that the change in the number n_2 of atoms in the state of energy E_2 ($E_2 < E_1$) in time dt due to absorption is $dn_2 = -pn_2 dt$. The quantity p is a constant which depends on the properties of the incident radiation. The change due to stimulated emission is $dn_2 = pn_1 dt$, where n_1 is the population of the energy level E_1. Because the total number of atoms must be conserved, we have

$$dn_2 = -dn_1 = p\,(n_1 - n_2)\,dt \tag{6.9}$$

We can thus see that the rate of change of n_2 increases with the population difference. Hence the changes in atomic properties are observed more easily, i.e., with an even larger signal to noise ratio, as the population difference between the two levels increases.

In thermodynamic equilibrium, the ratio of the level populations is given by Boltzmann's law:

$$\frac{n_1}{n_2} = \exp\left(-hf/kT\right) \tag{6.10}$$

where f is the transition frequency between the two levels, k is Boltzmann's constant and T the absolute temperature. For $f = 10\,\mathrm{GHz}$ and $T = 300\,\mathrm{K}$, we have $(n_1 - n_2)/n_2 \approx 1.5 \times 10^{-3}$. This shows that, in thermodynamic equilibrium, the population difference between levels linked by the clock transition is very small. The useful signal is very weak, so that steps must be taken to increase the population difference, which can be done by magnetic deflection or by optical pumping.

(b) Magnetic deflection.

We know that every physical system tends to minimise its potential energy. Figures 6.1 and 6.2 show that the atoms of hydrogen in the $F = 1, m_F = 0$ state, and of caesium in the $F = 4, m_F = 0$ state, are attracted to regions of low magnetic field. On the other hand, hydrogen atoms with $F = 0, m_F = 0$ and caesium atoms with $F = 3, m_F = 0$ are attracted by high-field zones. It can be shown that the force F acting on an atom is

$$\vec{F}(F, m_F) = -\frac{\partial E(F, m_F)}{\partial B}\vec{\nabla}B \qquad (6.11)$$

where $E(F, m_F)$ is given by (6.4) and $\vec{\nabla}B$ is the gradient of the magnetic field. The quantity $-\partial E(F, m_F)/\partial B$ is proportional to the magnetic moment. In high fields, its magnitude is very close to the Bohr magneton since the magnetic behaviour of the atom is then determined by its outermost electron. In high and inhomogeneous fields, the two states with $m_F = 0$ are thus subjected to forces pointing in opposite directions. When an atomic beam is used, the corresponding atoms are deflected in opposite directions, and are thus separated.

We recall that an experiment involving the magnetic deflection of an atomic beam of silver was performed by Stern and Gerlach, in 1921 and demonstrated the existence of atomic magnetism and its space quantisation.

(c) Optical pumping.

The optical pumping method, due to Kastler, is illustrated schematically in Fig. 6.3.

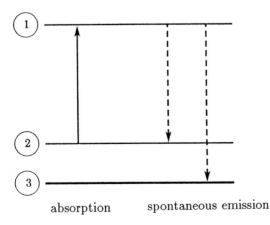

Fig. **6.3** Optical pumping.

Consider the special case of three atomic levels. For example, let us suppose that levels 2 and 3 are hyperfine structure levels in the ground state whilst level 1 is an excited state. The wavelengths λ emitted in transitions between ground-state and excited-state levels are listed in Table 6.1 for caesium and rubidium atoms. The corresponding frequency is of the order of 4×10^{14} Hz and is much higher than that of the hyperfine transition. According to Boltzmann's law, the population of level 1 is very small in thermodynamic equilibrium. If we suppose that the incident light is so monochromatic that only the 1-2 transition can take place, the interaction between the radiation and the atom is represented globally by the absorption of photons and the preferential population transfer from level 2 to level 1. However, at these wavelengths, the spontaneous emission probability is very high. In other words, the lifetime of level 1 is very short, i.e., of the order of 10 ns, and the atom returns very rapidly to the ground state, with equal probability of ending in levels 2 or 3. The net result of these absorption and emission processes is that the population of level 3 increases whilst that of level 2 decreases. This produces a population difference between levels 2 and 3 all the more efficiently as the intensity of radiation with the particular wavelength λ is allowed to increase. It is possible to raise the population of level 2 above that of level 3 if the wavelength λ corresponds to the energy difference between levels 1 and 3, and the corresponding lifetime ratio is right.

6.2.5 Correction for the Doppler effect

It is well known that the motion of a source of radiation relative to an observer produces an apparent change in signal frequency. In the case of an electromagnetic signal, the frequency shift caused by this so-called Doppler effect is $f_0 v/c$ where v is the component of the relative velocity in the direction of the observer and c the speed of light. The ratio v/c is usually of the order of 10^{-6}, so that it is obviously necessary to suppress the effect of this frequency shift as much as possible in the atomic standards of frequency and time. This can be done by standard spectroscopic techniques.

(a) The mean velocity of the atoms can be substantially reduced by mixing them with an inert gas known as a buffer gas. This method is used in rubidium clocks.

(b) When the atoms are confined to a volume whose linear dimensions are smaller than the radiation wavelengths, the absorption and emission spectra of the atoms consist of a central line that is not broadened by the Doppler effect and side lines which, in practice, do not interfere with the observation of the transition. This property is exploited in the mercury ion clock.

(c) If the interaction takes place in the central portion of a standing wave, in which the phase of the electromagnetic field is constant, the Doppler frequency shift is eliminated. This configuration is used in the caesium and hydrogen clocks.

The above Doppler effect is called the first-order Doppler effect as it assumes that $v/c \ll 1$. Relativistic analysis of the Doppler effect shows that there is also a second effect which for monoenergetic atoms is described by the relative frequency shift

$$\frac{f - f_0}{f_0} = -\frac{1}{2}\frac{v^2}{c^2} \tag{6.12}$$

where v is the velocity of atoms. For $v = 150\,\mathrm{m\,s^{-1}}$, a typical value in caesium clocks, we have $(f - f_0)/f_0 = -1.25 \times 10^{-13}$. Despite its small magnitude, this frequency shift must be accounted for in the caesium primary standard, as well as in the other frequency standards. The second-order Doppler effect can only be eliminated by slowing down the atoms (*see* Section 4.4.4).

6.2.6 Resonance width

Even after the physical causes of broadening such as collisions, the Doppler effect and so on have been minimised, there remains the natural line width due to the Heisenberg uncertainty principle which states that the uncertainty Δf in a measurement of the transition frequency is related to the observation time Δt by

$$\Delta f \, \Delta t \approx 1 \tag{6.13}$$

In atomic frequency standards, the observation times range between 1 ms and a fraction of a second. The width of an atomic resonance thus lies between, say, 1 kHz and 1 Hz. For the commonly used values of magnetic induction, the different possible transitions are well separated from each other, at least to a first approximation. Everything therefore happens as if, amongst all the hyperfine structure levels, only two levels interact with the microwave electromagnetic field. Atomic properties can therefore be considered in terms of the simplifying two-level approximation.

The Q factor of the atomic resonance, $f_0/\Delta f$, lies in the order of magnitude range of $10^7 - 10^9$. This high value is crucial for the frequency stability of atomic frequency standards. It also contributes to their accuracy.

6.3 THE CAESIUM BEAM CLOCK

The caesium atomic-beam clock is the outcome of experimental research into the magnetic properties and radio-frequency spectra of atoms and molecules that has continued since the 1920s. The idea of making an atomic frequency standard was first put forward by Rabi *et al.* in 1939, and the first caesium frequency standard was built in 1955 at the National Physical Laboratory near London. Industrial development, exploiting advances in high-frequency electronics, began in 1956 and led to the construction of the Atomichron. Since then, different commercial companies have improved the performance of these clocks, whilst research laboratories began studies of devices used for the definition of the second to a few parts in 10^{14}.

6.3.1 The caesium beam tube

(a) Description

Figure 6.4 shows a schematic diagram of the most common design of a caesium-beam resonator. The ribbon-shaped caesium beam is produced by a crucible heated to about 90°C. The sixteen hyperfine levels of the caesium atom are equally populated as the atoms leave the source. Unless we mentioned otherwise, we shall confine our attention to levels with $F = 4, m_F = 0$ (level 1) and $F = 3, m_F = 0$ (level 2) between which the clock transition takes place.

Magnet A produces a strong inhomogeneous field that deflects atoms occupying levels 1 and 2. This deflection is, say, towards the right for atoms in level 2 and therefore towards the left for atoms in level 1. The latter are stopped by a screen and their trajectory is not shown in Fig. 6.4. Let us suppose that atoms undergo the transition from level 2 to level 1 during their flight across the resonant cavity. Magnet B is arranged to deflect the atomic beam to the left, for example, and the atoms reach the detector. On the contrary, atoms that have not undergone the transition are deflected towards the right, and are not detected. The flux of caesium atoms reaching the detector is thus proportional to the $2 \to 1$ transition probability.

The detector of caesium atoms is a hot wire. It is made from a metal such as tantalum, or an alloy such as platinum-iridium, whose work function is higher than the ionisation energy of caesium. The wire therefore picks up the outermost electron of caesium and the positive ion Cs^+ is re-emitted. The caesium ions are separated from the ionised impurities, also emitted by the hot wire, by a mass spectrometer and finally reach a diode in which they eject secondary electrons whose flux is amplified by an electron multiplier.

The resonant cavity, tuned to the hyperfine transition frequency, has the unusual two-arm shape and is made from an X-band waveguide. Its shape allows the atoms to cross successively two regions of interaction with the microwave field (Ramsey's method of two separate oscillating fields). A standing wave is established in each interaction region.

A static magnetic field of $\sim 6 \times 10^{-6}$ T is applied parallel to the alternating field of the regions of interaction. It is as uniform as possible over

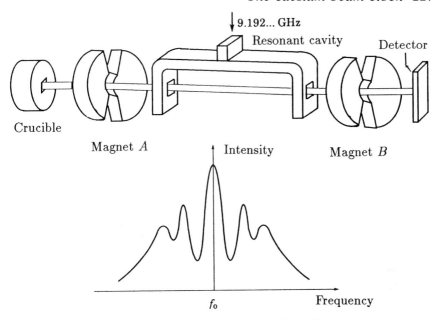

Fig. 6.4 Trajectory of atoms in a caesium beam tube and resonance curve.

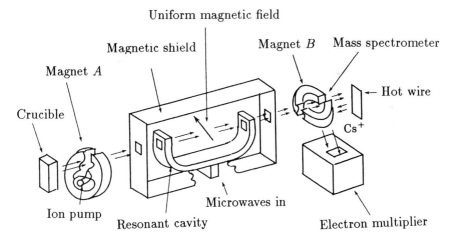

Fig. 6.5 Details of the caesium beam tube.

the region of interaction and in the intermediate space. This magnetic configuration allows the observation of transitions corresponding to $\Delta F = \pm 1$, $\Delta m_F = 0$, which include the clock transition $F = 3$, $m_F = 0 \longleftrightarrow F = 4$, $m_F = 0$. The volume occupied by the uniform static field is shielded from the ambient magnetic field and its variations.

Graphite getters (absorbers) incorporated in the caesium tube absorb unwanted caesium atoms arriving from the crucible. An ion pump maintains a high vacuum.

Figure 6.5 shows the main components of a commercial caesium beam tube. Other designs are also used. In one of them (Hewlett Packard 004), the caesium tube contains two separate atomic beams.

(b) Frequency response

The caesium beam tube behaves as a cavity resonator whose output current depends on the frequency of the applied microwave signal. The general shape of the resonance curve is shown in Fig. 6.6. We shall first give a qualitative interpretation of the shape of this resonance curve and then describe it more rigourously.

Qualitative interpretation. Let us sppose that in the first interaction region, near state selector magnet A, the role of the microwave field is simply to impose a precession on the magnetic moment of each atom. In the interval between the two interaction regions, the magnetic moment precesses at the natural angular frequency, equal to the transition angular frequency ω_0. If T is the time of flight between the two interaction regions, the phase difference between processional motion at the entrance to the second region and the microwave field in the first region is $\psi = (\omega - \omega_0)T$, where ω is the angular frequency of the microwave field. The standing waves in the resonant cavity are arranged so that their phases in the two interaction regions are the same. The quantity ψ therefore represents the phase difference between the processional motion at entrance to the second interaction region and the field in this region. If we accept that the strength of the interaction in the second region depends on ψ, we see that the probability of a transition from level 2 to level 1 is a maximum for $(\omega - \omega_0)T = 2n\pi$ where n is an integer or zero for $(\omega - \omega_0)T = (2n + 1)\pi$. So far, we have assumed that T was unique. However, the atomic beam has a velocity distribution, so that the transition probability has a well-defined maximum only for $n = 0$, i.e., for $\omega = \omega_0$. The full width at half height of the central peak, expressed in terms of the angular frequency, is $\Delta\omega = \pi/T$, where T is now the mean time of flight between the two interaction regions. We note that this result is in agreement with the uncertainty principle of equation (6.13). The corresponding width of the resonance peak is in the range 800–300 Hz for caesium tubes produced commercially, and can be as low as 20 Hz for certain laboratory models.

Proof by quantum theory. The interaction between atoms and the electromagnetic field can only be treated correctly in quantum mechanics. The most convenient approach is to use the density matrix. We shall adopt this method here.

For the two hyperfine structure levels $F = 4, m_F = 0$ and $F = 3, m_F = 0$, designated levels 1 and 2, respectively, the density matrix of this two-level system is

$$\rho = \begin{pmatrix} \rho_{11} & \rho_{12} \\ \rho_{21} & \rho_{22} \end{pmatrix} \qquad (6.14)$$

where ρ_{11} and ρ_{22} are real quantities and we have $\rho_{21} = \rho_{12}^*$, the asterix indicating a complex conjugate.

The rate of change of the density matrix is given by

$$i\hbar \frac{d\rho}{dt} = [H, \rho] \qquad (6.15)$$

where H is the Hamiltonian of the problem and the square brackets represent the commutator of H and ρ.

In the absence of electromagnetic perturbations, the Hamiltonian representing the hyperfine interaction has the eigenvalues E_1 and E_2, where $E_1 - E_2 = \hbar\omega_0$. It can be shown that the matrix element of the Hamiltonian describing the interaction with the electromagnetic field is

$$\hbar V_{12} = \hbar V_{21}^* = \mu_B B(t) \qquad (6.16)$$

where $B(t)$ is the component of the oscillating magnetic field parallel to the direction of the static field B_0. The Hamiltonian H is therefore given by

$$H = \hbar \begin{vmatrix} \omega_0/2 & V_{12} \\ V_{21} & -\omega_0/2 \end{vmatrix} \qquad (6.17)$$

If B is the amplitude of the magnetic field $B(t)$ and ω is the pulsatance, we have

$$V_{12} = b\cos(\omega t + \varphi) \qquad (6.18)$$

where φ represents a possible residual phase difference between the oscillating fields and

$$b = \mu_B B/\hbar \qquad (6.19)$$

The quantity b has the dimensions of pulsatance and is a measure of the amplitude of the microwave magnetic field acting on the atoms. It is often called the Rabi frequency. In the design shown in Figs. 6.4 and 6.5, the amplitude B is constant along each trajectory of an atom passing through the waveguide, so that b is a constant during the time of the interaction.

We can now write

$$V_{12} = \frac{1}{2}(b_1 + ib_2)e^{-i\omega t} + c.c \qquad (6.20)$$

where c.c. stands for the complex conjugate and b_1 and b_2 are real quantities given by

$$b_1 = b\cos\varphi \qquad b_2 = -b\sin\varphi \qquad (6.21)$$

The complex conjugate term lies outside the resonance and can be shown to be negligible.

The quantum state of the atoms depends both on the instantaneous value of the oscillating magnetic field and on the time interval ϑ during which the interaction takes place. We therefore seek solutions of the form

$$\rho_{12} = \frac{1}{2} \left[a_1 (\vartheta) + i a_2 (\vartheta) \right] e^{-i\omega t} \tag{6.22}$$

where

$$\rho_{11} - \rho_{22} = a_3 (\vartheta) \tag{6.23}$$

and $a_1(\vartheta), a_2(\vartheta), a_3(\vartheta)$ are real.

The sum $a_1(\vartheta) + i a_2(\vartheta)$ is the complex atomic coherence amplitude and $a_3(\vartheta)$ is the population difference between the two levels. Since $d/dt = \partial/\partial t + \partial/\partial\vartheta$ in equation (6.15), we obtain the following set of differential equations:

$$\frac{\partial a_1 (\vartheta)}{\partial \vartheta} + (\omega - \omega_0) a_2 (\vartheta) + b_2 a_3 (\vartheta) = 0$$

$$- (\omega - \omega_0) a_1 (\vartheta) + \frac{\partial a_2 (\vartheta)}{\partial \vartheta} - b_1 a_3 (\vartheta) = 0 \tag{6.24}$$

$$- b_2 a_1 (\vartheta) + b_1 a_2 (\vartheta) + \frac{\partial a_3 (\vartheta)}{\partial \vartheta} = 0$$

The solution can be expressed in matrix form:

$$\begin{vmatrix} a_1 (\vartheta) \\ a_2 (\vartheta) \\ a_3 (\vartheta) \end{vmatrix} = |R(\vartheta)| \begin{vmatrix} a_1 (0) \\ a_2 (0) \\ a_3 (0) \end{vmatrix} \tag{6.25}$$

where $a_1(0), a_2(0), a_3(0)$ are the parameter values at the beginning of the interaction.

We will write down the matrix $|R(\theta)|$ in full in two special cases. The first assumes that

$$|\omega - \omega_0| \ll b \tag{6.26}$$

which implies that we are concerned only with the central part of the resonance curve, as we shall show later. We then have $|R(\vartheta)|$ equal to

$$\begin{vmatrix} \cos b\vartheta + (1 - \cos b\vartheta) \cos^2 \varphi & -(1 - \cos b\vartheta) \sin \phi \cos \varphi & \sin b\vartheta \sin \varphi \\ -(1 - \cos b\vartheta) \sin \varphi \cos \varphi & \cos b\vartheta + (1 - \cos b\vartheta) \sin^2 \varphi & \sin b\vartheta \cos \varphi \\ - \sin b\vartheta \sin \varphi & - \sin b\vartheta \cos \varphi & \cos b\vartheta \end{vmatrix} \tag{6.27}$$

The second special case involves the region separating the two zones of interaction in which $b = 0$. Here

$$|R(\vartheta)| = \begin{vmatrix} \cos (\omega - \omega_0) \vartheta & - \sin (\omega - \omega_0) \vartheta & 0 \\ \sin (\omega - \omega_0) \vartheta & \cos (\omega - \omega_0) \vartheta & 0 \\ 0 & 0 & 1 \end{vmatrix} \tag{6.28}$$

To describe the flux of atoms reaching the detector, we have to find the probability of transition between levels $F = 3, m_F = 0$ and $F = 4, m_F = 0$ along the trajectory of an atom. We can easily show that

$$P(\tau) = \frac{1}{2}\left[1 - \frac{a_3(\tau, T, \tau)}{a_3(0)}\right] \qquad (6.29)$$

where the notation $P(\tau)$ indicates that we are assuming that all the atoms have the same velocity v, so that the time of interaction with each oscillating field is $\tau = l/v$, where l is the length of each interaction region; $a_3(\tau, T, \tau)$ is the population difference after the first interaction region, after transit through the region free of the oscillating field and after the second interaction region. We have $T = L/v = \tau L/l$, where L is the distance between the two arms of the resonant cavity.

If we choose the origin of the phase of the oscillating magnetic field in the first region, and if φ is the phase difference between the second and first oscillating magnetic fields, we obtain

$$P(\tau) = \frac{1}{2}\sin^2 b\tau \{1 + \cos[(\omega - \omega_0)T + \varphi]\} \qquad (6.30)$$

As we shall see, the phase difference φ is usually very small, and this expression then shows that for ω close to ω_0 the transition probability has a peak whose half width at half height is given by

$$\Delta f = 1/2T \qquad (6.31)$$

which agrees with the qualitative treatment. Moreover, the probability depends on the amplitude of the electromagnetic field through the term $\sin^2 b\tau$. It can therefore be optimised for $b\tau = \pi/2$ if all the atoms have the same velocity.

We note that the inequality given by (6.26) can be rewritten in the form

$$|f - f_0| \ll \frac{L}{l}\frac{b\tau}{\pi}\Delta f \qquad (6.32)$$

Since, in practice, $b\tau \approx \pi/2$ and $L/l > 10$, the above approximation is satisfactory for the central part of the resonance curve. It is useful in simplified analyses of the behaviour of the caesium beam resonator. The complete result, obtained without using this approximation, is

$$P(\tau) = \frac{4b^2}{\Omega^2}\sin^2\frac{\Omega\tau}{2}\left\{\cos\frac{\Omega\tau}{2}\cos\frac{1}{2}[(\omega - \omega_0)T + \varphi]\right.$$

$$\left. -\frac{\omega - \omega_0}{\Omega}\sin\frac{\Omega\tau}{2}\sin\frac{1}{2}[(\omega - \omega_0)T + \varphi]\right\} \qquad (6.33)$$

where

$$\Omega^2 = (\omega - \omega_0)^2 + b^2 \tag{6.34}$$

In practice, the atoms have a non-Maxwellian velocity distribution because the deflection produced by magnets A and B depends strongly on velocity. If we use $f(\tau)$ to represent the corresponding distribution of interaction times, the simplified expression for the transition probability becomes

$$P = \frac{1}{2} \int_0^\infty f(\tau) \sin^2 b\tau \left\{1 + \cos\left[(\omega - \omega_0)T + \varphi\right]\right\} d\tau \tag{6.35}$$

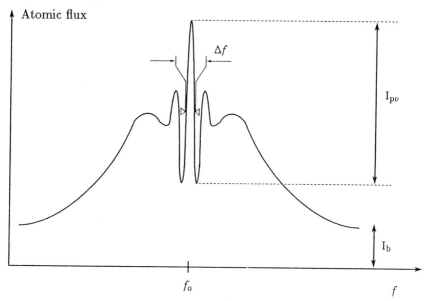

Fig. 6.6 Resonance in a caesium beam tube.

The velocity distribution introduces a dispersion in the position of the maxima of the function $\cos[(\omega - \omega_0)T + \varphi]$. This dispersion is all the more important as we move away from the central frequency ω_0, which reduces the amplitude of the side lines. When $|\omega - \omega_0|$ is much greater than the width of the central line, the response of the caesium beam tube has a symmetric profile on the ω_0 axis. Its full width at half height is then close to 20 kHz and is determined by the time taken to cross each interaction region. This profile is often called the Rabi pedestal. Figure 6.6 shows schematically the central resonance and the Rabi pedestal.

The flux I of detected atoms is a linear function of the probability P. It can also contain a contribution I_b that is independent of the interrogation frequency and is due to the parasitic caesium atoms. We then have

$$I = I_b + \frac{I_0}{2} \int_0^\infty f(\tau) \sin^2 b\tau \left\{1 + \cos\left[(\omega - \omega_0)T + \varphi\right]\right\} d\tau \tag{6.36}$$

where I_0 is a constant.

6.3.2 Electronics

(a) The error signal.

The caesium clock is a passive frequency standard: a signal has to be produced at the atomic transition frequency in order to generate the error signal necessary to control a quartz oscillator. Figure 6.7 illustrates the basic arrangement that exploits the principles covered in Chapter 2. The microwave signal at 9 193 GHz, which supplies the resonant cavity is obtained by frequency synthesis from a quartz oscillator, and its frequency is usually 5 or 10 MHz. In order to achieve this, an interrogation signal is generated by frequency synthesis from a quartz oscillator and is modulated around the value $f_0 + \Delta f_z$ where f_0 is the hyperfine transition frequency of the caesium atom in the ground state, which is equal to 9 192 631 770 Hz, and Δf_z is the frequency shift caused by the second-order Zeeman effect (about 1.5 Hz). We recall that Δf_z is related to the magnetic induction B_0 by equation (6.6). In primary laboratory standards, the frequency of the interrogation signal is modulated around the value $f_0 + \Delta f_z + \Delta f'$ where $\Delta f'$ is the algebraic sum of the frequency shifts other than that due to the second-order Zeeman effect.

The search for the resonant frequency of the caesium resonator is performed in a very classical manner. The frequency of the microwave signal is periodically modulated. The component of the caesium tube response with frequency equal to the modulation frequency is selectively amplified and demodulated in synchronism with the modulation signal. An error signal is generated when the mean frequency of the microwave signal is different from the resonant frequency of the caesium tube, i.e., from the hyperfine transition frequency of the ground-state caesium atom.

Let us examine this error signal in some detail, assuming that the frequency modulation is sinusoidal, i.e.,

$$\omega(t) = \omega_i + \omega_m \sin \omega_m t \qquad (6.37)$$

where ω_i is the interrogation angular frequency, ω_m is the modulation amplitude and ω_M the angular frequency of modulation. We assume that the modulation frequency is sufficiently low so that we can replace $\omega(t)$ by ω in equation (6.36) (we note that this implies that the modulation frequency is small compared to the width Δf of the atomic resonance curve). The atomic flux to be detected is periodic. Assuming that $\varphi = 0$, the fundamental component is

$$I_1(t) = -I_0 \int_0^\infty f(\tau) \sin^2 b\tau \left\{ \sin(\omega_i - \omega_0)T \right\} J_1(\omega_m T) \, d\tau \sin \omega_M t \quad (6.38)$$

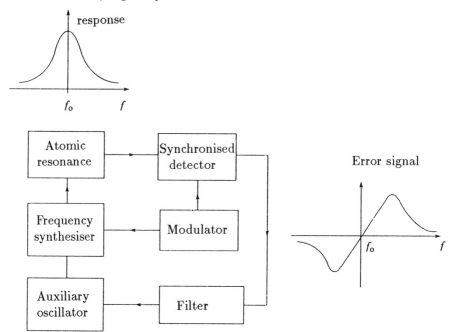

Fig. 6.7 Frequency locking of a quartz oscillator to the atomic resonance of caesium.

where J_1 is the Bessel function of order 1.

Synchronised detection involves multiplication of $I_1(t)$ by $+1$ during the first half-period and by -1 during the second. The error signal E is the low-frequency component of the synchronised detector output. For ω_i sufficiently close to ω_0, we then have

$$E = -\frac{2}{\pi} K I_0 (\omega_i - \omega_0) \int_0^\infty Tf(\tau) \sin^2 b\tau J_1(\omega_m T)\, d\tau \qquad (6.39)$$

where K is a constant representing the different amplification factors. The error signal is therefore zero for $\omega_i = \omega_0$ and is proportional to the difference between the interrogation frequency and the atomic resonance frequency when this difference is small. The slope of the error signal depends on the transit time T. It is therefore bigger for a narrower resonance peak, i.e., higher Q. The slope also depends on the amplitude of the microwave signal and on the modulation amplitude ω_m.

It can be shown that the second harmonic of the response of the caesium beam tube is a maximum for $\omega_i = \omega_0$. It can be used as a measure of the intensity of the atomic beam.

Other modulated waveforms are also used: square-wave frequency modulation and square-wave phase modulation. The slope of the error signal is not then markedly different from that obtained with sinusoidal frequency modulation.

(b) Locking of a quartz oscillator to the atomic resonance.

By using modulated interrogation frequency and synchronous-detection response functions we thus transform the caesium beam resonator into a frequency discriminator. It is then possible to lock the frequency of a quartz oscillator in such a way that $\omega_i = \omega_0$, i.e., the quartz-oscillator frequency becomes tied to that of the atomic resonance. (Fig. 6.8).

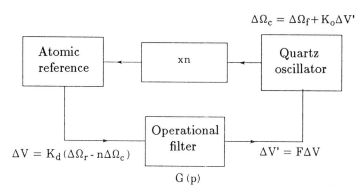

Fig. 6.8 Block diagram illustrating frequency locking.

We saw earlier that frequency control can be achieved by applying a command voltage to the varactor diode in the quartz oscillator. The operational filter is usually reduced to an integrator, so that

$$G(p) = A/RCp \qquad (6.40)$$

where A is a constant and RC is the time constant of the integrating circuit. Moreover

$$\Delta\Omega_c(p) = \frac{T_a p}{1 + T_a p} \Delta\Omega_f(p) + \frac{1}{1 + T_a p} \frac{\Delta\Omega_r(p)}{n} \qquad (6.41)$$

where $\Delta\Omega_c(p), \Delta\Omega_f(p)$ and $\Delta\Omega_r(p)$ represent, respectively, the Laplace transforms of the frequency fluctuations of the quartz-oscillator in a closed loop, of the quartz operator in an open loop and of the atomic reference frequency; n is the ratio of the atomic transition frequency to that of the quartz oscillator and T_a is the locking time constant, given by

$$T_a = RC/nAK_cK_d \qquad (6.42)$$

in which K_c and K_d are, respectively, the slopes of the quartz oscillator frequency control and of the frequency discriminator (*cf.* Section 2.3.2.)

Equation (6.41) shows that, when $G(p)$ is given by (6.40), we have first-order frequency locking. It is also possible to consider higher-order locking. The frequency fluctuations of the free quartz oscillator are filtered by a high-pass filter and those of the atomic reference by a low-pass filter. In other words, the locked quartz oscillator follows the slow fluctuations of the atomic reference frequency and the rapid fluctuations of the non-locked quartz oscillator. The cut-off angular frequency is $1/T_a$. Usually, we have $T_a \approx 1\,\mathrm{s}$.

We now turn to random frequency fluctuations. The spectral density S_y of relative fluctuations is

$$S_{y,c}(F) = \frac{(2\pi F T_a)^2}{1 + (2\pi F T_a)^2} S_{yf}(F) + \frac{1}{1 + (2\pi T_a)^2} \frac{S_{yf}(F)}{n^2} \qquad (6.43)$$

where F is the Fourier frequency and the indices have the same meaning (6.41).

6.3.3 Frequency stability

We will be interested more particularly in frequency stability which is set by the caesium beam resonator. Equation (6.43) shows that $S_{y,c}(F) = S_{y,r}(F)$ for $f \ll 1/2\pi T_a$. For sampling times $\tau > T_a$, the characteristic Allan drift of the locked quartz frequency fluctuations is therefore determined by the noise of the caesium beam resonator.

This noise is due to the peculiar nature of the atomic beam: the flux of atoms produces a granular noise in the detector, and the result is a fluctuation in the caesium resonator signal. For a constant interrogation frequency, the locking circuit interpretes this fluctuation as a variation in the resonance frequency of the caesium beam tube and, hence, in the reference frequency. If S_I is the power spectral density of the beam granular noise, we have

$$S_{y,r} = \frac{8}{\pi^2} \frac{K^2}{K_d^2 \omega_0^2} S_I \qquad (6.44)$$

where K_d is the slope of the frequency discriminator system in $\mathrm{rad\,s^{-1}}$. The factor K has the same meaning as in (6.39). The factor $8/\pi^2$ comes from the random part of the tube response recorded by a synchronous detector when this response is filtered near the modulation frequency. The noise of the reference frequency is a white noise.

The power spectral density S_I is related to the mean flux of atoms recorded by the detector. For a sinusoidal modulation, we find from (6.36) that

$$S_I = 2\left\{ I_b + \frac{I_0}{2} \int_0^\infty f(\tau) \sin^2 b\tau \left[1 + J_0(\omega_m T)\right] d\tau \right\} \qquad (6.45)$$

The value of K_d follows from equation (6.39):

$$K_d = \frac{2K}{\pi} I_0 \int_0^\infty T f(\tau) \sin^2 b\tau J_1(\omega_m T)\, d\tau \qquad (6.46)$$

The general expression for $S_{y,r}$ in the case of a slow sinusoidal modulation of the interrogation frequency is therefore

$$S_{y,r} = \frac{2\left\{2I_b + I_0 \int_0^\infty f(\tau)\sin^2 b\tau\left[1 + J_0(\omega_m T)\right]d\tau\right\}}{\omega_0^2 I_0^2 \left\{\int_0^\infty Tf(\tau)\sin^2 b\tau J_1(\omega_m T)d\tau\right\}^2} \tag{6.47}$$

This equation demonstrates the effect of the amplitude ω_m of the angular frequency modulation. There is an optimum value of ω_m for which $S_{y,r}$ is a minimum. This value is close to the half width of the resonance curve. It can be shown that power spectral density of the relative frequency fluctuation is then reasonably well described by

$$S_{y,r} \approx \frac{1}{2}\frac{2I_b + I_{pv}}{Q_\ell^2 I_{pv}^2} \tag{6.48}$$

where I_{pv}, defined in Fig. 6.6, is the of atomic flux corresponding to the height of the resonance curve between the peak at $\omega = \omega_0$ and the two adjacent minima. We also see the unwanted effect of the parasitic flux I_b which must obviously be reduced to a negligible value.

The numerator in (6.48) represents the granular-noise power P_B within a noise pass band of 1 Hz. The quantity I_{pv}^2 represents the power P_S of the useful signal. We can therefore write

$$S_{y,r} \approx \frac{1}{2Q_l^2\left(P_S/P_B\right)_{1\,\mathrm{Hz}}} \tag{6.49}$$

In accordance with Chapter 3, the characteristic Allan variance of the frequency fluctuations is then given by

$$\sigma_y(\tau) \approx \frac{1}{2Q_l\left(P_S/P_B\right)_{1Hz}^{1/2}} \tag{6.50}$$

which indicates, finally, that the relative frequency stability depends on the relative accuracy with which the signal to noise ratio allows us to detect the atomic resonance. In commercial caesium clocks, the width of the atomic resonance is of the order of 500 Hz and the amplitude signal to noise ratio in a band of 1 Hz is $(P_S/P_B)^{1/2} \sim 10^3$. We therefore have $\sigma_y(t) \simeq 3 \times 10^{11}\tau^{-1/2}$. The actual frequency stability depends on the particular model considered, as Table 6.2 shows. Figure 6.9 illustrates the frequency stability of commercial caesium clocks. Primary laboratory standards are usually assigned a frequency stability factor; it is equal to a few units of 10^{-12} for $\tau = 1\,\mathrm{s}$. This is achieved essentially by increasing the atomic-beam intensity leaving the crucible.

Long-term, for $\tau \geq 1$, frequency stability is no longer determined by the granular beam noise, and ambient conditions affect the operation of the

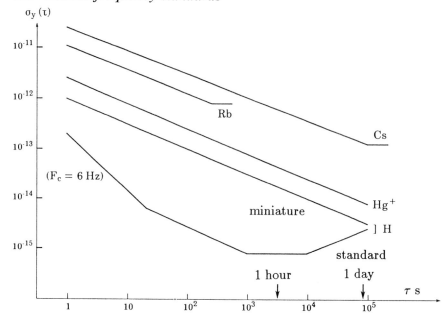

Fig. 6.9 Comparison of the stabilities of different atomic clocks.

clock. In a laboratory in which the temperature is held constant to better than 1°C, the relative frequency variations are of the order of a few parts in 10^{13} per annum.

6.3.4 Causes of clock frequency shifts

A number of physical effects, on the one hand, and of instrumental imperfections, on the other, affect the resonance frequency of a caesium beam tube. We must now briefly review the frequency shifts which can amount to approximately 10^{-13} in relative value. These shifts give rise to long-term frequency instability. They also determine the reproducibility and accuracy of the caesium clock.

(a) Shifts due to physical phenomena

Relativistic effects. The gravitational potential and the motion of a clock give rise to effects that do not depend on the nature of the clock. This has been analysed in detail by Ashby and Allan. We shall confine our attention to the gravitational effect. Near the Earth's surface, the frequency of any clock depends on its altitude as follows:

$$\frac{f_2 - f_1}{f_1} = \frac{g\,(h_2 - h_1)}{c^2} \tag{6.51}$$

Table 6.2. Numerical values of the parameters of frequency standards

Characteristic	CAESIUM		HYDROGEN		RUBIDIUM	
	High perform.	Standard perform.	Standard size	Small size	High perform.	Starndard perform.
Short-term stability: 1 s	5×10^{-12}	7×10^{-12} 6×10^{-11}	3×10^{-13}	2×10^{-12}	5×10^{-12}	3×10^{-11}
Medium term stability: 10^3s	3×10^{-13}	2×10^{-12} $2,5\times10^{-12}$	2×10^{-15}	6×10^{-14}	5×10^{-13}	1×10^{-12}
Long-term stability: 1 d	3×10^{-14}	3×10^{-13} 3×10^{-13}	$*<1\times10^{-14}$	$<1\times10^{-14}$		
Drift	$\pm2\times10^{-12}$ per tube life	$\pm2\times10^{-12}$ per tube life	$*<1\times10^{-15}$ per day	$<1\times10^{-15}$ per day $<1\times10^{-12}$ per device life	$<1\times10^{-11}$ per month	$<4\times10^{-11}$ per month
Accuracy	$\pm2\times10^{-12}$	$\pm7\times10^{-12}$ $\pm3\times10^{-12}$	$*2\times10^{-12}$			
Reproducibility	$\pm1,5\times10^{-12}$	$\pm3\times10^{-12}$ $\pm3\times10^{-12}$				
Temp. sensitivity	$<5\times10^{-12}$ /-28to61°C	$<5\times10^{-12}$ /-28 à 61°C	3×10^{-14} /1°C	$<5\times10^{-14}$ /1°C	$<4\times10^{-11}$ /0to50°C	$<3\times10^{-10}$ /-25to65°C
Magn. field sensitivity	$<2\times10^{-13}$ /2×10^{-4}T	$<2\times10^{-12}$ /2×10^{-4}T	$<2\times10^{-13}$ /1×10^{-4}T	$<1\times10^{-13}$ /1×10^{-4}T	$<5\times10^{-12}$ /1×10^{-4}T	$<3\times10^{-11}$ /1×10^{-4}T
Pressure sensitivity	$<2\times10^{-12}$ 0to13 000 m	$<2\times10^{-12}$ 0to13 000 m	$<3\times10^{-16}$ /1 mBar		$<5\times10^{-11}$ 0to13 000 m	$<1\times10^{-13}$ /1 mBar
Power consumption	43 W	43 W	100 W	70 W	50 W	13 W
Tot. volume	40 dm³	30 dm³ 40 dm³	650 dm³	65 dm³	25 dm³	1 dm³
Mass	32 kg	20 kg 30 kg	220 kg	35 kg	15 kg	1,5 kg
Guarantee	3 y tube	5 y tube	1 y	3 y maser	3 y lamp+cell	5 y lamp+cell

An empty box signifies that data are not available. Asterisks mark typical values obtained for active hydrogen masers obtained by the authors.
CAESIUM: high performance HP 5061 B opt. 004; standard performance FTS 4000 (left) and HP5061 B (right)
HYDROGEN: standard size EFOS; small size CPHM 100
RUBIDIUM: high performance HP 5065 A; standard performance FRK-L

where f_i $(i = 1 \text{ or } 2)$ is the clock frequency at the altitude h_i, g is the acceleration due to gravity and c the speed of light. The relative change in frequency is 1.09×10^{-13} per km. The frequency of the primary caesium standards is corrected for this effect (the NIST standard at Boulder is located at an altitude of 1.6 km). By convention, the surface of the geoid is taken as the altitude origin.

The second-order Doppler effect, mentioned in Section 6.2.5, is also a relativistic effect. The corresponding relative frequency shift is of the order of 10^{-13}. The exact value of the frequency shift depends on the effective velocity distribution of the atoms. Without going into details, we note from (6.35) that the effect of the velocity distribution is represented by the factor $f(\tau)\sin^2 b\tau$. Consequently, the exact value of the frequency shift due to the second-order Doppler effect depends on the amplitude of the microwave field. The long-term stability of the latter must therefore be excellent.

Zeeman effect. The $(F = 4, m_F = 0) \longleftrightarrow (F = 3, m_F = 0)$ clock frequency depends on the square of the magnetic induction B_0 experienced by the atoms. In practice, the corresponding frequency shift is large (its relative value is close to 1.7×10^{-10}). It must therefore be measured as accurately as possible. One method uses side bands in the spectrum of the interrogation signal separated, for example, by 42.82 kHz from the carrier in a field $B_0 = 6.12 \times 10^{-6}$T. The resulting $(F = 4, m_F = 1) \longleftrightarrow (F = 3, m_F = 1)$ and $(F = 4, m_F = -1) \longleftrightarrow (F = 3, m_F = -1)$ transition frequency varies linearly with B_0 and can be observed in the response of the caesium beam tube.

When the magnetic induction is inhomogeneous, the measurement gives $\overline{B}_0$, i.e., the mean value of B_0, and we use $(\overline{B}_0)^2$ to correct the clock transition frequency. The frequency shift Δf_z actually depends on $\overline{B_0^2}$ which can be different from $(\overline{B}_0)^2$. The magnetic induction must therefore be as uniform as possible inside the magnetic shield (*see* Fig. 6.5).

(b) Shifts due to the structure of the clock

Effect of a phase difference between the two microwave fields. The resonant cavity suffers losses due to the finite conductivity of its walls. There is therefore necessarily a travelling wave that transports energy lost through the Joule effect, and the standing wave is not entirely pure. If the two arms of the cavity do not have identical electric lengths, there is a phase difference φ between the oscillating fields in the two regions of interaction with the atoms. This phase difference is small (of the order of a fraction of a milliradian).

Equation (6.30) shows that if the atoms all have the same velocity, the transition probability maximum occurs at a frequency f that is different from f_0. We then have

$$f - f_0 = -\frac{\varphi v}{2\pi L} \qquad (6.52)$$

When a velocity distribution is present, it can be shown, using (6.35), that this frequency shift depends on the amplitude of the microwave field. Furthermore, the phase of the microwave field is a function of position in each interaction region and this leads to an additional frequency shift.

The relative frequency shift usually amounts to a few parts in 10^{12}. It can be considered as a residual first-order Doppler effect.

Frequency shift due to the cavity. If the resonant cavity is not tuned to the atomic transition frequency, the quantity b that represents the amplitude of the microwave field no longer varies symmetrically around ω_0. The result is a distortion of the atomic resonance and, hence, an apparent resonance frequency shift. By considering a variation of b in equation (6.35), we can show that the order of magnitude of the frequency shift is given by

$$\frac{f - f_0}{f_c - f_0} \approx \frac{Q_c^2}{Q_l^2} \qquad (6.53)$$

where $f_c - f_0$ is the detuning of the cavity and Q_c and Q_l the Q factors of the cavity and atomic resonance, respectively. When $f_c - f_0 = 1\,\mathrm{MHz}$ $Q_c = 500$ and $Q_l = 2 \times 10^7$, the relative frequency shift is close to 7×10^{-14}.

Effect of neighbouring transitions. Since the oscillating magnetic field is parallel to the main field, selection rules allow the excitation of $\Delta F = \pm 1$, $\Delta m_F = 0$ transitions. Amongst these, we will consider the effect of the transitions $(F = 3, m_F = -1) \longleftrightarrow (F = 4, m_F = -1)$ and $(F = 3, m_F = 1) \longleftrightarrow (F = 4, m_F = 1)$, which are nearest to the clock transition. When ω is close to ω_0, the probability of exciting these transitions is small, but not zero. It is given by (6.33), suitably averaged over the distribution of the interaction times, with ω_0 replaced by the neighbouring transition angular frequencies $\omega_1 = \omega_0 + 2\pi K_z' B_0$ and $\omega_{-1} = \omega_0 - 2\pi K_z' B_0$.

After magnet A, the atomic beam consists of atoms with an effective magnetic moment whose sign corresponds on the deflection. If this beam contains atoms in state $F = 3, m_F = 0$, it also contains, amongst others, atoms in the states $F = 3, m_F = -1$ and $F = 3, m_F = 1$. If these atoms undergo transitions with $\Delta F = \pm 1, \Delta m_F = 0$, their effective magnetic moment changes sign and magnet B deflects them towards the detector. This means that the tails of the neighbouring transitions that we are considering contribute, although in a small way, to the flux of atoms at the detector. This effect would be of no consequence if the side resonances were of the same intensity because they would then be symmetric relative to the clock resonance. However, in practice, the magnetic moments of the atoms in states with $F = 3, m_F = 1$ and $F = 3, m_F = -1$ are not exactly equal in the magnetic fields produced in the deflecting magnets. The side resonances do not therefore have the same intensity, and their tails distort the central resonance asymmetrically, so that, its frequency is shifted. The relative effect is of the order of 10^{-12} is commercially available clocks.

Effect of parasitic spectral components The spectrum of the interrogation signal can include unwanted components due to the electrical supply frequency and its harmonics. They can also originate from insufficient rejection of parasitic frequencies generated by the synthesis of the interrogation signal. The presence of such components disturbs the quantum system consisting of the two states $F = 3, m_F = 0$ and $F = 4, m_F = 0$ and causes a change in the energy difference between the levels and, hence, a shift of the transition frequency.

It can be shown that this effect cancels out when the parasitic spectral components are symmetric in position and height relative to the transition frequency. This is the case in practice for components due to the electrical supply frequency. On the other hand, the transition frequency is shifted by asymmetric spectral components separated from the transition frequency by 12.6 or 7.4 MHz and often present in the interrogation signal. For a single spectral component, the frequency shift is given by

$$f - f_0 = \frac{l}{L} \frac{(b')^2}{2\pi(\omega_0 - \omega')} \tag{6.54}$$

where ω_0 is the angular frequency of the transition and b' the Rabi frequency associated with the magnetic field component oscillating with pulsatance ω'. The quantity b' is defined by an expression similar to (6.19). For a parasitic line whose power is 10 dB below that of the interrogation power, we have $b' \approx 7 \times 10^3$ rad s^{-1}. When $\omega_0 - \omega' = 2\pi \times 12.6 \times 10^6$ rad s^{-1}, the relative frequency shift is 9×10^{-13}.

Effects of imperfections in electronics. We note that when the resonance curve at 9.2 GHz is located to within the nearest 10^{-13}, this corresponds to an error in the determination of the resonance frequency of approximately 10^{-3} Hz, i.e., two millionths of the resonance width if the latter is 500 Hz. The fact that this error is so small imposes stringent conditions on the interrogation signal and the treatment of the caesium beam response. To illustrate this point, consider the case of slow sinusoidal modulation. The first point is that the frequency modulation waveform can be distorted. If, for example, it contains a second harmonic impurity, we can write

$$\omega(t) = \omega_i + \omega_m(\sin\omega_M t + \delta_2 \sin 2\omega_M t + \delta_2' \cos 2\omega_M t) \tag{6.55}$$

where δ_2 and δ_2' describe the amounts of the second harmonic. We can then readily show that the fundamental $I_1(t)$ of the tube response contains an extra term as compared with (6.38). This term is proportional to δ_2. Since frequency locking tends to suppress $I_1(t)$, the result is that the interrogation frequency $\omega_i/2\pi$ is no longer centered on the shifted frequency of the atomic transition. The order of magnitude of the relative frequency error is

$$\frac{f - f_0}{f_0} \approx \frac{\delta_2'}{4Q_l} \tag{6.56}$$

For $(f - f_0)/f_0 = 10^{-13}$, we must have $\delta_2' \sim 8 \times 10^{-6}$ when $Q_l = 2 \times 10^7$. Generally, any contamination of the modulation waveform by even harmonics of the fundamental frequency is harmful. The effect is smaller for harmonics of order higher than 2. Frequency modulation therefore requires special precautions and is usually produced by a phase modulator whose linearity is as high as possible. The modulation is applied to a circuit tuned to the frequency of a quartz oscillator. It then has a small amplitude, which helps with linearity. The jump in phase and hence in frequency at 9.2 GHz is achieved by frequency multiplication. This type of shift is nowadays eliminated by using square-wave modulation of phase or frequency by digital techniques.

A frequency shift is also found to appear if the amplitude of the microwave field is modulated and contains odd harmonics of the modulation frequency. This can be described by replacing b in (6.36) by $b + \Delta b(t)$ where

$$\Delta b(t) = b\left(\xi_1 \cos\omega_M t + \xi_1' sin\omega_M t\right) \tag{6.57}$$

in which we confined our attention to the first harmonic. The fundamental of the response of the caesium beam tube again contains an extra term which causes a shift of the frequency of the locked quartz oscillator. The relative shift is given by

$$\frac{f - f_0}{f_0} \approx \frac{\xi_1}{Q_l} \tag{6.58}$$

A relative frequency shift of 10^{-13} implies $\xi_1 \approx 2 \times 10^{-6}$ for $Q_l = 2 \times 10^7$. This is also a stringent requirement and every effort must be made to keep the excitation level constant.

The bias voltage of the operational amplifier that performs the analogue integration described by (6.40) can also cause a shift in the frequency of the locked quartz oscillator. In the most recent models, this frequency shift is removed by numerical treatment of the response of the caesium beam tube.

6.3.5 Long-term frequency stability, reproducibility and accuracy

We have seen that there are many sources of frequency error. The associated frequency shifts are functions of ambient conditions and determine the long-term frequency stability and reproducibility of the frequency standard.

(a) Long-term frequency stability. The sensitivity of clock fluctuations to the magnetic field, temperature and even ambient humidity is largely responsible for the long-term frequency stability. Typical indicators of this sensitivity are indicated in Table 6.2. Variations in the ambient magnetic field influence the magnitude of B_0 and, hence, the frequency shift, through the second-order Zeeman effect (this despite the presence of the magnetic screen).

Temperature affects many of the frequency shifts, sometimes directly, as in the case of cavity frequency tuning, but most often indirectly through the amplitude of the microwave field in the interaction regions. The electronic circuit components depend on humidity, which can affect the apparent value of the atomic transition frequency. Aging effects, which influence some of the parameter values, must also be taken into account. The combined effect of these frequency shifts governs the long-term frequency stability, as indicated at the end of Section 6.3.3.

(b) Reproducibility. The reproducibility, also called intrinsic reproducibility, characterises the ability of an atomic frequency standard to reproduce its nominal frequency without calibration against another frequency reference. In the case of the caesium clock, this applies most particularly to commercial models. Reproducibility is measured by the relative difference between the output frequency of an instrument and another of a different type when the ambient conditions are within a specified range. Reproducibility indicates the quality of the design and manufacture. It ranges from 1.5×10^{-12} to 10^{-11} depending on the model.

Reproducibility must not be confused with repeatability. The latter characterises the spread in the output frequency of a given instrument between one operation and another under constant ambient conditions.

(c) Accuracy. The accuracy of a caesium beam frequency standard characterises its ability to deliver a frequency in agreement with the definition of the second. This implies the observation of isolated caesium atoms at rest, whilst in practice the atoms are moving and are subject to different perturbations. Accuracy is defined as the relative uncertainty in the output frequency.

In general, this uncertainty has two components. One of them is random, determined by fluctuations in the measured frequency shifts. The contribution due to random fluctuations in the measured magnetic field B_0 is an example. The other component is systematic and is due to insufficient knowledge of the operating conditions, e.g., magnetic-field inhomogeneities are not fully described, since we do not know exactly the difference between the measured $(\overline{B}_0)^2$ and the value of $\overline{B_0^2}$ necessary for the calculation of the frequency shift caused by the second-order Zeeman effect. In commercial atomic clocks, the second-order Doppler effect is another source of systematic uncertainty that is not corrected and can vary from one instrument to another, depending on the atomic velocity distribution. The quoted accuracy of commercial caesium beam clocks lies between 2×10^{-12} and 3×10^{-11}, depending on the model.

The accuracy of the primary caesium beam standards lies somewhere between 10^{-13} and a few units of 10^{-14}. This is achieved in long instruments, in which the distance between the two interaction regions is 1 to 2 m. This is the way to reduce the frequency shift which is inversely proportional to the Q of the atomic resonance. The frequency stability of primary standards is of the order of a few parts in 10^{12} for $\tau = 1\,\mathrm{s}$ and 10^{-14} for $\tau = 1\,h$.

6.3.6 Current and future research

Current research is partly aimed at improving the performance of caesium clocks of existing design. The laboratories that possess a primary standard are constantly trying to reduce and measure parasitic frequency shifts associated with the magnetic field, the second-order Doppler effect, the phase difference between the two arms of the resonant cavity, the imperfections in electronics and so on.

Commercial instruments employ digital electronics to generate the interrogation signal for atomic resonance, to process the signal produced by the caesium tube and to lock the quartz oscillator.

Several laboratories and companies have pursued research leading to a new class of caesium beam clocks in which the preparation and detection of the atoms are carried out optically. This has become possible since the advent of GaAlAs semiconductor lasers operating close to 0.85μm, which can be made to coincide with the caesium D_2 line.

The production of a population difference between these states $F = 3, m_F = 0$ and $F = 4, m_F = 0$ states is briefly explained in Section 6.2.4. Detection can also be carried out optically. Indeed, if we refer to Fig. 6.3, atoms prepared in state 3 are transferred to state 2 after crossing the two interaction regions in which they resonate with the microwave field. They can then again absorb photons whose wavelengths corresponds to the $1\rightarrow2$ transition, and emit fluorescence that is detected. If the $3 \rightarrow 2$ transition does not take place, the fluorescent light cannot be emitted. Figure 6.10 shows schematically the simplest caesium beam clock with optical pumping and detection. Other configurations are also possible. We can exploit the hyperfine structure of the D_2 line of caesium to prepare and detect at slightly different wavelengths. We can also use two lasers with suitably chosen wavelengths to improve the preparation of the atoms by transferring the population of the sixteen hyperfine sublevels to only one of the two states $F = 3, m_F = 0$ and $F = 4, m_F = 0$ and thus markedly increase the flux of atoms likely to be detected.

Optical methods offer the possibility of improved short-term and medium-term frequency stability as well as better long-term stability. The

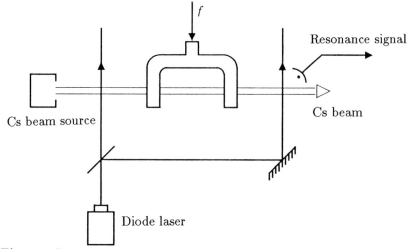

Fig. 6.10 Principle of an optically–pumped caesium beam tube.

deflection of atomic trajectories by magnets depends strongly on the velocity of the atoms. The result is a reduction by a factor >100 in the detected flux of atoms. There is no such reduction in the case of optical methods which have produced a measured short-term frequency stability characterised by $\sigma_y(\tau) = 2 \times 10^{-12}\tau^{-1/2}$. Moreover, plane-polarised light can be used to ensure that the amplitude of the $\Delta F = \pm 1, \Delta m_F = 0$ side lines varies symmetrically around the clock transition. This suppresses the frequency shift due to neighbouring lines. The magnetic field B_0 can then be reduced with the attendant reduction in the influence of its variation (the frequency shift due to the second-order Zeeman effect is proportional to B_0^2). We can also envisage the same medium-term and long-term stability as in magnetic-deflection caesium tubes, but in more compact instruments.

It will be possible to verify whether the realisation of the definition of the unit of time depends on the method of preparation and detection of the hyperfine states, which is very important for fundamental metrology. Moreover, the elimination or reduction of some of the frequency shifts will probably lead to more accurate primary caesium beam standards.

6.4 OTHER ATOMIC FREQUENCY STANDARDS

Hydrogen and rubidium clocks also play an important practical role in frequency metrology and are briefly described below. They constitute secondary standards.

6.4.1 The hydrogen clock

In around 1960, Ramsey demonstrated in Harvard that hydrogen atoms could undergo a very large number of collisions ($10^4 - 10^5$) with a surface such as that of Teflon before the relaxation of hyperfine states took place. Following this observation, it is now possible to confine such hydrogen atoms to a volume of the order of $1\,\mathrm{dm}^3$ for approximately $1\,\mathrm{s}$. The uncertainty principle given by (6.13) then shows that the hyperfine resonance has a width of the order of $1\,\mathrm{Hz}$, which corresponds to $Q \sim 10^9$. The ionisation energy of the hydrogen atom is too high for detection by the hot-filament method, so that the hyperfine resonance of the hydrogen atom is observed by using the maser effect (Microwave Amplification by Stimulated Emission of Radiation), which had already been used in 1955 to build the ammonia maser.

There are now two types of the hydrogen maser. In the first, the resonant cavity does not contain any material other than the storage cell and the hydrogen atoms. The resonant cavity, tuned to the hyperfine transition frequency, usually resonates in the TE_{011} mode. In the second type, invented in 1978, the dimensions of the resonating cavity are reduced by introducing dielectric or conducting media, which reduces the volume of the maser very markedly. Masers of the first category operate as self-oscillators, i.e., in an active mode. Those of the second category are either used as amplifiers, in a passive mode, or as oscillators. In the case of active operation, the associated electronic system locks the phase of the quartz oscillator to the maser oscillation. In the case of passive operation, the electronics locks the frequency of the quartz oscillator to the maximum of the amplifier gain. The hydrogen clock consists of a hydrogen maser and the associated electronics. The standard version is that in which the cavity resonates in the ordinary TE_{011} mode. The volume of the clock is then about $500\,\mathrm{dm}^3$. In the miniaturised version, the volume of the clock is of the order of $50\,\mathrm{dm}^3$.

(a) Principle of the hydrogen maser

General description. The hydrogen maser is illustrated schematically in Fig. 6.11. Molecular hydrogen is dissociated in a radio-frequency discharge and a beam of atomic hydrogen with a circular cross section is produced. The hyperfine structure states are selected by a magnet with four or six poles of alternate polarity, evenly distributed around the beam axis. The field in the magnet gap is perpendicular to the axis and its strength is independent of the polar angle; it increases uniformly with the distance from the axis, reaching $0.7 - 1\,\mathrm{T}$ on the poles. This field focusses atoms with $F = 1, m_F = 0$ and $F = 1, m_F = 1$, pulling them towards the axis, and defocusses atoms with $F = 0, m_F = 0$ and $F = 1, m_F = -1$, pushing them away from the axis. Atoms with $F = 1, m_F = 0$ and $F = 1, m_F = 1$ are collected in a quartz container of about $1\,\mathrm{dm}^3$, whose inner wall is coated with Teflon. The collector is placed in a cylindrical resonant cavity whose

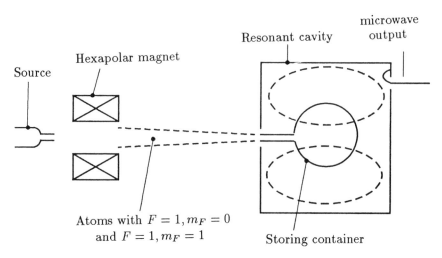

Fig. 6.11 Principle of a hydrogen maser.

axis is parallel to the atomic beam. The cavity resonates at 1420 MHz. The general shape of the microwave magnetic field lines of force is indicated in Fig. 6.11. The phase of the microwave magnetic field is constant within the volume occupied by the atoms, so that the transition frequency is not affected by the first-order Doppler effect and the width of the atomic resonance is effectively of the order of 1 Hz. A solenoid whose axis is parallel to that of the resonant cavity produces a magnetic induction of about 10^{-7} T. The microwave magnetic field being approximately parallel to the latter over the volume available to the atoms, the clock transitions $F = 1, m_F = 0 \longleftrightarrow F = 0, m_F = 0$ can take place.

The hydrogen maser also incorporates a magnetic screen consisting of four or five layers or materials with very high magnetic permeability around the resonant cavity. Hydrogen gas is removed by a pumping system.

Oscillation threshold. The population of atoms entering the quartz bulb with $F = 1, m_F = 0$ is very much greater than that of the atoms with $F = 0, m_F = 0$. The atomic beam therefore brings energy into the resonant cavity and the microwave field already present in the resonant cavity can be amplified if its frequency is very close to that of the hyperfine transition. A self-sustained oscillation can occur if a certain condition, which we will derive, is satisfied. The oscillation is then initiated by thermal noise at the transition frequency.

The two-level system that we have to consider here consists of the two hyperfine structure levels $F = 1, m_F = 0$ and $F = 0, m_F = 0$ of the ground-state atom. The density matrix of this system is given by (6.14) and its development is described by equation (6.15). The system is disturbed by the microwave magnetic induction, as in the case of the caesium clock, and also by relaxation mechanisms due to collisions of hydrogen atoms with the Teflon wall, collisions between hydrogen atoms and the loss of hydrogen through the bulb aperture. It can be shown that the coherence ρ_{12} and the population difference $\rho_{11} - \rho_{22}$ satisfy the following equations:

$$\frac{d}{dt}\rho_{12} = - \left[\frac{1}{T_2} + i\omega_0 \right] \rho_{12} + iV_{12}\left[\rho_{11} - \rho_{22} \right] \tag{6.59}$$

$$\frac{d}{dt}\left[\rho_{11} - \rho_{22} \right] = -\frac{1}{T_1}\left[\rho_{11} - \rho_{22} \right] + \frac{1}{T_b}\left[\rho_{11} - \rho_{22} \right]^0$$
$$+2i\left[V_{21}\rho_{12} - V_{12}\rho_{21} \right] \tag{6.60}$$

The time constants T_1, T_2 and T_b describe longitudinal relaxation, transverse relaxation and accumulation of the atoms, respectively; $[\rho_{11} - \rho_{22}]^0$ represents the population difference as the atoms enter the quartz collector.

If we suppose that there is a standing microwave magnetic field in the resonant cavity, we have, by neglecting the nonresonant component as before

$$V_{12} = \frac{b}{2}e^{-i\omega t} \tag{6.61}$$

where the phase of the microwave field is chosen as the origin of phase. The steady-state solution of (6.59) and (6.60) is

$$\rho_{12} = \frac{1}{2}\left(a_1 + ia_2 \right) e^{-i\omega t} \tag{6.62}$$

$$\rho_{11} - \rho_{22} = a_3 \tag{6.63}$$

where a_1, a_2 and a_3 are real constants.

At the resonance, i.e., for $\omega = \omega_0$, we have

$$a_3 = \frac{T_1 a_0^3}{T_b \left(1 + T_1 T_2 b^2 \right)} \tag{6.64}$$

As in the case of the caesium clock, we can now deduce the transition probability, using an equation similar to (6.29). This probability is found to be $T_1 T_2 b^2 (1 + T_1 T_2 b^2)$. Since each atom delivers energy hf_0 to the resonant cavity when a transition occurs, the power received by the cavity is

$$P = \frac{1}{2}I\hbar\omega_0 \frac{T_1 T_2 b^2}{1 + T_1 T_2 b^2} \tag{6.65}$$

where I is the flux of atoms entering in the cavity. Just above the oscillation threshold $T_1 T_2 b^2 \ll 1$ and (6.65) can be written in the form

$$P = I_s\omega_0 T_1 T_2 \mu_B^2 \langle B_z \rangle_b^2 / 2\hbar \tag{6.66}$$

where I_s is the intensity of the atomic beam at the oscillation threshold. In (6.66), the quantity $<B_z>_b$ is due to the random motion of atoms in the quartz collector, on the one hand, and the selection rules, on the other. The system behaves as if the atoms were in a mean field parallel to the maser axis, with magnitude equal to the amplitude of the microwave magnetic induction.

The power supplied by the atoms to the cavity maintains the electromagnetic field, so that, using the definition of the on-load Q, we have

$$P = \omega_0 W/Q_c \tag{6.67}$$

where W is the energy stored in the cavity. This energy is twice the energy stored in the form of magnetic energy.

$$W = V_c \langle B^2 \rangle_c / 2\mu_0 \tag{6.68}$$

where V_c is the volume of the cavity and $\langle B^2 \rangle_c$ is the mean square value microwave induction in the cavity. Hence, using (6.66)–(6.68), we have

$$I_s = \frac{\hbar V_c}{\mu_0 \mu_B^2 \eta Q_l T_1 T_2} \tag{6.69}$$

where $n = \langle B_z \rangle_b^2 / \langle B^2 \rangle_c$ is a measure of the coupling between the atomic medium and the electromagnetic field. In the standard hydrogen maser, $V_c = 15\,\mathrm{dm}^3, n = 2.5, Q_l = 3.5 \times 10^4$. T_1 and T_2 are approximately equal to $0.3\,\mathrm{s}$ and $I_s = 2 \times 10^{12}$ atoms per second. The maser operates in active or passive regime depending on whether we have $I > I_s$ or $I < I_s$. In the miniaturised maser, Q_l does not usually exceed 5000 or 10 000. One might think that the active mode operation could thus be obtained by increasing sufficiently the atomic flux. However, this is not so because of the broadening of the atomic resonance by collisions between the hydrogen atoms and by the resulting reduction in T_1 and T_2. Active-mode operation cannot usually be achieved for Q_l lower than about 1.5×10^4.

The order of magnitude of the oscillation power can be obtained by neglecting collisional broadening. We then have, using equation (6.67) and (6.69),

$$P \approx \frac{1}{2} \hbar \omega_0 (I - I_s) \tag{6.70}$$

Assuming that $I = 2I_s$, we obtain $P = 9 \times 10^{-13}\,\mathrm{W}$. The power coupled to the receiver is a fraction of this and amounts to about $10^{-13}\,\mathrm{W}$.

(b) Radio-frequency model of the maser

The behaviour of the maser can be described with the help of the radio-frequency model. One of the advantages of this model is that it shows that, despite the quantum-mechanical origin of the amplification mechanism, the behaviour of the maser can be described by the general theory of oscillators. We refer the reader to the bibliography for the derivation of the model.

Let us take the microwave magnetic induction in the form

$$B_1(t) = b \cos(\omega t + \varphi) \tag{6.71}$$

where B is defined by (6.19), ω is the pulsatance of this induction and φ is the phase. The microwave magnetic moment of the atoms is

$$M_a(t) = -\mu_B N (\rho_{12} + \rho_{21}) \tag{6.72}$$

where N is the total number of hydrogen atoms interacting with the electromagnetic field. We now define the reduced magnetic moment:

$$M_1(t) = -M_a(t)/\mu_B \tag{6.73}$$

and put

$$M_1(t) = m \sin(\omega t + \psi) \tag{6.74}$$

After some algebra, equations (6.59) and (6.60) then lead to the following expressions:

$$\frac{d^2}{dt^2} M_1(t) + \frac{\omega_0}{Q_l} \frac{d}{dt} M_1(t) + \omega_0^2 M_1(t) = 2\omega_0 B_1(t) M_3(t) \tag{6.75}$$

$$\frac{d}{dt} M_3(t) + \frac{1}{T_1} M_3(t) = -bm \cos(\varphi - \psi) + I \tag{6.76}$$

where M_3 is the population difference of the ensemble of atoms:

$$M_3(t) = N (\rho_{11} - \rho_{22}) \tag{6.77}$$

Equation (6.75) shows that the atomic magnetic moment resonates at the angular frequency ω_0 of the atomic transitions and that the Q factor of this resonant system is given by

$$Q_l = \omega_0 T_2/2 \tag{6.78}$$

The system is excited by the microwave magnetic field, but proportionally to the population difference $M_3(t)$.

The equation for the magnetic induction in the cavity is

$$\frac{d^2 B_1(t)}{dt^2} + \frac{\omega_c}{Q_c} \frac{dB_1(t)}{dt} + \omega_c^2 B_1(t) = K \frac{d^2 M_1(t)}{dt^2} - \frac{\omega_c^2}{Q_c} B_e(t) \tag{6.79}$$

where ω_c is the angular frequency of the cavity resonance and Q_c its on-load Q factor; K is a coefficient that depends on fundamental constants and n and $B_e(t)$ represens the possible injection of a signal into the resonant cavity, e.g., the interrogation signal in the passive mode.

(c) Standard hydrogen clock

Figure 6.12 shows in greater detail a possible design of a standard hydrogen maser. The influx of molecular hydrogen into the HF discharge separator is regulated by taking advantage of the selective hydrogen permeability of palladium, which depends on temperature. An oscillator supplies a few watts to the discharge, at about 100 MHz, and the diameter of the quadrupole or hexapole magnet piece is about 1 mm. The internal diameter and the height of the resonant cavity are both about 27 cm. The cavity is self-compensating for thermal expansion, and an evacuated thermal screen ensures temperature uniformity. The on load Q factor is usually close to 35 000 and a varactor coupled to the cavity is used to fine tune the resonant frequency. A low noise amplifier precedes the microwave receiver. A solenoid produces a magnetic induction of about 10^{-7} T inside the shield, with fluctuations below 10^{-13} T. Usually, a three-stage thermostat maintains the cavity temperature constant to within a millikelvin or better.

The standard hydrogen maser is generally used as a self-oscillator. Its oscillation frequency is mainly determined by the atomic transition frequency, but nevertheless depends on the resonance frequency of the resonant cavity. This frequency pulling by the cavity can be easily worked out and is given by

$$\omega - \omega_0 = \frac{Q_c}{Q_l}\left(\omega_c - \omega_0\right) \tag{6.80}$$

With the usual values of Q_c and Q_l we deduce from this expression that $(\omega - \omega_0)/\omega_0 = 2.5 \times 10^{-15}$ for $(\omega_c - \omega_0)/2\pi = 1$ Hz. It is therefore necessary to tune the resonant cavity very carefully.

The atomic oscillation is used to control the phase of a quartz oscillator at 5 or 10 MHz. Figure 6.13 shows a block diagram of the necessary electronics. The quartz oscillator to be controlled drives a frequency synthesiser which produces the local oscillator signal. After mixing, the signal delivered by the maser is amplified by an intermediate-frequency amplifier whose output signal is compared in phase with a second signal of different frequency, delivered by the frequency synthesiser. An error signal is generated if the phases are not identical and acts on the frequency of the oscillator to be locked, after passing through an operational filter whose role is to adjust the transfer function of the phase locked loop.

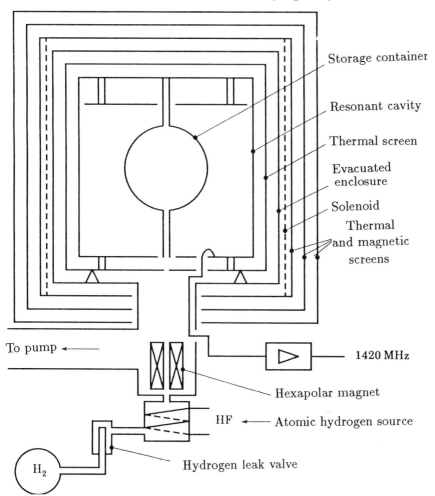

Storage container

Resonant cavity

Thermal screen

Evacuated enclosure

Solenoid

Thermal and magnetic screens

To pump

1420 MHz

Hexapolar magnet

HF ← Atomic hydrogen source

Hydrogen leak valve

H_2

Fig. 6.12 The classical hydrogen clock.

(d) Miniaturised hydrogen clock

The size of the hydrogen maser is largely dictated by the size of the resonant cavity. Its dimensions can nevertheless be reduced while keeping the resonance frequency at 1420 MHz and maintaining the magnetic field configuration within the volume available for the atoms. This is done by inserting dielectric or conducting materials as shown in Fig. 6.14.

The result is a cavity whose diameter D and height H are both about 15 cm when the dielectric is alumina. The dimensions of the cavity can be reduced even further when conductors are employed. A hydrogen clock has been built with $D = H \approx 7.5$ cm. In all cases, the atoms are confined to the central zone of the cavity by using Teflon-coated walls. The distance

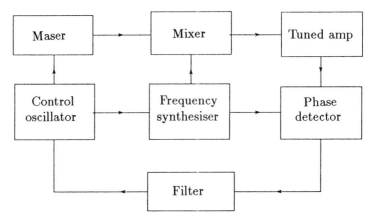

Fig. 6.13 Electronics for the hydrogen clock.

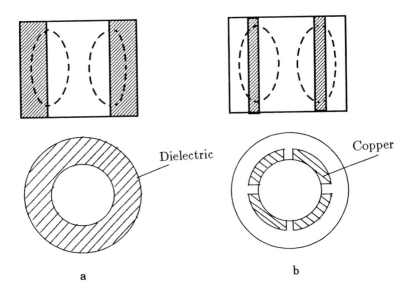

Fig. 6.14 Miniaturised cavity of the hydrogen clock.

between the source of atomic hydrogen and the cavity can be reduced and very compact electronics can be constructed. The overall clock volume is between 25 and 60 dm^3, depending on the design. Nevertheless, the insertion of materials into the resonant cavity gives rise to additional losses and Q falls to less than 10 000. Two approaches are then possible. We can use a Q factor for which self-oscillation can no longer take place, so that an interrogation signal has to be generated. The maser then functions as a highly tuned amplifier with a pass band of the order of 1 Hz and a resonance curve similar to that of the caesium tube. This is the passive mode of operation in which a quartz oscillator is controlled by the atomic resonance and the principle is similar to that illustrated in Fig. 6.7. The other approach is to increase Q by electronic means to a value for which the self-oscillations appear. This is active mode operation.

(e) Frequency stability

The short-term and medium-term frequency stability of hydrogen masers is limited by random frequency fluctuations produced by the thermal noise in the resonant cavity and the additional noise of the input stage of the receiver.

Active masers: the effects of thermal noise in the resonant cavity. When thermal noise is present in the resonant cavity, the magnetic induction has a quasi-sinusoidal part described by (6.71) in which φ is now random, and an additional part $B_b(t)$ that represents thermal noise centred on the resonant frequency of the cavity, i.e., in practice, the oscillation frequency. Let

$$B_b(t) = p_c(t)\cos\omega t + p_s(t)\sin\omega t \tag{6.81}$$

where $p_c(t)$ and $p_s(t)$ have the same dimensions as the amplitude of $B_1(t)$. Moreover, these are slowly-varying random variables with identical spectral densities. In practice, the maser noise effect is measured over observation times longer than 0.1 s. The corresponding pass band is therefore much smaller than the pass band of the resonant cavity, and we can assume that the spectra of $B_b(t)$, $p_c(t)$ and $p_s(t)$ are broad. The one-sided power spectral densities of these quantities are then related by

$$S_{B_b} = \frac{1}{2}S_{p_c} = \frac{1}{2}S_{p_s} \tag{6.82}$$

where the factor $1/2$ comes from the fact that the width of the low-frequency spectrum of $p_c(t)$ and $p_s(t)$ has a range that is half that of the microwave spectrum of $B_b(t)$ (see Section 3.1.2).

In addition, we have the usual conditions

$$p_c(t)/b \le 1 \quad \text{and} \quad p_s(t)/b \le 1 \tag{6.83}$$

The energy of the thermal noise contained in the useful mode is equal to kT where T is the thermodynamic temperature of the cavity. Moreover, the noise pass band of the resonant cavity being equal to $\omega_c/4Q_c$, this energy is proportional to $\omega_c S_{B_b}/4Q_c$. Hence we have

$$\frac{\omega_c}{4Q_c}S_{B_b} = C\frac{kT}{2} \tag{6.84}$$

where C is a constant. The factor $1/2$ reminds us that half the energy is contained in the magnetic component of the microwave field.

The energy W associated with the quasi-sinusoidal magnetic induction described by (6.71) is such that

$$\overline{[B_1(t)]^2} = \frac{b^2}{2} = C\frac{W}{2} \tag{6.85}$$

where the constant is the same as in (6.84). This energy is related by (6.67) to the power P supplied by the atoms to the cavity. From (6.82), (6.84) and (6.85) we then obtain

$$\frac{S_{p_c}}{b^2} = \frac{S_{p_s}}{b^2} = \frac{4kT}{p} \tag{6.86}$$

which gives the power spectral density of the thermal noise component.

We will now use the radio-frequency model to determine in a simplified manner the frequency stability of active models. By superimposing in the resonant cavity a purely sinusoidal field of the form $b\cos\omega t$ on the random field described by (6.81), we obtain a quasi-sinusoidal field with fluctuating phase $\varphi(t)$. When $p_s/b \ll 1$, the phase variation is given by

$$\varphi(t) = -p_s(t)/b \tag{6.87}$$

If we consider first of all the rapid variation of $p_s(t)$ with respect to the pass band F_c of the atomic medium, these fluctuations are not transmitted by the latter and we find ourselves in the situation just described where the phase fluctuations of the oscillator are due to the additive noise of the cavity. We then have

$$S_\varphi(F > F_c) = \frac{4kT}{P} \tag{6.88}$$

and hence

$$S_y(F > F_c) = \frac{4kT}{P}\frac{F^2}{f_0^2} \tag{6.89}$$

On the other hand, if the phase variations are slow, the effect of the feedback loop must be taken into account. The selectivity of the atomic medium introduces the following relation between the variation in the angular frequency and phase:

$$\Delta\varphi = 2Q_l\Delta\omega/\omega_0 \tag{6.90}$$

where $\Delta\varphi$ is due to the noise in the cavity, and we have

$$S_y(F < F_c) = \frac{kT}{PQ_l^2} \tag{6.91}$$

We will assume that the two parts of $S_y(F)$ given by (6.89) and (6.91) are additive.

Active masers: the effect of receiver noise. Receiver noise is superimposed on the signal from the maser cavity. The result is an additional phase fluctuation, given by an expression similar to (6.87):

$$\varphi'(t) = -\frac{p'_s(t)}{b'} \tag{6.92}$$

where $p'_s(t)/b'$ represents the amplitude of the quadrature component of noise fed back to the amplifier input, relative to the amplitude of the signal from the maser. We have

$$\frac{S_{p'_s}}{(b')^2} = \frac{F_b k T Q_{\text{ext}}}{P Q_c} \tag{6.93}$$

where F_b is the receiver noise factor at the same temperature as the maser cavity and Q_{ext} is the Q of the coupling loop to the receiver. The corresponding spectral density of the relative frequency fluctuations is therefore

$$S_y(F) = \frac{F_b k T Q_{\text{ext}}}{P f_0^2 Q_c} F^2 \tag{6.94}$$

Active masers: frequency stability. The two-sample variance of an active maser follows from the results of Chapter 3 and it is given by

$$\sigma_y^2(\tau) = K_{-1}\tau^{-1} + K_{-2}\tau^{-2} \tag{6.95}$$

where

$$K_{-1} = \frac{kT}{2PQ_l^2} \tag{6.96}$$

and

$$K_{-2} = \frac{3kTF_c}{2\pi f_0^2 P}\left[1 + F_b\frac{Q_{\text{ext}}}{4Q_c}\right] \tag{6.97}$$

in which F_c is the threshold frequency of a low-pass filter with a simple pole. These results typically apply to the standard maser.

For $T = 313\,\text{K}, P = 5 \times 10^{-13}W, Q_l = 1.5 \times 10^9, F_c = 6\text{Hz}, F_b = 1.5$ and $Q_{\text{ext}}/Q_c = 6$ we obtain

$$\sigma_y^2(\tau) = [6.6 \times 10^{-14}\tau^{-1/2}]^2 + [2.0 \times 10^{-13}\tau^{-1}]^2 \tag{6.98}$$

which is in agreement with experimental results for $\tau \leq 10^3\,\text{s}$. The function $\sigma_y(\tau)$ is shown in Fig. 6.9. Long-term frequency stability is affected by systematic effects that must be minimised by protecting the frequency standard from temperature fluctuations by thermostats and from magnetic-field fluctuations by magnetic shields. Some hydrogen clocks equipped with automatic resonant cavity tuning systems have frequency stability of a few parts in 10^{14} over several years.

In the miniaturised version operating in the active mode, the electronic feedback loop, which increases the apparent Q of the cavity, has the effect of increasing the noise temperature of the cavity. The result is a degradation of frequency stability. In practice, $\sigma_y(\tau) \approx 10^{-12}/\sqrt{\tau}$ for $\tau > 1\,\text{s}$.

Passive masers: frequency stability. In miniaturised hydrogen clocks operating in the passive mode, the locking of the quartz oscillator is similar to that in the caesium clock. Consequently, and even though the origin of the noise may be different, the short-term and medium-term stability is described by an equation similar to (6.49) in which fluctuations are determined by the thermal noise of the cavity and by receiver noise. In practice, the frequency stability is very similar to that of the miniaturised version used in the active mode. The latter is illustrated in Fig. 6.9.

(f) Accuracy

In the hydrogen maser, the atomic transition frequency is shifted by the applied static magnetic field, the second-order Doppler effect, frequency pulling by the cavity and the collisions of the hydrogen atoms with the wall of the quartz collector collector. The last of these is due to the deformation of the atom as it collides with wall, which modifies the interaction energy between the proton and the electron.

The frequency shift corresponding to the lack of reproducibility of the properties of the Teflon coating is least accurately known, but it is this shift that limits the accuracy of the hydrogen maser.

The frequency shift due to collisions with the wall of the collector can be determined by measuring the frequency of a hydrogen clock for two or more values of the mean collector diameter, other things being kept constant, by extrapolating to an infinite diameter. This gives oscillation frequency in free space and the frequency correction due to this wall effect for given collector diameter and given Teflon coating. The relative uncertainty in the clock frequency is usually close to 2×10^{-12}.

By using a double collector, it has been possible to attain an uncertainty of 6×10^{-13} and measure the hyperfine transition frequency of atomic hydrogen to within 7×10^{-13}. This is the most accurate physical measurement currently available.

(g) Current and future research

Since short-term and medium term frequency instability is mainly due to the thermal noise of the resonant cavity, it is reasonable to consider cooling of the complete system consisting of the cavity and collector. Moreover, research in fundamental physics aimed at studying Bose-Einstein condensation of atomic hydrogen has shown that this gas can be confined for several seconds in a vessel whose walls are coated with liquid helium. For helium-4, the most favourable temperature which produces the smallest frequency shifts is about 0.5 K. At very low temperatures, the broadening of the hyperfine resonance by collisions between the hydrogen atoms is considerably reduced. Hence, it seems possible advantageously to combine the reduction in thermal noise, reduction in the resonance and improved reproducibility of wall properties. Consequently, an improvement in the frequency stability of the hydrogen maser by two or three orders of magnitude seems possible with stability of about 10^{-18} for $\tau = 10^3$ s. Several laboratories have built and operated cryogenic hydrogen masers.

6.4.2 The rubidium clock

The rubidium clock is an application of the double-resonance methods in which an atom is subjected simultaneously to a light wave and microwave field. The light wave is used for optical pumping of one of the two levels between which the microwave resonance occurs, and also to detect this resonance. The first experiments were concerned with sodium and caesium, but the development of the method of hyperfine filtering has enabled, since 1958, the efficient pumping of the rubidium-87 atom, so that most of the subsequent development has been concerned with this atom.

The rubidium clock is the most widely used atomic frequency standard. Its cost/performance ratio makes it suitable for a wide range of applications e.g., in navigation.

(a) Optical and microwave configurations

Figure 6.15 shows a schematic diagram of the rubidium clock. The operation can be described with the help of Fig. 6.16 which employs principles that we have already covered.

In the lamp containing the rubidium-87 isotope, the atoms are excited by a radio-frequency discharge and emit D_1 and D_2 resonance lines (Fig. 6.16a) with wavelengths of 795 and 780 nm, respectively. These exhibit the hyperfine structure that is mainly due to the two ground-state hyperfine levels with $F = 1$ and $F = 2$.

The isotope filter cell contains the rubidium-85 isotope whose atoms absorb one of the hyperfine components of the D_1 and D_2 lines, so that the filter is opaque to this component in the line spectrum emitted by the lamp, but is transparent to the other component. The filter contains a gas, generally argon at a pressure about 10 kPa, which shifts and broadens the resonance lines of rubidium-85 in order to improve filtering. Figure 6.16b illustrates schematically the spectrum of light transmitted by the filter.

The filter cell is placed in the resonant cavity. The incident light is selectively absorbed by the atoms occupying the $F = 1$ level of the ground state. When the rubidium atoms have been excited and raised to one of the P states, they relax back to the ground-state $F = 1$ and $F = 2$ levels by spontaneous emission or as a result of collisions with the gas molecules. The absorption/relaxation cycle ensures that the $F = 1$ level becomes less

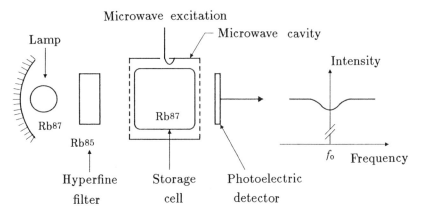

Fig. 6.15 Schematic diagram of the rubidium clock.

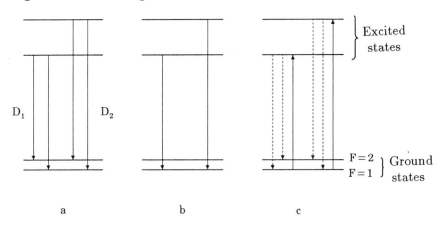

Fig. 6.16 Energy levels used in the rubidium clock.

populated than the $F = 2$ level (Fig. 6.16c). The cell is then transparent to the incident light because the number of atoms likely to absorb the light is now smaller.

So far, we have ignored the microwave field of the cavity; moreover, what has just been described is valid when the microwave field is not in resonance with the transition between the two ground-state hyperfine levels. Let us now assume that the resonance condition is satisfied. The $F = 2 \rightarrow F = 1$ transition is stimulated and the $F = 1$ level is thus re-populated. Light transmission by the cell is now altered and the intensity reaching the silicon photocell is reduced slightly. Figure 6.15 shows the resonance profile obtained by sweeping the microwave frequency across the resonance frequency at approximately 6 835 MHz. This signal can be described by the expression

$$I = I_b \left(1 - \frac{k}{1 + 4(f - f_0)^2/(\Delta f)^2} \right) \qquad (6.99)$$

where I is the collected photon flux, I_b is the flux outside the resonance, Δf is the width of the resonance, usually of the order of 500 Hz, and k is a parameter equal to a few thousandths.

The resonant cell contains a buffer gas, often nitrogen, at a pressure of a few kPa, which has several functions, namely, elimination of the first-order Doppler effect by reducing the velocity of the atoms to a few centimetres per second, reducing the effect of collisions with the cell walls, adjustment of the temperature coefficient of the clock and so on.

In the simplified and compact versions of the rubidium clock, the hyperfine filter is removed and the resonant cell contains natural rubidium (Rb ^{87} + Rb ^{85}). In this configuration, the latter also has the function of the hyperfine filter.

In high-performance versions, the cavity resonates in the TE_{011} mode. The TE_{111} mode is used in compact versions because the cavity can then be smaller.

Magnetic shields protect the cell from the ambient magnetic field and from its variations. A coil is placed inside the magnetic shield and produces a static magnetic field of around 10^{-5}T. The volume of the compact version of the rubidium clock is of the order of 1 dm^3.

(b) Electronics

Since the rubidium clock is a passive atomic frequency standard, the frequency of the quartz oscillator is controlled by the atomic resonance as shown in Fig. 6.7. Other circuits generate the radio-frequency signal necessary for the excitation and thermal regulation of the discharge lamp, the resonant cell and the hyperfine filter (if any). The density of rubidium and of the gases contained in the different elements has to be controlled to ensure satisfactory operation of the lamp, adequate signal to noise ratio in the detection of the hyperfine resonance and control of parasitic frequency shifts.

(c) Short-term and medium-term frequency stability

As in all passive clocks, a good approximation to short-term and medium-term frequency stability is provided by (6.50). Here the noise is the granular noise of the current supplied by the photocell. In practice, we have $Q_l = 10^7, (P_S/P_B)_{1\,\text{Hz}}^{1/2} \approx 4 \times 10^3$, which leads to $\sigma_y(t) \approx 10^{-11}\sqrt{\tau}$.

Actually, frequency stability depends on the model considered, as shown in Table 6.2.

(d) Frequency shifts and long-term frequency stability

In addition to the frequency shifts common to all atomic clocks, there are two frequency shifts that are specific to the rubidium clock. One of them is specific to double-resonance experiments in which two transitions with a common energy level are excited simultaneously. It is called the optical shift because it is the result of optical pumping. It amounts to a few hertz under normal conditions. The other frequency shift is due to collisions between rubidium atoms and buffer-gas molecules, in which the interaction forces induce a modification of the hyperfine gap. This shift amounts to several kilohertz.

It is clear that the resonance frequency of the rubidium clock must be calibrated because of the parasitic frequency shifts. It is very hard to make them reproducible because the filling pressure of the buffer gas cannot be controlled with sufficient accuracy. The frequency of the oscillator locked at 5 MHz is regulated by the synthesiser in the feedback loop and by the magnetic field applied to the atoms.

The optical and collision shifts are relatively large, and frequency fluctuations will occur due to, for example, variations in ambient conditions and aging. Actually, for $\tau \leq 2 \times 10^2$ s, the frequency stability deviates from the value set by granular noise of the photodetector current (*cf*. Fig. 6.9). The rubidium clock exhibits a relative frequency drift of the order of 10^{-11} per month. When this drift is taken into account, the frequency stability is between 10^{-12} and 10^{-13}, depending on the model, for sampling times τ between 10^3 and several days.

(e) Current and future research

The primary aim of research is to reduce the sensitivity of the rubidium clock to external perturbations. This is accomplished by a judicious choice of the buffer-gas mixture that ensures that the shift of the transition frequency and its temperature coefficients are as small as possible. The second aim is to reduce further the volume of the rubidium clock. An overall volume close to 100 cm^3 seems feasible at present.

Optical pumping by a low-efficiency rubidium lamp and a semiconductor lasers is being studied. First results show a very clear increase of the strength of the resonance signal. However, problems connected, for example, with the frequency instability of these lasers have to be solved.

We note that laboratory studies have led to the development of rubidium-87 masers operating in the active mode. This is done by optimising population inversion between the $F = 2, m_F = 0$ and $F = 1, m_F = 0$

levels, and by placing the resonant cell inside the cavity with best possible Q. The main interest in the rubidium maser rests on its excellent short-term frequency stability which is of the order of 10^{-13} for $\tau = 1\,\mathrm{s}$. This stability tends to deteriorate for $\tau > 1\,\mathrm{s}$ due to the optical and collision shifts in the buffer gas.

6.4.3 Trapped mercury-ion clock

It has been known for several decades that charged particles can be suspended in vacuum by static electric and magnetic fields (Penning trap) or a radio-frequency electric field (Paul trap). Ions can be held in this way for several days in confined space without walls. A clock using mercury ions confined in a radio-frequency trap has been built. At the time of writing, only a few prototypes are available.

(a) Confinement of ions in a radio-frequency trap

Consider a trap with electrodes in the form of hyperboloids of revolution (Fig. 6.17).

A potential difference is applied between the ring, on the one hand, and the two lids, on the other. The axial and radial components of the electric field in this structure are proportional to the distance from the centre of the trap but have opposite signs. If the potential difference is constant, a charged particle placed in the trap will experience a force whose axial and radial components have opposite effects. If, for example, the axial component pulls the particle towards the centre, the radial component is centrifugal.

The situation changes when the potential difference between the electrodes is alternating with frequency f_c. The alternating electric field forces the particles into periodic motion of frequency f_c and amplitude that is small under normal conditions. This so-called microperiodic motion is then approximately sinusoidal. To analyse the favourable consequences of this motion, let us examine its axial and radial components with the help of Fig. 6.18.

The two components have opposite phases because, as mentioned earlier, they have opposite signs. The microperiodic motion allows the charged particle to explore the electric field. Since the field is non-uniform, the electric force acting on the particle during its microperiodic motion is not exactly sinusoidal. For a positively charged particle, for example, the variation in one period is illustrated in Fig. 6.18. We note that the mean electric force is not zero and is directed towards the centre of the trap. This conclusion is valid for both axial and radial components. This restoring force impresses on the particle a harmonic motion in the neighbourhood of the trap centre,

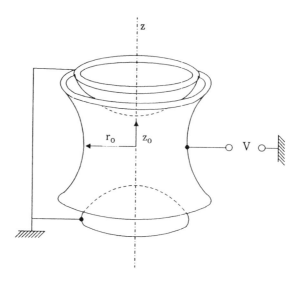

Fig. 6.17 Radio-frequency trap.

which is superimposed on the microperiodic motion. Its frequency f_s is generally small compared to f_c. It is called *secular* motion.

An alternating potential difference applied to the electrodes of the trap will therefore confine the particles to the neighbourhood of its centre. The ions become suspended in vacuum under the action of the inhomogeneous alternating electric field. They describe sweeping secular motion with the higher-frequency microperiodic motion superimposed upon it with a smaller amplitude. In a trap whose characteristic dimensions are of the order of 1 cm, and the applied potential difference has $f_c = 200\,\text{kHz}$ and amplitude of the order 100 V, the secular frequency f_s for mercury ions is of the order of 20 kHz and the amplitude is typically of the order of a third of the internal dimensions of the trap.

The density of the ion cloud that can be confined in this way is very low, mostly because of the repulsive electrostatic forces between the charged particles. This density is of the order of 10^6 ions per cubic centimetre, which corresponds to a partial pressure of the order of 10^{-7} Pa, comparable to that of the residual-gas pressure in a high vacuum. The ionic medium is therefore very diluted.

Let us now suppose that the ion cloud is exposed to an electromagnetic wave of fixed frequency f'. As a consequence of the Doppler effect, the ions 'feel' a frequency-modulated wave. The signal spectrum seen by the ions is therefore made up of a carrier of frequency f' and side lines of which the closest ones are separated by f_s. If the ions have a transition frequency very close to f', and if the resonance at f' is very sharp, they are sensitive

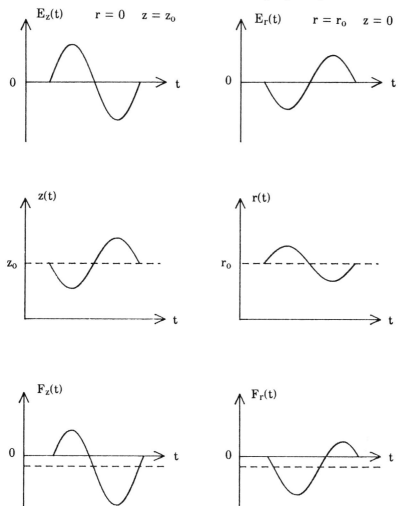

Fig. 6.18 Motion of an ion in a radio-frequency trap.

only to the latter frequency, and the side components do not affect them. Everything then happens as if the Doppler effect did not occur. This is observed in hyperfine transitions.

(b) ^{199}Hg$^+$ frequency standard.

The nuclear spin of the mercury-199 isotope is 1/2. Moreover, singly-ionised mercury has an electronic structure similar to that of the alkali metals, with one electron in the outmost shell, so that ^{199}Hg$^+$ has a hyperfine structure as simple as that of the hydrogen atom. However, as indicated in Table 6.2, the hyperfine spacing is much larger and amounts to approximately 40.5 GHz.

The transition between the $F = 1, m_F = 0$ and $F = 0, m_F = 0$ levels is observed by a method similar to that used in the rubidium clock.

A lamp containing ionised mercury-202 emits ultraviolet radiation whose wavelength is about 194 nm. As a consequence of a favourable isotopic coincidence, it is selectively absorbed by trapped $^{199}\mathrm{Hg^+}$ ions in the $F = 1$ level of the hyperfine structure. These ions are raised to an excited state and drop back almost immediately, and with equal probabilities, to the $F = 1$ and $F = 0$ ground-state levels by emitting 194-nm radiation. The number of ions in the $F = 1$ state therefore decreases, to the advantage of those in the $F = 0$ state. This eventually leads to a state of equilibrium in which the population of the $F = 1$ state is small. Few ions are then available for raising to the excited state and the radiation emitted by them is of low intensity when this equilibrium is reached. The trapped ions are now exposed to electromagnetic radiation whose frequency is close to 40.5 GHz and is adjusted to stimulate transitions between the $F = 0$ and $F = 1$ levels, so that the $F = 1$ level finds itself re-populated and the intensity of the light fluorescing at 40.5 GHz increases. The intensity of this light thus acts as an indicator of the hyperfine structure transition. The ionic medium is very diluted, so that the intensity of the fluorescent light is considerably lower than that of the incident optical pump radiation. The fluorescent light is therefore observed at right angles to the direction of the pump radiation, taking care to avoid light scattered by the apparatus.

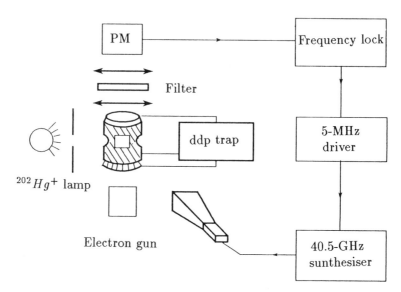

Fig. 6.19 Schematic diagram of a mercury ion clock.

Figure 6.19 illustrates schematically the mercury ion clock. The trap has a simple cylindrical shape and, from the point of view of confinement, behaves very much like that shown in Fig. 6.17. The upper lid is a loose metal grid that acts like an equipotential surface, but allows 90% of the fluorescent light through. The mercury-199 atoms are ionised by bombarding them with electrons in the trap. The trap and the electron gun are placed in vacuum. The pump radiation is produced by a lamp containing mercury-202 which is ionised and excited by an intense 100-MHz electric field. The fluorescent light is filtered and detected by a photomultiplier. The 40.5-GHz wave is produced electronically by frequency synthesis from a quartz oscillator. A horn allows it to be aimed at the ion cloud.

The resonance line obtained in this way is very narrow. Its width is of the order of 1 Hz which corresponds to $Q \sim 5 \times 10^{10}$.

This is a passive frequency standard that can be locked to a quartz oscillator using the hyperfine resonance of $^{199}Hg^+$ and the general method shown schematically in Fig. 6.7. Its frequency stability is excellent, with $\sigma_y(\tau) = 2 \times 10^{12}\sqrt{\tau}$ where τ is of the order of 1d.

The kinetic energy of the stored ions is of the order of 1 eV, which corresponds to a temperature of 10^4 K. The second-order Doppler effect amounts to approximately 5×10^{-12}. This relative frequency shift can be reduced to a value close to 10^{-13}, by cooling the ions by collisions in the buffer gas, e.g., helium. This is effective at very low pressures, close to 0.01 Pa. The corresponding relative frequency shift is very small, i.e., of the order of 10^{-14}. The optical shift can be eliminated by cutting off the ultraviolet illumination when the 40.5-GHz interrogation signal is applied. The second-order Zeeman effect is intrinsically small, the constant K_0 in of equation (6.6) having the value $9.70 \times 10^8 Hz\,T^2$ for $^{199}Hg^+$. The parasitic frequency shifts are small, so that the long-term frequency stability is excellent. The frequency drift amounts to a few parts in 10^{15} per day. Moreover, an accuracy of the order of 10^{-13} seems possible.

6.4.4 Slowing down ions and atoms: a clock with a single ion at rest?

We have mentioned the thermalisation of confined ions by collisions with a light buffer gas. Much lower temperatures can be obtained by using laser radiation whose frequency is lower than that of one of the absorption lines of the ion under consideration. As in the case of hyperfine structure transitions, the Doppler effect ensures that the spectrum optical transitions of confined ions consists (in the laboratory frame of reference) of a component at the resonance frequency f_0 of the ion at rest and the side bands. The latter are generally not resolved in the optical spectrum. However, an ion can absorb a photon in one of the side bands of frequency f_L lower than the transition frequency at rest f_0. The ion then re-emits this photon by isotropic spontaneous emission. Its mean frequency is f_0. This means that, in each absorption-emission cycle, the ion loses on average energy of $h(f_0 - f_L)$. This is taken from the kinetic energy, so that the ion slows down. These principles have been verified experimentally and temperatures of the order of 10 mK have been achieved. Theoretical predictions set the limiting temperature at about 1 mK. The method can also be used with atomic and laser beams travelling in opposite directions. It is even possible to exploit stimulated emission to obtain even more efficient slowing down. An ion or atom at rest is thus a real possibility. Very interesting avenues are thus revealed for spectroscopy and for the development of frequency standards in the microwave and optical ranges.

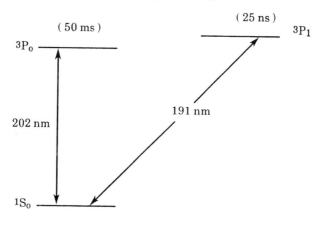

Ion 205Tl +

Fig. 6.20 Energy diagram of the thallium ion.

The various proposals that have been made include the suggestion by Dehmelt for a frequency standard using a single ion at rest. We note, first, that it is indeed possible to observe a single ion confined to a trap. When the rate of creation of ions by ionisation is very small, it is possible to fill a trap very slowly, say, at the rate of one ion per minute. Moreover, an ion (or an atom) is able to absorb photons at a rate whose order of magnitude is the reciprocal of the lifetime of the excited state. When this lifetime is 30 ns, the ion can absorb 3×10^7 photons per second. A laser beam can supply such a flux. The ion re-emits the same photon flux when, by spontaneous emission, it returns from the excited state to the ground state. The flux of fluorescent photons is high and it has been reported that, when the number of ions contained in the trap changes by one, a jump in the fluorescent light intensity can be observed. The fluorescence of a single confined ion has effectively been observed. This method can be used to detect a different transition. In the proposed scheme, a ^{205}Tl$^+$ ion, whose simplified energy diagram is shown in Fig. 6.20, is held in a radio-frequency trap with an electrode separation of a few hundred micrometers. It absorbs a photon of wavelength 202 nm and undergoes the $6^1S_0 \rightarrow 6^3P_0$ transition. The lifetime of the excited state is approximately 50 ms, so that the relative width of the resonance is very small, i.e., 10^{-14}. The resolution of this method compares very favourably with that obtained by the Mössbauer effect. The resonance is detected by means of the intermediate $^1S_0 \longleftrightarrow ^3P_1$ transition. The lifetime of the $^3P_1$ state is short, of the order of 1 ns, so that a large number of fluorescent photons, of the order of $10^8 \, \text{s}^{-1}$, is emitted if the ions are in the $^1S_0$ ground state. On the other hand, no fluorescent photons are emitted if the ion has been raised to the $^3P_0$ state by the 202-nm radiation. The corresponding transition is thus detected with a very good signal to noise ratio. Moreover, the $^1S_0 \longleftrightarrow ^3P_1$ transition can be used to cool the ions down to about 1 mK. The ion is then suspended at the centre of the trap, in the neighbourhood of the point with zero electric field, and moves around within a volume of approximately $1 \, \mu\text{m}^3$. This takes us very close to the ideal conditions for the observation of a single, motionless and isolated particle. This scheme should produce very good frequency stability, very high accuracy and improved performance (by several orders of magnitude). However, a working clock will require some advances in the generation of coherent radiation in the ultraviolet, on the one hand, and in the measurement of the corresponding frequency, on the other. Meanwhile, current research is focussed on the simultaneous utilisation of two species of ion. One of them is Mg^+, cooled by laser radiation; the other ion offers a favourable spectrum, with, for example, ground-state hyperfine structure, e.g., the mercury-199 ion or the yttrium-171 ion.

6.5 SOME APPLICATIONS OF ATOMIC CLOCKS

The range of application of atomic clocks is not confined to quantities such as frequency and time. Accurate distance measurements are carried out by determining the propagation time or the phase shift of radio-frequency signals. Velocity measurements are carried out by exploiting the Doppler effect, and voltage measurements employ the Josephson effect.

Purely scientific applications usually require the most stable and/or accurate frequency standards. Technical applications involve the largest number of atomic frequency standards, selected according to their performances and operational and financial constraints.

6.5.1 Fundamental metrology

Atomic frequency standards obviously provide absolute frequency and time standards when they run continuously as clocks.

In the 70s, measurements of the frequency and wavelength of radiation, made with the aid of infrared gas lasers, produced a very accurate value for the velocity of light. This has led to the re-definition of the metre, whereby the speed of light is *fixed* at $c = 299\,792\,458\,\mathrm{m\,s^{-1}}$ and the metre is the *defined* as the distance travelled in vacuum by light in a time interval of $1/299\,792\,458$ seconds.

The impact of frequency standards on electric metrology has already been mentioned.

6.5.2 Astronomy

The distance between the Earth and the Moon has been determined to within few centimetres by measuring the time taken by a laser pulse to travel to the Moon and back.

The international atomic time scale (TAI) is used in studies of the motion of celestial bodies. It is particularly important because of its uniformity and its rapid access to the ephemeris time.

The rotational period of certain pulsars appears to be particularly stable. Its measurement requires long-term frequency stability better than the stability achieved to date.

6.5.3 Radio astronomy

Radio-interferometry techniques with a very long baseline allow the measurement of the angular diameter of radio sources with a resolution of 10^{-3} seconds of arc. On the other hand, strong radio sources, natural or artificial, allow the determination of the position of geodesic points to the nearest centimetre as well as very accurate geodynamic studies of the drift of tectonic plates and of polar movements. The rubidium clock, and especially the hydrogen clock, provide the frequency sources necessary for very long baseline radio-interferometry because of their remarkable frequency stability over $1\,\mathrm{s} < \tau < 1\,\mathrm{h}$.

6.5.4 Test of general relativity

The effect of the gravitational potential on the frequency of a clock has been examined using a caesium clock, on the one hand, and a hydrogen clock, on the other. In a particular experiment, a hydrogen maser was sent up in 1976 on a ballistic trajectory with an apogee of $10\,000\,\mathrm{km}$. The experiment has produced the most accurate measurement of the gravitational effect. Projects for the detection of gravitational waves using hydrogen clocks carried by satellites have been proposed.

6.5.5 The metrology of frequencies: dissemination of frequencies and time

The atomic frequency standards associated with frequency synthesisers are widespread in metrological services. The frequency stability of hydrogen clocks is useful for the characterisation of very high stability quartz oscillators. Some radio stations transmit carrier frequencies and/or time signals driven by atomic clocks. Certain public services have their own time broadcasting networks controlled by atomic clocks. For example, certain electrical utilities in North America use such facilities in order to be able to reconstruct the chronology of events that can occur in power distribution lines.

6.5.6 Telecommunications

Rubidium clocks are used in the USA to generate the sub-carrier frequency for colour television broadcasting.

Caesium clocks ensure quasi-synchronism between different national transmission networks of high density information.

Transmissions of diplomatic or strategic interest are sent using frequential or temporal codes requiring the use of rubidium or caesium clocks.

6.5.7 Navigation and radar

Numerous short-range and long-range navigation systems rely on atomic frequency standards. Some of them use on-board caesium or rubidium clocks.

Studies of spacecraft or missile trajectories rely on rubidium clocks.

The Loran C and Omega systems, for example, are driven by caesium clocks located in the transmitting stations. In the NAVSTAR-GPS system, quartz, rubidium and caesium clocks are carried on board satellites. This system allows the determination of position to within a few metres, and time comparisons to within a few nanoseconds, as long as the transmission code is not voluntarily degraded.

The tracking of the space probes exploring the solar system requires very accurate Doppler measurements of velocity and of position by long baseline radio-interferometry, using hydrogen clocks.

7
Atomic time scales

7.1 INTRODUCTION

In previous chapters we described the operation of oscillators and the development of frequency standards. We now turn to one further application, namely, that of standards of time.

A familiar way of becoming aware of the passage of time is to observe change, e.g., change of position or change of concept. A change of this kind involves a duration. Another approach is to perform continuous observations on a process that evolves and passes through a sequence of identifiable states. By assigning a date to each state we effectively define a time reference generator. The simplest to use is a periodic process characterised by a *period* and a *frequency*. We start from a chosen origin and obtain a time reference – we speak of a *time scale* – by counting the periods. If we adopt the period of the oscillation as our unit of time, the difference between two dates is equivalent to the measurement of duration.

The combination of a generator of units of duration, i.e., a frequency standard, and a counter of units constitutes a clock – the basic instrument for the setting up of a time and/or frequency standard. An important feature of this arrangement is that a clock must be able to repeat the unit of duration without dead time. Atomic frequency standards that generate a constant frequency (to a first approximation) are remarkable generators of units of duration; they are the basis of atomic time and frequency standards.

We saw in previous chapters that there are now many devices that rely on the physical properties of atoms, e.g., the atoms of rubidium, caesium, hydrogen and so on. Caesium standards are now particularly important because, since 1967, the unit of time – the second – has been defined as the duration of 9 192 631 770 periods of the radiation corresponding to the transition between the two hyperfine levels of the ground state of the caesium-133 atom.

The caesium beam standard has thus become the official standard of time interval. The number 9 192 631 770 was originally proposed by Markowitz *et al* in 1958, following a comparison between astronomical time (more precisely, the ephemeris time) and atomic time based on the caesium beam standard. It was confirmed later by other experiments, and finally adopted in 1967. This definition of the second leads to that of the hertz – the unit of frequency – since the frequency f and the repetition period T of an event are related by $fT = 1$.

The introduction of this definition of the *second* represents a fundamental change. Up to middle of the twentieth century, the concept of time in science was the exclusive province of astronomers as far as development, measurement and utilisation were concerned. However, in the second half of the twentieth century, the establishment of the standard of time and its measurement became the concern of physicists and engineers.

Since 1955, atomic standards have enabled the evaluation of a semi-official atomic time reference. The first such reference was was established by using continuously operating quartz clocks that were calibrated from time to time against caesium standards. The time derived from caesium atomic clocks is by defintion accurate in frequency if we assume that the caesium clock gives the *second* perfectly.

An atomic standard can, of course, break down or its operation may deteriorate. The integrity of an atomic time reference must therefore be assured by running several instruments simultaneously and comparing them with each other. The metrological institutions of industrialised nations have therefore set up national time/frequency references based on a network of atomic standards. For example, in France, the Bureau National de Métrologie, has appointed the Laboratoire Primaire du Temps et des Fréquences (LPTF) of the Observatoire de Paris as the laboratory responsible for the setting up and distribution of national references of time and frequency. This laboratory sets this reference of scientific time, known as $TA(F)$ (Temps Atomique Français) and based (1988) on data taken from the caesium atomic clocks operating in nine French institutions. This reference requires a certain amount of time for its evaluation (one month in 1988). LPTF also offers to users a reference labelled UTC (OP) (Universal Time Coordinated of the Observatoire de Paris) which is available in real time. The differences between these two atomic time references are distinguished by delays in publication and their scientific quality, and are published regularly.

At the international level, an agreement has been reached to consider the time references produced by the Bureau International de l'Heure, and from the 1st January 1988 by the Bureau International des Poids et Mesures (BIPM), as world references. This is the International Atomic Time (TAI) which is the time reference coordinate defined relative to a geocentric reference and using the SI second as the unit of scale realised by the rotating geoid. Consequently, the TAI can be widely distributed in the current state of the art, and with a sufficient precision at points fixed or mobile in the neighbourhood of the geoid, by applying the first-order corrections of general relativity, i.e., corrections for the gravitational potential and velocity differences as well as for the rotation of the Earth. The relationships that can be used with this definition of TAI were given by the Ninth Comité Consultatif pour la Définition de la Seconde. They are published in Report 439-3 of the Comité Consultatif International pour les Radiocommunications (CCIR). From a practical point of view, TAI is calculated from readings taken with clocks operating in different institutions (about 150 clocks in 38 laboratories were included in this calculation in 1987). Moreover, the International Earth Rotation Service (created in 1987 and official since the 1st January 1988 with central office at the Observatoire de Paris) sets up the UTC (Coordinated Universal Time) that is the basis of worldwide coordination of the time signals and frequency standards. The two references TAI and UTC have the same step (the same frequency) but differ by an integral number of seconds, the relationship between them being

$$\mathrm{TAI} - \mathrm{UTC} = m(t)$$

where $m(t)$ is an integer that varies with time so that the difference between UTC and the time calculated from the Earth's rotation (UT1) is less than 0.9 s:

$$|\mathrm{UTC} - \mathrm{UT1}| < 0.9\,\mathrm{s}$$

On the 1st January 1988, $m(t)$ was equal to 24 s. We see that UTC is a compromise between the astronomical time based on Earth's rotation and physical time based on atomic properties.

The atomic times TAI and TA(F) are established so as to satisfy criteria for metrological quality: they are the best current approximation to the physical ideal. Is it necessary to have several atomic time such as the TAI, TA(F) and so on? A single world reference, for example, the TAI, might be thought sufficient, but there is a number of arguments against this. First, the availability of several time references is a guarantee of security. Second, some local references are believed to be of higher quality than TAI, e.g., in respect of frequency precision or accessibility. Third, the existence of several atomic time references based on different networks of clocks and on different evaluation algorithms tends to engender improvements and serves as a guarantee of scientific quality.

There is a considerable and extensive need for a time/frequency reference of high metrological quality, e.g., as we saw earlier, different events frequently have to be referred back to the same time scale. Another example is provided by the physicists working in cosmology. Experiments performed with atomic clocks, e.g., caesium clocks transported around the Earth and rocket-borne hydrogen masers, have provided reliable verifications of the theory of relativity. For astronomers interested in the signals from pulsars, the accessment of the feasibility of pulsars as generators of time references must be based on comparisons with metrological time references. A high-quality time reference is also essential in modeling the operation of clocks. A model of this kind can be used to predict the clock signal, which is essential for clocks isolated in space or under the sea, so that the corresponding data can be related to the outside world. Finally, a time reference also provides a mean frequency reference over time intervals of the order of a day to several months. Time scales driven by primary laboratory standards provide the best realisation of the second.

7.2 PROPERTIES OF AN ATOMIC TIME REFERENCE

7.2.1 Durability and accessibility

Durability and accessibility are the essential qualities of astronomical time references. The durability of an atomic time reference is assured by continuously running several clocks in parallel and regularly comparing them among themselves. The question that then arises is: how many clocks are needed? Three seems to be a minimum for the detection of defective operation of one of them. We saw earlier that, in practice, the TAI reference is calculated from the date derived from 150 clocks (in 1987) and the French TA(F) relies on 18 caesium clocks, which ensures a very high statistical durability of these references. Changes sometimes intervene in the establishing of a time scale: the higher performance clocks and new methods of evaluation are introduced from time to time, so that durability must be considered in a wide sense: the time reference must belong to a family whose durability is assured.

The purpose of a time reference is to provide a basis for the dating of events, which implies that it must be accessible to all potential users. If the events of interest occur at the same place, they can be dated on a local scale, e.g., the scale of a laboratory. However, in general, events occur at several locations and must be dated by reference to a scale adopted by the scientific community. Each country has its own national time references for this purpose. At the international level, the TAI and UTC references have the properties of universality and accessibility, but both are *paper references* and time data of the highest accuracy are derived from the clocks taking part in the TAI. Radio transmissions of signals in the UTC system provide access to the TAI for a wide range of users but with a lower accuracy than previously. Moreover, we note that references like TAI and TA(F) are only known after a delay required for their evaluation. In 1987, the delay amounted to one month. However, for real-time applications, it is possible to use a prediction based on a previous reference.

Durability, accessibility and universality are the essential general properties of a time reference, but scientists are also interested in its stability and accuracy.

7.2.2 Stability

A quantity is said to be stable if it is independent of time, which implies constant generation quantity and significance. Measurements of instability employ comparisons between the present and the past, the latter being taken as the reference. For a time reference, we measure the time interval between two states of the process involved in the generation of the reference, i.e., we determine the frequency stability of the reference. It is clear that a perfectly periodic process provides a stable reference.

The instability of a time reference is examined by first assuming that there is a perfectly periodic process that goes through states $A, B, C, D, ...$ (after n periods) that have associated them dates $t_A, t_B, t_C, t_D, ...$, respectively, on the scale being examined. Comparison of the time intervals $t_B - t_A, t_C - t_B, t_D - t_C, ...$ then gives a measure of the instability of the reference. The role of the periodic process is thus to give a meaning to the measured quantities; it is an implicit reference. The measure of instability may vary with the period of the process, so that it is necessary to specify this period, which is also called the standard duration.

We note that the concept of stability involves a present/past comparison that is meaningful only if it relies throughout on the same mathematical entity obtained by using a time-invariant mathematical process; this means in practice that the reference being examined is derived by using the same algorithm.

The notion of stability is often described in terms of the uniformity of a time reference. Uniformity is defined in the following manner: a physics experiment that is repeated under identical conditions is always found to have the same duration if the time reference used to measure this duration is uniform, whatever the actual duration of the experiment happens to be. The periodic process used above to measure instability was a physics experiment of set duration. We thus see that the nonuniformity of the time reference as revealed by duration corresponding to the period of a process is intimately linked to its instability.

The mathematical tools used to characterise the frequency stability of oscillators were developed earlier in this book. They can be applied directly to clocks and time/frequency references because a clock is a continuously-operating step-counting frequency generator. The frequency stability of a time reference (the term stability is used much more frequently than instability) is measured in terms of the two-sample (or Allan) variance, estimated as the average of the square of the second difference of time errors divided by twice the square of the sample time, or by the corresponding power spectral density.

7.2.3 Precision

The precision of a quantity is measured by comparing it with a standard.

For a time/frequency standard, there are two types of precision. The precision of time is obtained by comparing the dates of an event with two time scales, one of which is used as standard.

As there is no absolute time reference, it is impossible to measure the absolute precision of time; the only possibility is the relative precision of two references. This is the case, for example, with scales such as UTC(OP) which are coordinated with the UTC ; in other words, the difference |UTC−UTC(OP)| is kept below m microseconds in order to have a permanently satisfactory approximation to UTC in France.

The precision of the step (or frequency) of a time reference is determined by comparing the unit time interval produced by the reference with the primary standard. By 1988 there were several laboratories with caesium frequency standards, i.e., primary laboratory standards, which they designed and built, with known uncertainty in the definition of the second. At the practical level, the result of the calibration of a time/frequency reference against a primary standard is expressed by a number that is a measure of the frequency precision of the reference at a given date and an uncertainty due to the different sources of errors and the noise in the comparison between the standard and the time/frequency reference. This uncertainty is usually taken as the square root of the sum of the squares of the uncertainties associated with the different possible errors. Important improvements have been achieved in the precision of primary laboratory standards over the last twenty years. Figure 7.1 illustrates this improvement which amounts to an order of magnitude between 1965 and 1975.

There has been no overall progress in the primary standards since 1975. In 1987, there were eight primary standards with relative precision equal to or better than 1×10^{-13}.

7.3 EVALUATION OF AN ATOMIC TIME REFERENCE

7.3.1 General aspects

Atomic time references are evaluated from the readings of simultaneously operating atomic clocks. By 1988, the laboratories for time/frequency metrology were equipped with caesium and rubidium clocks, hydrogen masers (active and passive) and mercury-ion devices (prototypes). The first of these were much more numerous and consisted of commercial and laboratory clocks. Their relative precision is typically of 5×10^{-14}, which means that their long-term stability is better than 7×10^{-14}.

Let us suppose (to simplify matters) that several caesium clocks are operating at the same location, and that we wish to produce a time/frequency reference as close as possible to the ideal time. Each clock provides time information that differs from the others and is also a function of time. The question is: what is the best time reference that can be obtained with this group? There is no absolute time reference, so that we cannot judge the quality of the absolute dating on the time reference that we evaluate. On the other hand, we can calculate the frequency stability of the reference and estimate its frequency precision in relation to laboratory standards.

Let E be the resulting time reference; this implies that an logarithm has been chosen for the evaluation, and also an agreed time origin. The reference consists of a series $X(t)$ in which the parameter t varies between an arbitrary origin and infinity. Each value of $X(t)$ can be considered as coming from a single clock running continuously and representing all the clocks in operation. $X(t)$ indicates the time difference between the reference E and an ideal reference produced by an ideal oscillator at the date t marked on the ideal scale, so that $X(t)$ represents the error in the reference E at the date t. The quality of atomic oscillators is such that the (average) variation of $X(t)$ is slow. The date t does not therefore need to be known with high precision: one hour is often acceptable. It is customary to measure t on the UTC scale. On the other hand, one tries to obtain $X(t)$ with the highest possible accuracy .

The time reference E is regarded as satisfactory for the user if two conditions are satisfied: (1) there is no interruption of the reference and (2) the error $X(t)$ is as close as possible to the initial error $X(0)$.

We will see how these two conditions are satisfied in a particular implementation . In what follows, $X(t)$ indicates, in a very general manner, a realisation of an atomic time reference at a date t. Other designations will be used for particular references.

7.3.2 Data and results

Suppose that the available data take the form of differences $H(L,t) - H(i,t), H(M,t) - H(j,t)..., H(L,t) - H(M,t)...$ between the references at laboratories $L, M, \ldots$ and clocks $i, j, \ldots$, and between the references themselves (in general, the clocks $i, j, \ldots$ belong to different laboratories and take part in establishment of a time reference). The quantities $H(i,t), H(j,t)$ are the time errors of clocks i, j apart from a constant term. The result takes the form $X(t) - H(L,t), X(t) - H(M,t), \ldots$. If A represents the algorithm used in the calculation, then:

$$X(t) - H(L,t) = A\{H(L,t) - H(i,t), H(L,t) - H(M,t), ...\}$$

where $i = i(t), j = j(t), \ldots$ represent the different clocks. The series $X(t)$, which represents the scale E, depends on
 • each quantity $H(i,t)$, i.e., the error in each clock defined to the nearest constant
 • the comparison between the clocks L and i, L, and M, $\ldots$
 • the changes in clocks i and j with time.

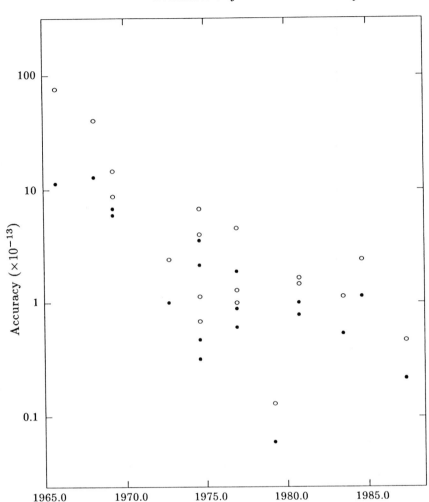

Fig. 7.1 Estimated precision of primary laboratory caesium standards for 1965 - 1986. These values are taken from the lists of uncertainties produced by the respective laboratories. Open circles represent the sum uncertainties and full points show the square root of the sum of the squares of the uncertainties. The data come from measurements performed at the Conseil National des Recherches du Canada, NIST in USA, the Physikalisch Technische Bundesanstalt in Germany and the Laboratory for Time and Frequency Standards in the former USSR.

The aim of the calculation and algorithm is to minimise the difference $|X(t) - X(0)|$ whilst the above three factors tend to increase this quantity.

The result $X(t)$ drifts away from $X(0)$ because of errors $H(i,t)$ in each clock i. These variations can be quite large (relative to the initial value

$H(i, 0))$ and are due to systematic terms and the noise that affects the running of clocks. Clocks are usually selected by examining the variation of these errors relative to a reference that is supposed to provide a scale closer to the ideal scale than each clock i. For example, we can take the average scale $X(t)$ as the reference. When the variations are irregular, the readings of the corresponding clock are given a low weight. If they are regular, the clock is assigned a greater weight. The relationship between the weight and the variations of $H(i, t)$ is generally different for different scales. It is specific to each algorithm.

Comparisons of clocks (clocks L and i, M and j, L and M in our example) contribute to the error $X(t) - X(0)$; their results depend on the apparatus used, the propagation of the comparison signal and so on. More generally, there is also the comparison noise. It is clear that it is essential to use the best possible communication links that ensure the highest precision. Several cases can arise. For example, two clocks or two time scales can be simultaneously linked two or three times a day; the optimum result is achieved by weighting the links. Another possibility is to have a regular link and, from time to time, to perform more accurate comparisons. The function of the evaluation algorithm is then to achieve optimum utilisation of the data.

The third possible cause of an increase in $|X(t) - X(0)|$ is a change in the clocks with increasing t. The limited life of atomic clocks makes such changes inevitable. The general rule adopted for the computation of atomic time scales is that clock changes do not modify the value of $X(t)$ or of the scale step represented by $X(t)$ at the date at which they take place. It is commonly said that the scale is continuous in time and in step; or, in oscillator language, phase and frequency are continuous around a given date. Consider the simple example of a clock introduced at a date t. By saying that the scale step is unaffected by the introduction of the clock we mean that the mean scale step remains the same over the interval $[t, t + \Delta t]$ with or without the new clock. This involves the parameter Δt, i.e., the time taken by the measurement of the mean frequency. Ideally, this measuring time must be long enough to give a meaningful estimate of the scale of step, i.e., it must give both the signal and the noise. This point of view is valid if we consider that the noise associated with the scale step is stationary in the first order. In the more general case of nonstationary noise, confusion between signal and noise may arise.

We shall now examine the case of a time scale calculated as a simple mean of the readings of clocks and the implications of the rule governing changes in clocks.

7.3.3 Time reference calculated as a simple mean

Let $\mathrm{ECH}(t)$ be the mathematical expression for the reference E, calculated as a mean of the readings of N clocks:

$$ECH\,(t) = \frac{1}{N}\sum_{i=1}^{N} H\,(i,t) \qquad (7.1)$$

where $H(i,t)$ is the reading of clock i at date t.

Let us suppose that p new clocks are introduced and q taken out at the date t_1 (see Fig. 7.2), which means that t_1 is the first date of use of the p new clocks and the last date for the q clocks. We calculate $ECH(t_1)$ from the above relationship for the N clocks. For t_1 and subsequent dates, the mean of the data for the $N+p-q$ clocks is given by

$$ECI\,(t) = \frac{1}{N+p-q}\sum_{i=1}^{N+p-q} H\,(i,t) \qquad t \geq t_1 \qquad (7.2)$$

where we use the p new clocks and reject the old q clocks. The quantity $ECI(t_1)$ is usually different from $ECH(t_1)$, and $[(d/dt)EC1(t)]_{t=t_1}$ is different from $[(d/dt)ECH(t)]_{t=t_1}$.

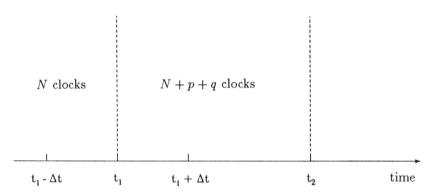

Fig. 7.2 Groups of clocks participating in the calculation of the time scale.

The scale E is said to be continuous in time and in step if it is possible to find A and B such that the mathematical expression for E for $t \geq t_1$ is

$$ECH'\,(t) = ECI\,(t) + A + B\,(t - t_1) \qquad (7.3)$$

where

$$ECH'\,(t_1) = ECH\,(t_1) \qquad (7.4)$$

$$\left(\frac{d}{dt}ECH'\,(t)\right)_{t=t_1} = \left(\frac{d}{dt}ECH\,(t)\right)_{t=t_1} \qquad (7.5)$$

From the relationship that defines $ECH(t)$ and the expression for $ECI(t)$, we obtain

$$A = \frac{1}{N+p-q} \sum_{i=1}^{N+p-q} [ECH'(t_1) - H(i,t_1)] \qquad (7.6)$$

The value of A that ensures continuity in time is readily obtained by replacing $ECH'(t_1)$ by $ECH(t_1)$ which is known.

The derivation of $ECH'(t)$ for $t = t_1$ leads to

$$B = \frac{1}{N+p-q} \sum_{i=1}^{N+p-q} \left(\frac{d}{dt}[ECH'(t) - H(i,t)] \right)_{t=t_1} \qquad (7.7)$$

If we know how to calculate the derivatives $(d/dt)[ECH'(t) - H(i,t)]_{t=t_1}$ for each clock i, we obtain B. The latter can be estimated from the sample data available for $t \leq t_1$. They are used in the calculation of $ECH'(t)$ for $t \geq t_1$. In other words, we suppose that the estimates are valid for $t \geq t_1$, i.e., we make a prediction. It is clear that the continuity of the scale step calculated in this way is only approximately secured.

Replacing A and B by the values obtained above, we obtain the following expression for $ECH'(t)$ in which ˆ indicates a prediction:

$$ECH'(t) = \frac{1}{N+p-q} \sum_{i=1}^{N+p-q} [H(i,t) + ECH(t_1) - H(i,t_1)$$

$$+ (t-t_1) \frac{\hat{d}}{dt} (ECH'(t) - H(i,t))_{t=t_1}] \qquad (7.8)$$

The quantity $ECH(t_1) - H(i,t_1) + (t-t_1) \times (\hat{d}/dt)[ECH'(t) - H(i,t)]_{t=t_1}$ constitutes a linear prediction for $ECH'(t) - H(i,t)$ at the date t.

The approach to the problem of continuity used in the above simple example can be generalised. Let $p(i,t)$ be the normalised weight of clock i for $t \geq t_1$. The mathematical expression for the time scale for $t \geq t_1$ then becomes

$$X(t) = \sum_i p(i,t) [H(i,t) + \hat{c}(i,t,t_1)] \qquad (7.9)$$

where the values of $\hat{c}(i,t,t_1)$ are the predictions for $X(t) - H(i,t)$ obtained from the past scale (before t_1). The required equation is

$$\sum_i p(i,t) \left[X(t) - H(i,t) - \left(\widehat{X(t) - H(i,t)} \right) \right] = 0 \qquad (7.10)$$

The resulting continuity depends on the quality of the prediction for each clock i.

If $\hat{c}(i,t,t_1)$ is a first order polynomial in $t - t_1$, we can again use the approach to continuity developed earlier. Four remarks are now in order.

(a) The above scale calculations generally employ discrete sampled data. The clock step is estimated for intervals of duration Δt that are multiples of the sampling period. The average step within the interval $[t_1, t_1 + \Delta t]$ can be calculated in different ways. The first estimate is obtained by using the values of $X(t) - H(i,t)$ for t_1 and $t_1 + \Delta t$. The step is usually obtained as the slope of the linear least squares fit to the values of $X(t) - H(i,t)$.

(b) The problem of prediction is more general than that of changing clocks. It is perfectly acceptable to use the technique of prediction at a date t_1 when there is no change of clocks; the prediction remains valid up to a date t_2 when a new prediction is made. The interval $t_2 - t_1$ is a multiple of the sampling period. In the particular case where there is no change of clocks at date t_1, and no change of clock weighting , it can be shown that the result of the calculation, $X(t)$, is independent of the algorithm used.

(c) It is interesting to note that, in the above calculations, the introduction and removal of clocks are considered in the same manner. Actually, there is an asymmetry. A device as sophisticated as a caesium standard can suddenly cease running. In contrast, the introduction of a new clock in a group is left to the discretion of the user: it is perfectly normal for a clock to run for a certain time before it is introduced into a group.

(d) The expression for $X(t)$ is written in terms of the values of A and B:

$$X(t) = X(t_1) + \frac{1}{N - p - q} \sum_{i=1}^{N+p-q} [H(i,t) - H(i,t_1)$$

$$+ (t - t_1) \times \frac{\hat{d}}{dt}(X(t) - H(i,t)_{t=t_1})] \tag{7.11}$$

which involves the primary differences $H(i,t) - H(i,t_1)$ and where the symbol ̂ indicates a prediction. If clock i has a normalised weight $p(i,t)$ at date t, the expression for $X(t)$ becomes

$$X(t) = X(t_1) + \sum_{i=1}^{N+p-q} p(i,t)[H(i,t) - H(i,t_1)$$

$$+ (t - t_1)\frac{\hat{d}}{dt}(X(t) - H(i,t))_{t=t_1}] \tag{7.12}$$

We note that instead of introducing the weights $p(i,t)$ of the actual data $H(i,t) - H(i,t_1)$ and the prediction of the step $(\hat{d}/dt)[X(t) - H(i,t)]_{t=t_1}$, we can use $p(i,t)$ and the step-prediction weights $p'(i,t)$. These weights characterise short-term and long-term behaviour, respectively.

In this first approach to the calculation of an atomic time scale, the algorithm has three rôles that can be classified in order of importance as follows. The first concerns the introduction and withdrawal of clocks. Then comes the use of clock readings, their weighting and the use of the clock comparisons. The procedures and parameters that are associated with these three functions involve

- the method of prediction
- the duration of the interval in which the prediction is used
- the duration of the past interval used for the prediction
- the method of assigning the weights to the clocks
- the duration of the interval used for each weight
- the comparison of clocks; filtering and weighting

7.4 SPECIFICITIES OF AN ATOMIC TIME REFERENCE

In this section, we will specify the distinctive features of atomic time references, and will examine their stability and the comparison of clocks.

The modelling of a time reference is an exceedingly important problem that provides the basis for any prediction.

Consider the example of a submerged submarine with its own time reference, completely isolated from the outside world. To enable the submarine to use its time scale correctly, it is vital that the latter should be well modelled in order to maintain an acceptable synchronism between the reference of the submarine and, say, the naval base or another ship. This relies on a long-term modelling, with the periods of observation ranging from one to dozens of days. This long-term aspect is the first characteristic feature of time references: we have to be sure that they are stable over very long periods of time. This means the signal from an atomic clock has to be modelled over a time interval corresponding to the period of operation of the instrument.

Moreover, we saw earlier that the calculation of an atomic time reference takes into account the readings of several clocks and the comparisons between them. We shall now examine the methods of comparing time, and the instability they introduce. Finally, we will refer to the renewing of the clocks in a time reference and will address the question of the precision.

7.4.1 Modelling of a clock signal

The signal generated by a clock can be described mathematically by the sum of a deterministic term $D(t)$ and a random term $z(t)$:

$$H(i,t) \equiv D(t) + z(t) \tag{7.13}$$

Statistical entities such as the two-sample variance have been developed to characterise the random terms, but we still have to be able to separate the contributions to $D(t)$ from those that belong to $z(t)$. If we have access to the data from a primary laboratory clock that is running simultaneously with the clock under investigation, we obtain estimates of $D(t)$ as a result of comparisons with this standard. The uncertainty in these estimates is a combination of the primary clock uncertainty, i.e., at best, a few parts in 10^{14}, and the noise inherent in the comparison. We can also obtain an estimate of $D(t)$ by comparing the clock under investigation with several other clocks running in the same laboratory.

From the practical point of view, it is also interesting to examine the behaviour of the two-sample variance in the presence of deterministic terms. The two-sample variance corresponding to $D(t) = at^2$ is $\sigma_D{}^2(\tau) = 2a^2\tau^2$ which is independent of t, and the result for $D(t) = b\cos\omega t$ is found to be $\sigma_D{}^2(\tau) = (4b^2\sin^4\omega\tau/2)/\omega^2\tau^2$. Figure 7.3 illustrates the square root of the two-sample variance for $H(i,t) = D(t) = at^2 + b\cos\omega t$ with $a = 10^{-13}$ per annum, i.e., $\sim 0.3 \times 10^{-15}$ per day, $b = 500$ ns and $\omega = 2\pi/T$ where $T = 1$ year. It is clear that the two-sample variance – a statistical tool – can be used in the presence of certain deterministic variations and that the presence of, say, terms that are quadratic in t (frequency drift) becomes evident for large values of the sampling interval τ. However, in practice, there is an advantage in applying statistical analysis to data that have been freed from deterministic terms.

We have estimated of the different terms in the model of a clock signal by investigating a group of about 30 commercial caesium clocks distributed among different laboratories. These clocks had functioned for more than two years without interruption, and belonged to groups of at least 3 clocks running simultaneously in each laboratory. The comparison at a given instant t of at least three clocks running in a laboratory is sufficient for the identification of the characteristic variation of each clock by evaluating the three pairs of differences. Table 7.1 shows the results obtained for the deterministic variations.

Periodic variations of a set of clocks have been studied by comparing them with two primary laboratory clocks (one in a German laboratory and the other in a Canadian laboratory). We found that certain clocks display variations with a periodicity of one year, as indicated in Table 7.1. It is interesting to note that two OSA 3200 clocks were found to exhibit annual variations that were in antiphase with those of HP clocks. The origin of these periodic variations is not clear, especially since metrological labora-tories impose strict controls on the environment of their clocks, inncluding their temperature which generally varies by less than 1°C per annum. Hu-midity variations, usually less strictly controlled, have been suggested as a possible cause.

After detecting and estimating the systematic term, we calculated the two-sample variance for each clock. This yielded estimates of the instability of sets of clocks HP 5061A and HP 5061A opt 4. The results (mean and extremal values) are shown in Figs 7.4 and 7.5. The noise level was $\sim$

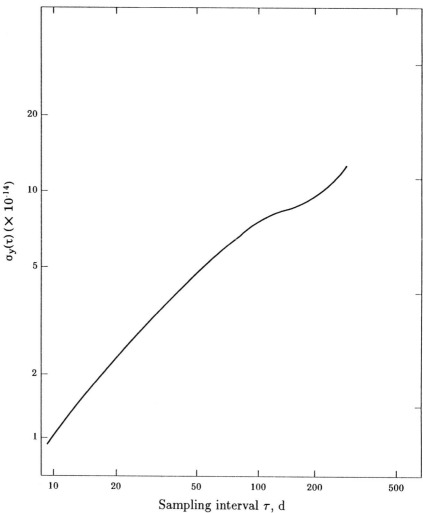

Fig. 7.3 Instability, measured by the square root of the two-sample variance, of a fictitious clock with a frequency drift of 10^{-13} per annum and an annual amplitude variation of 1 000 ns.

8×10^{-14} for the HP5061 clocks and 4×10^{-14} for the HP 5061A opt 4; however, for the latter, random noise rapidly becomes the predominant contribution. The stability of the three OSA 3200 clocks was found to be close to the mean obtained for the HP 5061A clocks.

In addition to the special caesium clocks used in metrology, there are a few primary laboratory clocks: in 1988, two instruments of this type were in operation at the Physikalisch Technische Bundesanstalt in Germany and four at the Conseil National des Recherches in Canada. The instability of

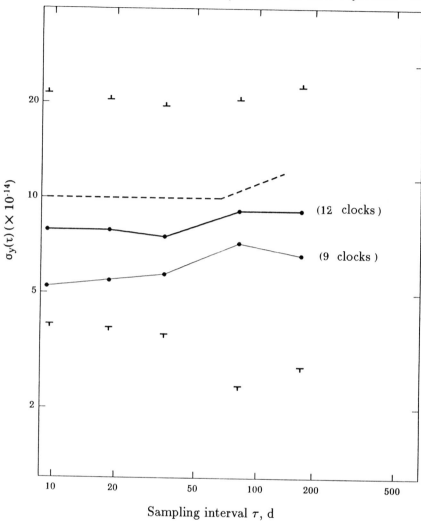

Fig. 7.4 Instability, measured by the square root of the two-sample variance, of HP 5061A caesium clocks as a function of sampling time: points – mean values, dashed line – published data (the symbols ⊥ and T indicate the extreme values of instability).

the best of these clocks is less than 2×10^{-14} over one to two months.

7.4.2 Comparisons of clocks

Comparisons of clocks are an essential feature of atomic time. Local comparisons in a laboratory and comparisons between reference clocks (one speaks of *master clocks*) at several laboratories have two aims, namely, to control the operation of each clock and to allow the calculation of a

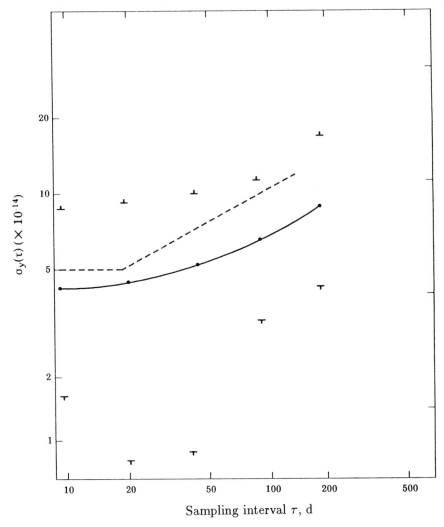

Fig. 7.5 Instability, measured by the square root of the two-sample variance, of caesium clocks HP 5061A opt. 4: points – mean values, dashed line – published data (the symbols T and ⊥ indicate the extreme values of the instability).

reference that integrates the data produced by standards at one or more laboratories so that local or international time can be established.

Local comparisons are readily made: the clocks are connected by cables to measuring equipment and the associated uncertainties are very small (≤ 1 ns). When long distances are involved, an interface becomes necessary. Before we describe the methods of comparison, we note that there are relativistic corrections that apply to such comparisons. Each clock supplies its own time at a particular reference point. Relativistic corrections allow for the effect of the gravitational potential, the rotation of the Earth and the possible motion of the clocks. Expressions for these corrections are published in CCIR report 439-3.

(a) The transport of clocks

The actual physical transport of a clock from one laboratory to another is the simplest method of comparing two clocks separated by a distance of one to several thousand kilometers. It reduces the measurement chain to a minimum as we simply have two local measures. The uncertainty of this method comes from the uncontrollable behaviour of the clock during transport. It is therefore desirable to minimise changes of environment during transport.

Caesium clocks were first transported in 1959. Since that date, a very large number of atomic clocks (both caesium and rubidium) have been transported and used for the calibration of other methods of comparison. The uncertainties currently attained lie in the range $10 - 100$ ns. A a modified version of clock transportation was tried by ONERA in 1970: an aircraft equipped with an atomic clock and a transmitter-receiver set was flown over two laboratories equipped with similar communication set. Since 1985, there has been a reduction in the transport of clocks between metrological laboratories. There are essentially two reasons: the transport of a clock imposes severe constraints and the advent of satellites will soon allow us to enter the 1 ns range.

(b) The 'long-range navigation system' (LORAN)

LORAN operates in the hyperbolic or circular mode. LORAN stations are equipped with caesium standards that transmit a $100-$kHz ground wave whose delay varies very little. Simultaneous reception of these signals by two laboratories and differential calibration of the corresponding delays can be used to compare two clocks in an arrangement employing LORAN. The system has been used operationally (in daily comparisons) for international links between Europe and the USA between 1968 and 1983.

(c) Television

The idea of comparing clocks by quasi-simultaneous reception of a pulse from a television raster was proposed by J. Tolman in 1965. Developments have taken place in numerous countries and in particular in France where special receivers have been developed. An uncertainty of a few nanoseconds can be regularly attained when two laboratories receive directly from the same transmitter. When the laboratories are separated by several hundred kilometers, the uncertainty rises to a few dozen nanoseconds. Since 1968, several French laboratories have regularly used television as a mean of comparison.

Since 1988, the advent of direct television satellites (TVSAT, TDF1 *etc.*) has opened new possibilities for time and frequency comparisons.

(d) Satellites

Satellites offer new and interesting possibilities for the terrestrial comparison of atomic clocks (the same time comparison techniques can be used to determine the location of satellites). The two essential advantages of this method are: the radio signals transmitted by a satellite reach a very large area on the Earth's surface (one third in the case of geostationary satellites) and the associated delays which come into the time comparison by satellite can either be eliminated or accurately estimated. The relevant data on the use of satellites for the dissemination of time and the principal results obtained since 1962 are presented in CCIR report 518.3.

There are two general methods for comparing time. The first (known as the one-way method) is perfectly illustrated by the global positioning system (GPS) [GPS data are published in the annual Precise Time and Time Interval (PTTI) proceedings and the Forum Européen Temps-Fréquence (EFTF)]. Each satellite in the GPS constellation (18 satellites) has its own clock with a known time delay relative to the reference time of the system (the GPS time). The arrival of the signal from each satellite (1.5 GHz in C/A code) is dated on the local scale of the user. After applying the necessary corrections (for ionospheric, tropospheric, relativistic, transmission time and satellite clock effects) one obtains the time difference between the local reference and the GPS time; the uncertainty in the results is associated with these corrections. The most important one, i.e., the signal propagation time, depends directly on the position of the satellite. The National Institute of Standards and Technology has developed a common sighting technique that reduces this uncertainty. The transfer of time through GPS began in 1981 between the NIST and USNO; by 1988, this was being used operationally by more than twenty metrological laboratories (of which approximately 15 are national laboratories).

Figure 7.6 shows the results obtained by two American laboratories, using GPS, LORAN-C and clock transport. It is clear that the GPS links assured a precision of 10 - 20 ns. Recent work at BIPM has shown that it was important to know the position of the GPS aerials to better than 1 m. Time comparisons with an accuracy better than 10 ns are readily attained with GPS.

The second method of time transfer (known as the two-way method) relies on the transmission and reception of signals by the user. This reduces or even eliminates some of the corrections mentioned above. The method was used with the Symphonie satellite between 1976 and 1982 for time transfers between three stations: one in France (l'Observatoire de Paris), another in Germany and the third in Canada.

The principle of the measurement was as follows. Pulses referenced to the times of stations A and B were sent quasi-simultaneously and were transmitted back by the satellite. The time difference between the departure of a pulse from A and the return of the pulse from B was measured and vice-a-versa. The time difference between the references at A and B were obtained as half the difference of the measurements at A and B after corrections (for instrumental delays, ionospheric, tropospheric, and relativistic effects, and the motion of the satellite). These corrections appear only in differences; the main uncertainty comes from the estimated instrumental delays at A and B. Uncertainties of a few nanoseconds (and even 1 ns in 1982) have readily been attained, and the uncertainty in time has been limited to some 50 ns. Tests were made during the last few years by using an INTELSAT channel and MITREX modems. The precision of these comparisons is better than 1 ns. Finally, a high-precision time comparison experiment (1 ns or better) was carried out in 1988, using laser beams aimed at a satellite equipped with reflectors.

Up to 1983, the problem of maintaining precision between two time references (for example, two national references) was solved by means of clock transport and the corresponding uncertainty was in the range 10 - 200 ns. With the expansion of GPS (uncertainties of 10 ns), the transport of clocks has become inadequate as a calibration method. Under the aegis of NBS and then BIPM, several transports of GPS receivers have been carried out, thus enabling the time calibration with an uncertainty of 10 ns.

7.4.3 Renewal of clocks

According to CCIR report 898, the average time between breakdowns of commercial caesium clocks is between 2.5 and 4.5 years. We have obtained similar results for all the caesium clocks operating in French laboratories (20 clocks in 1988). Frequent clock replacements are therefore necessary amongst the standards used to calculate time references. For example, in the case of TAI, as calculated by BIPM, the mathematical weight of clocks entering and leaving at the beginning of the calculation interval was about 10% of the total weight in 1986. The question is: what is the impact of such changes on the stability of the time reference? The answer depends on the manner in which these changes are introduced.

We have used a simulation to estimate the noise introduced by clock replacement . An initial time reference was calculated as the simple mean of the readings of 15 clocks running continuously. The dates on which clocks were stopped and introduced were obtained by random sampling, subject

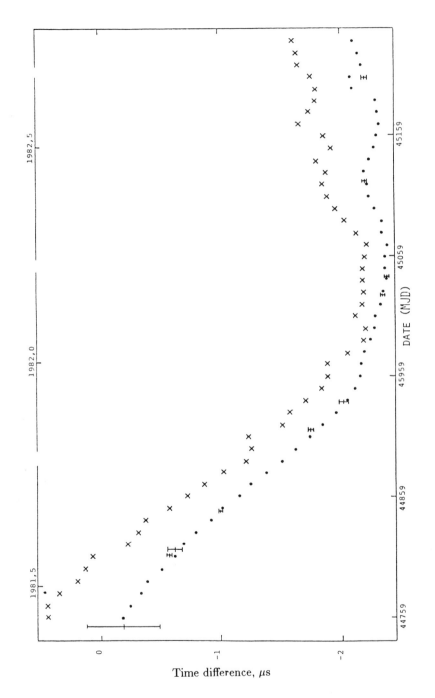

Fig. 7.6 Time differences between the two master clocks at the Naval Observatory in Washington and NIST (USA) obtained by GPS (●), LORAN-C (×) and clock transport (+). The LORAN-C results are given to within a constant.

to the average time of 2.7 years between breakdowns. The time references were then calculated as simple means of the available data, with allowance for these clock replacements, the clock step being taken as equal to the past step. We compared the references obtained in this way with the initial reference according to two criteria: the deterioration was calculated from the relative stability of the final reference relative to the initial reference and, secondly, we calculated the stability of the differences between the initial and final references. Several simulations have shown that there is a probability of deterioration (up to 60% for a sampling time of 640 days), but also a probability of improvement (up to 10% in our simulations). Moreover, the differences between the initial and final references indicate a noise of a frequency random walk for sampling times longer than 40 days. For a set of 100 clocks, the estimated relative noise level is 0.5×10^{-14} for an internal of 80 days.

In conclusion, Table 7.1 shows the order of magnitude of the instabilities of the three noise components of a time reference.

7.4.4 Precision of a time reference

We now turn our attention to the other important quality of a time/frequency reference, namely, its precision In what follows, the term precision is used for frequency precision. We note that the notions of stability and precision meet in the sense that the very long term stability of a time reference is obtained by assuring its precision.

The most precise time references are supplied by primary laboratory standards which establish the unit interval (the second) with uncertainties associated with measurements and/or estimates, which are likely to decrease as further advances are made. The precision of the second achieved with a laboratory standard is at least ten times better than that of a commercial standard; it is defined by the resultant uncertainty obtained by the summation of uncertainties (usually, of their squares) associated with the different departures from the ideal performance due to cavity detuning, the Doppler effect, magnetic effects *etc.* Table 7.3 lists the uncertainties involved in the definition of the second. The performance of laboratory caesium standards has substantiallytially improved over the last twenty years (Fig. 7.1) because of advances in the physics of atomic oscillators and in measuring techniques.

Continuous operation was an important advance from the point of view of time references. By 1988, there were four such instruments at the Conseil National des Recherches du Canada and two at the Physikalisch Technische Bundesanstalt. The standards are high stability clocks and high precision references; they are called primary clocks.

Five laboratories have primary standards with relative precision of 10^{-13}. Apart from the two just mentioned, they are at the NIST in the

USA (one standard), the Radio Research Laboratories of Japan (one standard) and the State Laboratory for Time and Frequency Standards of the former USSR (two standards). Several other laboratories have lower performance instruments; several are currently engaged in the construction of new devices (e.g., the Laboratoire Primaire du Temps et des Fréquences de l'Observatoire de Paris).

Calibrations confirm the time reference calculated by the laboratory. It is also possible to use the calibration results to drive a time reference other that of the laboratory by using a filter simulating particular models that rely on stationarity, linearity and statistical independence.

7.5 ATOMIC TIME REFERENCES

There are several types of atomic time reference. They can be local, national or international. Some are based on a criterion of stability whilst others rest on a stability/precision compromise. Each relies on a more or less sophisticated algorithm that satisfies the conditions considered earlier. We shall illustrate this by four examples.

7.5.1 The atomic time scale of the National Institute of Standards and Technology

The NIST has a set of caesium clocks, hydrogen masers and two primary laboratory standards of which one provides the definition of the second with a precision of 10^{-13}. The atomic time scale TA(NIST) is a local reference (all the sources operate in Boulder, Colorado) and a national reference (NIST and the US Naval Observatory share the national responsibility for time references). It has been calculated since 1982, using an algorithm that relies on Kalman filtering and takes into account the criteria of stability and precision.

Kalman filtering is very well suited to the establishment of a time scale. Starting with the state vector $\underline{Z}(t_0)$ with three components (time error, frequency error and frequency drift) and the covariance matrix $P(t_0)$ at the date of the origin t_0, we obtain $\underline{Z}(t_0 + 1)$ and $P(t_0 + 1)$ as filter output. The state vector is initialised by choosing as initial errors of time and frequency for each clock the existing differences between the clock (time and frequency means) and a reference scale; the covariance matrix is initialised by estimating the uncertainties of these time and frequency errors. The time scale algorithm usually includes checking for errors (errors in time and frequency), so that problematic clocks can be removed from the calculation and very rapidly reintroduced. The absence of data at one or several dates does not give rise to special difficulties: the procedure is to use the extrapolated value of the state vector $\underline{Z}(t/t - 1)$ instead of $\underline{Z}(t/t)$.

Moreover, it is relatively easy to introduce calibrations or other corrections aimed at improving the scales. Indeed, time-scale algorithms based on Kalman filtering are very flexible.

Figure 7.7 shows for 1984, 1985, 1986 and 1987 the time difference between the American reference TA(NIST) and the TA(PTB), TA(F) and TAI references considered below.

7.5.2 The atomic time scale of the Physikalisch Technische Bundesanstalt

The PTB operates a set of caesium clocks and, in 1988, two primary laboratory clocks. As in the case of NIST, the atomic time reference TA(PTB) is a local and national reference. Since 1980, the time TA(PTB) is derived directly from a local oscillator, e.g., a commercial caesium clock locked to the primary clock PTB-CS1. The following relationship is satisfied, usually to better than the nearest 10 ns:

$$TA(PTB) = T(PTB - CS1) + constant$$

where $T(PTB - CS1)$ is the characteristic time of the PTB-CS1 clock. It follows that the precision and long-term stability of TA(PTB) is very close to those of the primary clock, i.e., 3×10^{-14}.

7.5.3 The TAI scale calculated by the Bureau International des Poids et Mesures

The BIPM is the official custodian since January 1988 of the Temps Atomique International (TAI) which is the world reference of scientific time [this was previously done by Bureau International de l'Heure (BIH)]. The TAI is produced by an algorithm that ensures its stability and precision (Fig. 7.8). Stability is obtained by connecting together a large group of atomic clocks running at metrological laboratories. In 1988, approximately 150 clocks in 38 laboratories were taking part in the TAI, mostly commercial caesium standards, a few hydrogen masers and some laboratory caesium standards. The precision of TAI is achieved by locking the unit time interval to the best realisations of the second by laboratory standards and clocks. The reader will find all the information on TAI in BIH and BITM reports and elsewhere in the literature (see the references at the end of this book). The basic ideas of the TAI alogorithm are summarised below.

The essential quality characteristic of TAI is its long and very long term stability (frequency), by which we mean periods of one month or more. Very long term stability (one year or beyond) is assured by maintaining the precision of the step (frequency) as indicated above. Long term stability is achieved by using the readings of a large number of clocks operating under metrological conditions (controlled environment, monitoring of working parameters, multiple comparisons *etc.*).

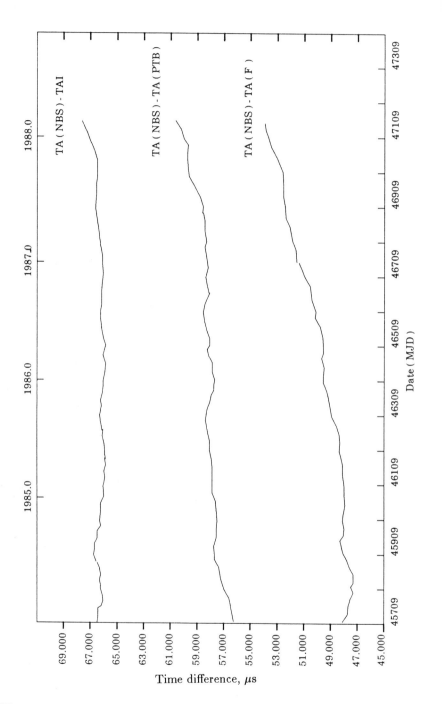

Fig. 7.7 Time differences for 1984, 1985, 1986 and 1987 between TA(NIST) and TA(PTB), TA(F) and TAI.

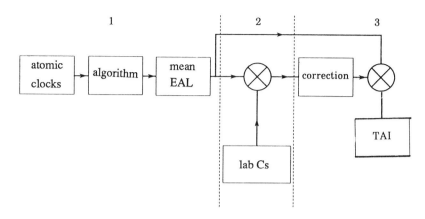

Fig. 7.8 The three stages of TAI calculation for 1988 (EAL stands for Echelle Atomique Libre - independent atomic standard).

The TAI algorithm involves the following steps: prediction of the running of each clock from the data for previous ten months, detection of abnormal clock behaviour and calculation of the characteristic weight of each clock. Until 1988, two weights were calculated: one relating to the prediction and the other to the frequency stability in terms of the stability of the mean frequencies over two months of operation of each clock. The final weight was taken as the smaller of the two. Since 1988, only the weight based on the variance of the mean frequencies is used. Present (as opposed to past) data, i.e. clock readings for the TAI calculation interval, are used to establish the weight; this implies that the calculation procedure is iterative. In practice, the weights converge after three iterations. The instability of the TAI over intervals of two months is estimated as lying in the range $5 \times 10^{-15} - 7 \times 10^{-15}$; its precision and its very long term instability has been calculated to be 5×10^{-14}.

7.5.4 Atomic time scale of Laboratoire Primaire du Temps et des Fréquences (LPTF) de l'Observatoire de Paris

This laboratory calculates the atomic time TA(F). In 1988, it was based on the reading of a set of commercial caesium clocks at the Paris Observatory and in about 10 French laboratories. All these clocks are compared daily with the master clock of the Paris Observatory, i.e., UTC(OP), either via GPS satellites or television.

The TA(F) algorithm conforms to the same long term stability criterion as TAI. It does not, however, guarantee the precision of frequency because, in 1988, a primary time/frequency laboratory standard that could be used as the ultimate frequency reference was not available in France. The algorithm assigns a weight to each participating clock, as in the case of TAI.

Table 7.1. Estimated deterministic terms obtained for a group of commercial caesium clocks (OSA–Oscilloquartz; HP–Hewlett Packard)

Cs clock	Frequency drift $D(t) \sim t^2$		Quadr. freq. variation $D(t) \sim t^3$		Annual variation	
	Nb.	Value per day	No.	Value per day	No.	Value per annum
HP 5061A	3/12	-0,6 to 1,4 10^{-15}	1/12	-3.10^{-18}	6/15	0,4 to 2,5 10^{-13}
HP 5061A opt. 4	8/16	-1,6 to 1,2 10^{-15}	1/16		2/17	0,2 to 0,8 10^{-13}
OSA 3200	Three clocks examined				2/4	2.10^{-13}

Table 7.2. Stability ($\times 10^{-14}$ as a function of sampling time in days for three noise components

Sampling time, d	10	20	40	80	160	320
• HP 5061A OSA 3200	8	8	8	9,5	11	
	Frequency scintillation					
HP 5061A opt. 4	4,3	4,6	5,2	6,6	9	
			frequency random walk			
• Comparison of clocks (LORAN C)	18	11	7	5	3,5	2,5
	white, phase and frequency					
GPS	1 à 2	0,5 à 1 white, phase				
• Replacement of clocks (100 clocks)				0,5	0,7	1

The instability of TA(F) over one month in 1988 was of the order of 2×10^{-14}. A guarantee of the accuracy of the TA(F) is planned for the time when LPTF will have a laboratory standard (in principle, 1992 onward).

Users have access to the TA(F) reference either through a monthly bulletin, or through a link with the computer files of LPTF.

Table 7.3. Uncertainties associated with perturbations (NIST, USA)

	Uncertainty ($\times 10^{-13}$)		
	NBS-4(1974)	NBS-5(1974)	NBS-6(1976)
Magnetic field mean value	0,08	0,08	0,03
inhomogeneity	0,4	0,4	0,02
Majorana transitions	0,05	0,05	0,03
Doppler effect (2nd order)	2,8	1,5	0,1
Resonator dephasing			0,8
Detuning of cavity	0,5	0,03	0,01
Undesirable transitions	0,1	0,05	0,2
Spectral impurities	0,8	0,04	0,02
Locking Amplifier	1	1	0,02
Distortion			0,15
Lowest estimated uncertainty	3,1	1,8	0,85
Highest estimated uncertainty	5,73	3,15	1,38

7.6 CONCLUSION

We saw on the previous pages that the setting up of time/frequency atomic references rests on the operation and frequent comparison of atomic standards.

Since 1967, the year of the change in the definition of the second and the official start of atomic time, important improvements have been made in the two domains of frequency standards and time/frequency transfers; current research suggests that further advances will be made. In 1988, caesium laboratory standards were better than $\sim 10^{-13}$; moreover, the instability of hydrogen masers amounted to a few 10^{-15} over periods of several hours. Projects aimed at improving these figures have been carried out in several areas, including optical pumping of caesium, cryogenic hydrogen maser, ion storage, cooling of atomic beams and stored ions *etc.*. It is not unrealistic to expect an improvement by an order in magnitude in the precision and stability by the year 2 000. In addition, the millisecond pulsar presents a remarkable temptation to the time/frequency metrologist. Indeed, comparison of signals from pulsar 1937 + 21 with the caesium caesium clock indicate excellent relative stability over periods of one year or more.

There have also been remarkable advances in time/frequency transfer, mostly due to the availability of artificial satellites. There is an order of magnitude improvement between the uncertainties of LORAN time links between Paris and Washington and those involved in time/frequency comparisons by GPS. We have progressed from 100 ns to 10 ns and 1 ns seems within reach albeit with great precautions. The two-way system by satellite offers interesting possibilities through better control of certain parameters as compared with GPS. We note that, when satellite techniques of time/frequency transfer are used, it is just as simple and nearly as accurate to compare two clocks separated by 10 km or by 5 000 km on the Earth.

Time/frequency metrology has extensive possible applications in many domains. The definition of the metre was changed in 1983 (the velocity of light in empty space was fixed) to take into account the results obtained with stabilised lasers. The development of improved frequency standards also presents promising possibilities, e.g., for the verification of gravitational effects in relativity and the detection of gravity waves in the universe.

8

A few ideas on 1/f noise

8.1 A FEW EXAMPLES

The most natural approach to noise is to describe it as a succession of impulses that are very short, independent of one another and produced randomly in time (Poisson distribution). In 1918, Schottky used this to describe the noise observed in a hot-filament diode. If n electrons are emitted, on average, in a time τ, the mean current is

$$\bar{i} = \frac{\bar{n}e}{\tau} \tag{8.1}$$

where e is the charge of the electron. If the phenomenon is Poissonian, the mean square difference between n and $\bar{n}$ is equal to $\bar{n}$ (Section 3.2). The current fluctuation

$$\Delta i = \frac{(n - \bar{n})e}{\tau} \tag{8.2}$$

then has the mean square value

$$\overline{\Delta i^2} = \frac{\bar{n}e^2}{\tau^2} = \frac{\bar{i}e}{\tau} \tag{8.3}$$

A variance that is inversely proportional to τ is associated with a *white* spectrum with spectral density given by

$$\frac{\Delta P}{\Delta F} = 2e\bar{i} \tag{8.4}.$$

By convention, this is the power dissipated in unit resistance.

This description is meaningful only if the noise has a sharp high-frequency cut-off at $F = F_M$. According to Section 3.1.2 this means that the impulses have finite duration of the order of $1/F_M$, and that the instant of the impulse was held in memory for a limited time. Physically this is readily understood since an electron emitted by the hot filament is perfectly identified whilst it is moving in the gap between the filament and the plate. However, once it is captured by the plate, the memory of the emission is lost. The corresponding lifetime of the electron is the transit time between filament and plate, which determines F_M.

In 1925, Williams and Hughes measured $\bar{i}$ with a highly damped galvanometer and $\overline{\Delta i^2}$ in a narrow filter with ΔF centred on $F \sim 700\,\text{kHz}$. They found that the elementary charge was $e = 1.59 \times 10^{-19}$, in good agreement with Millikan's $e = (1.59 \pm 0.002)10^{-19}$, obtained by the oil drop method. However, at much lower frequencies, the values of e obtained by the noise method are found to be very different, and no longer agree with Millikan's result: they are too high by three orders of magnitude. This is observed by going down to a few dozen hertz for a tungsten filament or a few dozen kilohertz for an oxide cathode. Schottky suggested that this could be explained by assuming that thousands of electrons were emitted simultaneously and that this should produce to a scintillation of light on the oxide cathode (flicker effect). This explanation is not really compatible with the assumed independent behaviour of individual electrons which underlies the Poissonian phenomenon. Moreover, it was already known in 1926 that the noise level tended to increase at low frequencies, going well beyond the values calculated from classical theory (Fig. 8.1).

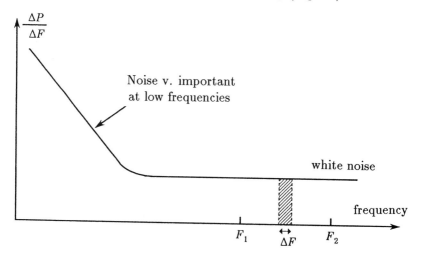

Fig. 8.1 White noise and excess noise (logarithmic scales).

L.B. Johnson in 1927 and Nyquist in 1928 put forward a theory of noise due to the thermal agitation of electrons in a conductance G at temperature T. This predicts white noise with power spectral density

$$\frac{\Delta P}{\Delta F} = kT \tag{8.5}$$

corresponding to intensity fluctuations

$$\overline{\Delta i^2} = 4GkT\Delta F \tag{8.6}$$

within a band ΔF.

The noise is cut off at high frequencies by a mechanism comparable to that already mentioned. It relies on jumps of free electrons, each of average length equal to the *mean free path*. The frequency F_M is then of the order of the inverse of the time taken to traverse one mean free path.

These formulas were verified by Bernamont in 1937, who observed that although the theoretical model was satisfactory at high frequencies, the noise level was higher than expected at low frequencies. The measurements were accurate enough to show a variation of the form $1/F$. This appears to have been the first time that such a law was observed experimentally (Fig. 8.1).

Since the end of the Second World War, the techniques for measuring random phenomena have improved considerably. The noise current in transistors has been measured down to 10^{-5} Hz and, within the measurement range, is described by the $1/F$ law. The same noise spectrum has been obtained for the Hall effect, for Josephson junctions and for electrochemical phenomena (ion currents through membranes, electrochemical potentials and so on) As a general rule, the noise spectral density increases with decreasing frequency. A law of the form F^{-a} is found to represent experimental phenomena for a very close to 1. White noise is again observed at high frequencies, but there is no plateau at low frequencies. Measurements carried out at 10^{-6} Hz show that the noise level continues to increase with decreasing F.

The advent of the fast Fourier transform has facilitated calculations of the spectral density of a numerical signal, sampled with a period θ and measured over a time interval T. The procedure has been applied to biological data, e.g., to calculate the daily insulin injection for an unstable diabetic. It has also been used to calculate the spectrum of the recording of a concerto and to determine the spectrum of traffic fluctuations on a motorway. The result is always a $1/F$ polychromatic noise at low frequencies. Of course, this method of calculation does not allow us to go down in frequency beyond $1/T$ as mentioned in Chapter 1. To obtain meaningful results at 10^{-6} Hz one has to record over a period of the order of two weeks and millions of data have to be saved in memory. Moreover, there are practical limits in exploration towards zero frequency.

A reasonable conclusion seems to be that the $1/F$ law is confirmed by the frequency and time domain analyses performed so far. There seems to be no limit at low frequencies, but experiments are possible only over limited periods of time and the exploration of very low frequencies is meaningful only for experiments performed over extremely long periods of time.

Signals can be constructed by a computer and their spectra can be calculated. By throwing five dice and adding the resulting five numbers we obtain a signal (number) lying between 5 and 30; if we continue after having shaken the dice well, we obtain a sequence of random numbers. FFT calculations then show that the corresponding spectrum is white. We can also obtain the successive sets of numbers in a more subtle way. For example, we can write the order of throwing in binary form and identify each dice with a binary element. For the 24th throw, we only throw the dice assigned to a binary element which has changed in going from 23 to 24 (the first four dice; the fifth still indicates the same value). This yields a sequence of numbers that are *more or less random*, since there is at least *one dice* that has been thrown again, but there are also several dice that have not been touched and allow the system to maintain partial memory. The fast Fourier transform applied to such data (ten dice with ten possible values, yield 1 024 random numbers between 10 and 100) gives a $1/F$ spectrum.

The above system can also produce $1/F^2$ spectra. The procedure is to throw again all the dice together, after shaking, when only one dice, well identified (e.g., by the assignment of the third binary element), will give a definite result (say, 4, but other criteria can be chosen). The memory of the system is considerable because, statistically, it is complete in five cases out of six; in the sixth, the randomness is complete since all the dice are thrown again. This is a *random walk* (a perfectly systematic operation, interrupted by a totally random change).

With artificially produced signals, we can synthesise physical signals, e.g., sound (using a digital to analogue converter, amplifier, loudspeaker and so on). The experiment has been done and shows that the *musical* sensation is different. The signal with the $1/F$ spectrum is quite easily distinguished from the others and gives a more pleasant sensation.

8.2 DIFFICULTIES WITH PHYSICAL MODELS

We conclude that, in general, noise is described by a *coloured* (i.e., not flat) spectrum. This is hardly surprising. The most *classical* spectrum, i.e., that of sunlight, is presented by the rainbow in which red is stronger than blue. A flat spectrum (wrongly called white noise, since white light does not have a flat spectrum) is only found in limited spectral bands. Formulae such as (8.4) and (8.6), reproduced in technical books, tend to propagate wrong ideas. They are valid only for the input noise of a circuit whose pass band lies in a region in which the noise is white (flat spectrum). They are excellent for a filter working between F_1 and F_2 in Fig. 8.1, but they are meaningless at very low or very high frequencies. There is nothing abnormal with the noise spectrum of a natural system being coloured rather than flat, but it has to be measured with care and then used. In most cases, a formula such as

$$A(F) = A_0 + \frac{A_1}{F} \quad (F < F_M) \tag{8.7}$$

describes experimental results very adequately, but there are some difficulties.

The first difficulty is that measurements cannot be extrapolated to $F = 0$ in order to use them simply, e.g., in the Wiener-Khinchin theorem. The $1/F$ law seems to agree well with observations down to very low frequencies, beyond which it is almost impossible to continue such measurements (Section 3.1.) and we have to resign ourselves to the fact that we will never reach $F = 0$. The theory can therefore be approached in two ways. We can assume, as we did earlier, that the phenomena are stationary (once the drift has been measured separately and then corrected). We can also try a combined theory of noise and drift, and construct a fundamentally nonstationary theory. This is a very difficult option that is currently at an exploratory stage. These difficult problems are outside the scope of this monograph.

On the other hand, it is proper to enquire why the $1/F$ variation, perfectly in agreement with experimental data, poses such curious problems. Actually, the most natural *model* of noise observed in a system is a combination of two factors. First, a *Poissonian* source produces extremely short pulses, randomly distributed in time. The system then reacts to these pulses and produces responses, often complicated, which superimpose to give the observed noise. The white *source* noise is unobservable, and what we see is the noise filtered by the system, which is be coloured. We can readily show that the output spectrum is the product of the original flat spectrum and the square of the modulus of the transfer function describing the system. Thus, in order to obtain the $1/F$ spectrum we need the transfer function $1/\sqrt{F}$. On the other hand, a linear differential equation, analysed by the Fourier transform, always leads to a rational transfer function of F (which is what we saw in the case of the $1/F^2$ noise, the random walk and so on). In order to obtain more complicated transfer functions, we need transmission systems such as that shown in Figure 8.2.

This is an RC transmission line, assumed infinite. It is sufficiently damped to allow us to make this hypothesis even for short line lengths. It is readily shown that the transfer function for the input current and the measured voltage is

$$\frac{v}{i} = \sqrt{\frac{d}{4\pi\sigma e \varepsilon_0 \varepsilon W^2}} \frac{1}{\sqrt{jF}} \tag{8.8}$$

so that white current noise due to Poissonian phenomena gives rise to a $1/F$ voltage noise.

This simple model allows us to demonstrate the special nature of $1/F$ noise. The extremely short pulses due to the Poissonian source change shape in the course of propagation. They retain their memory throughout the propagation process so long as the damping is too high. We know from experiment that low velocity propagation (which is the case for the RC

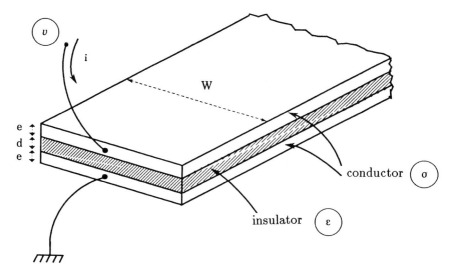

Fig. 8.2 A system equivalent to a transmission line with an infinite number of equivalent RC cells.

line) is accompanied by persistent preservation of the memory of a short initial pulse. We can also imagine that systems with more complicated propagation (e.g., propagation involving several guided modes) can produce memory durations distributed according to the traversed paths. This is the hypothesis underlying the model of Section 3.1.2.

Before we close this brief overview of the difficulties encountered in the elaboration of a model of $1/F$ noise, we note one further important property. We first calculate the variance for the $1/F$ spectrum, using one of the formulae of Chapter 3. Apart from the problems of convergence associated with the limit at $F = 0$ (which are not specific to $1/F$ noise), we see that the term to be integrated can be expressed in terms of the single variable $x = F\tau$ (because the kernel depends on $F\tau$ and the spectral density term involves $dF/F = dx/x$). It follows that the integral, i.e., the variance, is independent of τ. The $1/F$ noise gives rise to a horizontal plateau for the variance, and vice versa. This apparently simple result, conceals a difficulty if we wish to describe the random signal by the method of Section 3.2.1. Indeed, it implies that the result is independent of τ, i.e., the time over which the mean is evaluated. We have to determine all the properties of the signal for a sampling time τ whatever the value chosen. If we start with, say, τ_1, we obtain a value for the variance, and if we repeat this with $\tau_1/10$ we obtain the same result. This means that any part of the signal contains as much information as the complete signal. This strange property is typical of a specific category of mathematical entities known as *fractals*, which are of undoubted interest for the description of $1/F$ noise.

8.3 OSCILLATOR NOISE AND THE PHYSICS OF $1/F$ NOISE

The techniques available for the measurement of noise present in an oscillator output are now well established. They are described in Chapters 2, 3 and 4 and can be used to perform accurate and reliable measurements. This allows us not only to characterise oscillators, but also to trace the causes of noise if a suitable model of the circuit is available. Everything thus depends on the validity of the model, which has to be carefully verified. We conclude with an example.

A lossy resonator produces thermal noise and we can consider measuring its spectrum. This is a very complicated operation that involves the use of a carefully balanced passive bridge (which implies that we have two very similar resonators). We observe the noise spectrum and note, for example, an interesting empirical relationship between the noise spectral density and the Q-factor of the resonator in the $1/F$ region (at around, say, 1 Hz). When plotted on double logarithmic scale, the experimental points lie on a straight line with slope equal to -4. It would therefore appear that the characteristic noise of the resonator in the zone of interest can be described by

$$\frac{dP}{dF} = \frac{K}{Q^4 F} \tag{8.9}$$

By introducing a single resonator of this kind into an oscillator circuit, we can readily and accurately measure the phase noise. The experiment shows that the above empirical law is preserved: the noise at a given frequency is inversely proportional to Q^4. We can then envisage a comparison of the data for the resonator on its own and those obtained when the resonator is part of the oscillator circuit, and we can then attempt to verify simplistic theories such as those described in Section 3.4.

The techniques for measuring $1/F$ noise are well established for oscillators, but there is still a lot of work to be done before we can go back to the $1/F$ noise of each component. We will then have to attempt physical interpretations that we will be able to justify provided we have precise numerical data at our disposal. Currently, schematic descriptions abound, and are sometimes contradictory. They are not generally supported by undisputed numerical data.

9

Natural oscillators

9.1 SOME EXAMPLES

Observations of the motion of celestial bodies reveal periodic phenomena that are sometimes simple but often highly complex and due to the superposition of a very large number of sinusoidal processes that have different periods and are difficult to isolate, especially if the amplitudes are comparable. These periods can provide standards for the measurement of frequency and time. For example, the rotation of the Earth on its axis can be used to define the sidereal day. The rotation of the Earth around the Sun gives us the tropical year, and the Moon orbits the Earth in a complicated motion that involves several periods, including the 28-day period that defines the month. All these processes have a very clear origin: they are projections of rotational motion on to an axis (the sine is the projection of the radius of a circle on to a diameter). The difficulty lies in the decomposition of the apparent motion into a series of rotations with different periods and amplitudes. Other periodic astronomical phenomena (pulsars, variable stars, cepheids and so on) have a more complicated origin and have given rise to numerous studies.

Natural oscillators include musical instruments that are as old as civilisation. It is useful to distinguish between damped instruments that emit sound within a limited coherence time, and undamped instruments that emit continuously. The former are excited resonators and are easy to describe; the latter are oscillators whose operation is often difficult to grasp. Resonators that are made to vibrate and respond with damped motion, include the vibrating string (hit by a hammer in the piano or pulled away from its initial position by pinching and then released in the harp or the harpsichord), the membrane (in the drum), the plate (in the cymbal, the xylophone, the bell and so on) and, finally, the tuning fork which is widely used as a frequency standard for musical instrument tuning. Oscillators include the violin (friction between string and bow creates negative resistance that supports the oscillation), the reed mouthpiece (described by a model similar to that of the rubbed string) and, finally, the flute mouthpiece that is responsible for the working of the whistle which has a very complex mechanism that relies on the phenomenon of turbulence (we shall return to this later).

Nature offers us a huge variety of periodic phenomena. For example, intermittent fountains can act as sources of periodic flow that in turn excite the sound produced by geysers. They are an interesting example because they suggest the simple model shown in Fig. 9.1. A permanent source of water producing a flow rate D_0 feeds a tank in which the level varies between A and B; the volume of the tank between A and B is V. Initially, the level is at A and the water rises in the ascending part of the syphon I. When it reaches B, a small amount of water escapes into the descending branch II of the syphon, thus initiating the flow of water into II. If the corresponding flow rate D is greater than D_0, the tank will empty itself. As soon as the level reaches A, a small amount of air enters branch I , the syphon switches itself off and the entire cycle starts again.

The figure shows the form of the oscillations, whose period is given by

$$T = V\left|\frac{1}{D - D_0} + \frac{1}{D_0}\right| \tag{9.1}$$

This assumes that switching is instantaneous, which is not the case in practice.

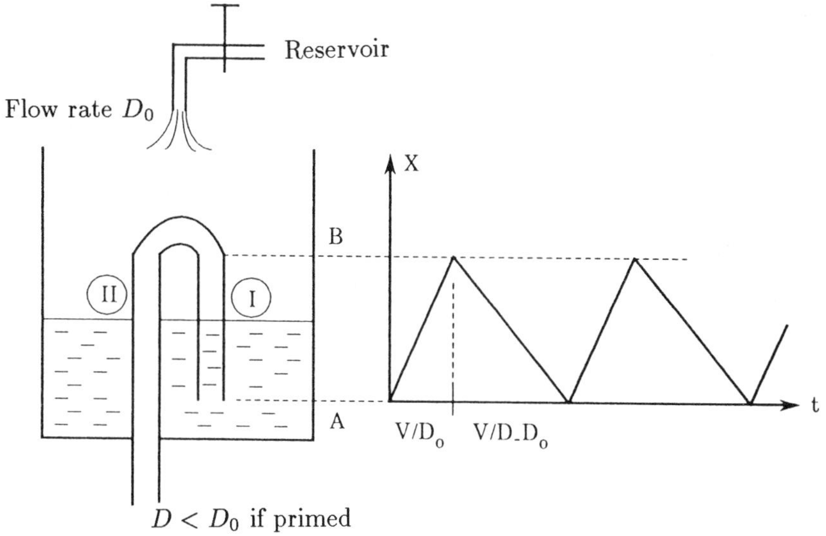
Reservoir
Flow rate D_0
II
I
B
A
X
t
V/D$_0$
V/D-D$_0$
$D < D_0$ if primed

Fig. 9.1 The syphon tank is a simple model of a relaxation oscillator.

The above relaxation oscillator is a bistable system, with two well-defined states. Each of them is unstable and is succeeded by the beginning of the other, so that the system oscillates between the two states. There is no resonator to impose the frequency, i.e., the frequency is largely determined by the energy-supplying system. Relaxation phenomena are quite common. The screeching of chalk on a blackboard is an example. A large number of electrical circuits can be described by this model (e.g., the multi-vibrator). The main biological oscillator, i.e., the heart, can be described in the first approximation by the model (filling and draining of the ventricle). The switching is performed by a nerve signal along the Hiss bundle which, roughly speaking, acts as a delay line, which complicates the operation.

More complex oscillations are produced when the two regimes just described are combined. This occurs, for example, in the case of oscillations of population which we shall now consider. Let us place X mice in an isolated enclosure (say, an island) in which there is plenty of vegetation so that they never run out of food. They will multiply and their population X will increase exponentially. If instead of mice we place Y wolves in the same enclosure, a reduction in Y will be observed because wolves are carnivorous, so that they can only fight among themselves and devour the resulting carcases. However, if we mix wolves and mice, the phenomenon becomes more complicated and two regimes are found to coexist. In one, the wolves devour the mice and, being well fed, cease to fight and proliferate. Of course, the mice population declines, but soon there are not enough mice left and the starving wolves begin to fight again; this gives a break to the mice which proliferate again, and Y decreases. The cycle starts again when there are enough mice to feed the surviving wolves. The oscillations of population modelled by Volterra depend on this type of bistable system in which the two states exist permanently. A mathematical analysis of the model, to which we will come back, demonstrates the complexity of the phenomena (see Section 9.2.).

A similar model applies to chemical oscillators. The oxidation-reduction reaction that takes cerium Ce^{++} to oxidised cerium in the form Ce^{+++} is an example. A colour indicator (ferroin) turns red when Ce^{++} predominates, and blue in the opposite case. We place in the solution of cerium ions a mixture of an oxidising agent (potassium bromate) and a reducing agent (malonic acid which gives the bromomalonate). The reduction reaction produces Br^- ions that are slowly absorbed by the malonic acid and disappear from solution, but they instantaneously block the oxidisation reaction by catalysis. Thus, if we start with Ce^{+++} (blue indicator), the reducing agent will begin to act and give a few reduced Ce^{++} ions. These will not be oxidised because the oxidisation reaction is brought about by the Br^- poison that has only just appeared. Everything will then be rapidly reduced and the coloured indicator will change to red. The poisonous Br^- ions will be absorbed and, from then on, nothing is opposed to the oxidisation that will take place. The coloured indicator changes back to blue and the Ce^{+++} ions reappear. The reduction reaction can then start again, produce the poison, block oxidisation and so on, and we observe oscillations of colour, between blue and red, for a very long time. The reaction has been studied by Belousov and Zhabotinskii and bears their names (Fig. 9.2).

Biological oscillators may operate according to this model. They regulate the unfolding of life (alternation of waking and sleeping periods, periodic return of the sensation of hunger, menstruation and so on). Chronobiology is an expanding subject and the description of signals has advanced considerably, but we still know relatively little about the elementary phenomena on which any model must be based. Certain economic phenomena appear to exhibit periodicities (Kondratev cycles), but specific conclusions are difficult to come by because the necessary observations are made over limited periods of time, the variables that have to be measured are poorly defined and our knowledge of economic phenomena is still a matter of some dispute.

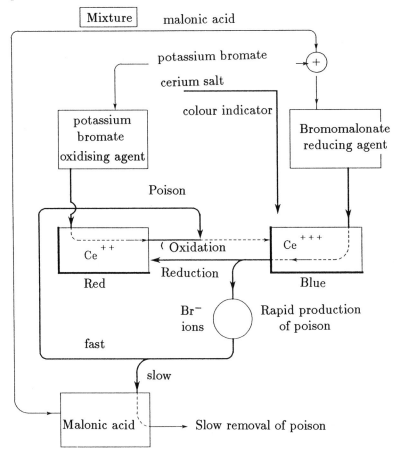

Fig. 9.2 A simplistic model of an oscillatory chemical reaction.

Considerable difficulties are encountered when we attempt to analyse phenomena occuring in inhomogeneous media. In the oscillatory chemical reaction, a simple scheme that can be described is achieved if we vigorously stir the medium so as to establish perfect homogeneity. Unless this is done, each microcell of space produces its own reaction, without synchronicity with the others. All the intermediate colours between red and blue are then present and we observe very complicated patterns that evolve in time. Everything depends on the diffusion of reaction products within the available volume.

Another example of oscillation in time and space was described by Bénard in 1900 and involved the frying pan problem. Suppose that we heat a container full of oil by placing it on a hot plate. Hot oil has lower density and must rise whilst colder oil sinks. Obviously, cold oil reaching the bottom of the pan is heated and hot oil at the top cools down. The process continues and we observe a structure that depends on the temperature difference and also on the shape of the container. Periodic motion occurs as time goes by, and regular patterns are established in space (Fig. 9.3). Bifurcations, and even chaos, may be observed in the spatial oscillations.

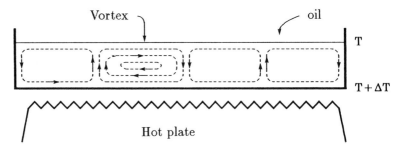

Fig. 9.3 Periodicity in time and space in a frying pan

Turbulence in outflowing fluid is a similar phenomenon. By reducing the cross section of the nozzle available for the outflow of a viscous fluid, we produce pressure and density gradients that act in a similar manner to that of the temperature gradient in the previous example. Vortical motion can result and, hence, oscillations in time. At the same time, in a given vortex (Fig. 9.4), the velocity increases from the centre outwards, but it cannot increase beyond the limit imposed by the initial outflow velocity. The size of a vortex is limited, and when the limit is reached, the whirl breaks off and moves in the direction of the drainage, and another whirl appears. All this results in a series of vortices that progress in the direction of the sink. A periodic signal is produced by the successive formation of vortices. The mouthpiece of a flute works in this manner, and the pitch of the emitted sound can be altered by changing the velocity of the flowing air.

Ionised media (in gas discharges, the ionosphere, stars, thermonuclear reactions and so on) can behave as oscillators emitting electromagnetic waves. The associated phenomena can be periodic in both space and time simultaneously. The oscillations can be maintained by turbulent phenomena in the flow of neutral fluids or in the interaction of a beam of charged particles with the internal electromagnetic field of a plasma. The principle is similar to that of a microwave tube or even that of a rubbed string.

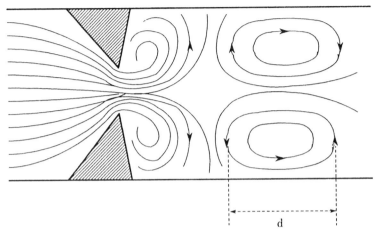

Fig. 9.4 Vortices produced during the drainage of a fluid (the whirls move with velocity v along the tube and the particles describe epicyclodal trajectories).

We must now stop our necessarily brief review of natural periodic phenomena. Advances in measuring techniques and signal processing (e.g., the advent of the fast Fourier transform) have led to excellent descriptions. We now turn to models and their numerical application as a test of their validity.

9.2 A FEW MODELS OF NATURAL OSCILLATORS

The periodic phenomena observed in the motion of celestial bodies were described by a mechanical model as early as the eighteenth century. Two fundamental laws were involved in this, namely, the law of universal graviation and the law relating force, mass and acceleration. The mathematical formulation of these laws culminates in the equations of celestial mechanics. Many nineteenth century mathematicians were therefore engaged in trying to find ways of obtaining the solutions of these equations. On the other hand, numerical values had to be found for these solutions and courageous mathematicians such as Bessel calculated *by hand* the data for tables giving numerical values for the solutions of certain equations that they encountered. It was then possible to compare the results of these calculations with measurements and thus check the validity of the underlying model.

The difficulty with the numerical calculations was that they were tedious (but indispensable), and this stimulated attempts by theoreticians to transform the equations analytically in order to reduce to a minimum the volume of computation. The result was a particularly successful form of the equations, namely, analytical mechanics. This was followed in the nineteenth century by the advent of the equations of state, limit cycles, Poincaré sections and Lyapunov stability conditions. These concepts are still used in the study of oscillators and were introduced in Chapter 5.

The situation has changed radically with the advent of the ubiquitous computer. The analytical methods of the nineteenth century often provided a demonstration of the existence of solutions, but were not readily amenable to numerical evaluation.

Nevertheless, the computer has its limitations. It runs for a limited time and has a limited number of digits to express the result. With eight digits, we cannot distinguish between, say, 203.3 and 203.4. We have to round off the number to 203. The computer is thus handicapped when detecting limit cycles. All it can then do is to show that we pass through the same point (to within the rounding off error). To be more sure of the result, we have to consider a large number of successive cycles, but we can only do this for a finite number of such cycles. When the frequency is low, the computing time soon becomes prohibitive. Computers are thus able to give us, with ease, the solution of certain problems encountered in the study of oscillators, but they introduce other problems (before a numerical calculation is started, it is often useful to exploit the techniques of analytical mechanics to the full in order to gain an insight into the *behaviour* of the solutions).

The intermittent fountain can be modelled by the simple electrical circuit indicated in Fig. 9.5 in which the source of emf E charges a capacitance C via a resistance R. The syphon is represented by the gas discharge between two electrodes. The corresponding characteristic curve is shown in the same figure and is obtained by introducing a slowly-varying current and measuring the voltage v. An acceptable analytical description is

$$v = V_0 - r(i - I_0) + \alpha(i - I_0)^3 \tag{9.2}$$

The discharge tube is connected to the terminals of the capacitance and is controlled by the voltage across it. As v increases, the current remains quite low up to V_a. Physically, this means that the discharge tube is turned off. Once V_a is reached, the current rises abruptly by Δi at constant voltage V_a, and the tube fires. If we now reduce the voltage, the current drops gradually until we reach V_e, at which point there is another jump in current, this time in the opposite direction. In the electrical model, such jumps represent the priming and unpriming of the syphon. These operations take place for different values of the control variable, depending on whether it increases or decreases. However, an abrupt change in the current during the short time interval τ in which the voltage remains constant is only possible if the

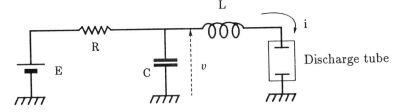

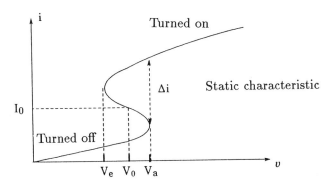

Fig. 9.5 Model of a relaxation oscillator.

model includes a stray self-inductance given by

$$L = \frac{V_a \tau}{\Delta i} \tag{9.3}$$

The circuit of Fig. 9.5 can then be described by the equation

$$\frac{d^2 x}{dt^2} - \gamma \left(1 - \varepsilon x^2\right) \frac{dx}{dt} + \omega_0^2 x + \beta x^3 = 0 \quad (x = i - I_0) \tag{9.4}$$

where

$$\gamma = \frac{r}{R} - \frac{1}{RC} \quad \varepsilon = \frac{3\alpha RC}{rRC - L} \quad \omega_0^2 = \frac{R - r}{LCR} \quad \beta = \frac{\alpha}{RCL} \tag{9.5}$$

This is the van der Pol equation augmented by a nonlinear term of the type that appears in Duffing's equation.

The system has two variables of state. The phase trajectory can be described step by step, which presents no difficulty on a computer, but requires a lot of time if we have to perform the calculation *by hand*, which is what van der Pol did in 1930 for different values of the parameters. He demonstrated that there was a smooth transition from large-amplitude quasi-elliptical cycles corresponding to sinusoidal oscillations and small-amplitude cycles corresponding to periodic phenomena with numerous harmonics. This is an interesting result because it suggests that the quasi-sinusoidal oscillator and the relaxation oscillator are in fact two limiting cases of a more general model. Figure 9.6. shows the shape of the cycle in the limiting case of a small stray self-inductance. We see that this is a relaxation circuit with instantaneous switching. As soon as the self-inductance increases, the amplitude of the cycle increases and we approach an elliptic cycle. In the case of the discharge tube, we have practically complete relaxation, the current and voltage being limited by $v = i = 0$. If we add a self-inductance, we obtain an oscillator with fewer harmonics. This is what happens in the singing arc whose nonlinear characteristic is the result of a discharge between two carbon electrodes. Such oscillators were employed in the transmitters used in the early days of radio broadcasting.

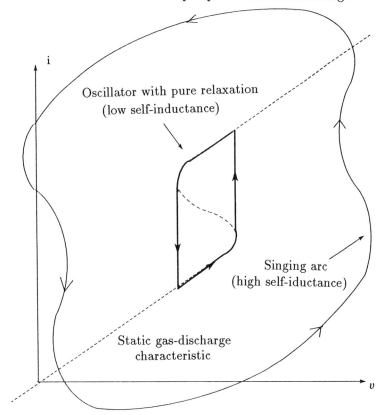

Fig. 9.6 Pure relaxation cycle evolving towards a more or less sinusoidal cycle as the parameters are modified.

Oscillations in population can be described, in the first approximation, by a model with two variables of state. If X mice are alone on an island with plenty of food, they proliferate. This situation is described by an equation of the form

$$\frac{dX}{dt} = \alpha X \qquad \alpha > 0 \tag{9.6}$$

If Y wolves are on their own, they fight and eat each other. The corresponding equation is

$$\frac{dY}{dt} = -\beta Y \qquad \beta > 0 \tag{9.7}$$

and describes the decline in their population. If the wolves and the mice are on the island at the same time, the processes become coupled, but must still be described as simply as possible. We may suppose that the growth rate of the population of mice is reduced (or even changed in sign) by the hunting wolves. A simple representation is

$$\alpha = \alpha_0 \left(1 - \frac{Y}{Y_0}\right) \tag{9.8}$$

Similarly, the rate of decline in the number of wolves will be lower (or even changed in sign) if the number of mice (the food of the wolves) is high. We can then try

$$\beta = \beta_0 \left(1 - \frac{X}{X_0}\right) \tag{9.9}$$

This leads to the Volterra model with two parameters of state:

$$\dot{X} = \alpha_0 \left(1 - \frac{Y}{Y_0}\right) X \qquad \dot{Y} = -\beta_0 \left(1 - \frac{X}{X_0}\right) Y \tag{9.10}$$

The corresponding analysis is simple. First, we notice that there is a biological equilibrium:

$$X = X_0 \qquad Y = Y_0 \tag{9.11}$$

Outside this equilibrium, the system finds itself on a cycle whose equation is

$$\frac{1}{\alpha_0} \left(\frac{X}{X_0} - \log \frac{X}{X_0}\right) + \frac{1}{\beta_0} \left(\frac{Y}{Y_0} - \log \frac{Y}{Y_0}\right) = \text{constant} \tag{9.12}$$

This relation is readily obtained by integrating the Volterra equations. Figure 9.7 shows this family of cycles, which are not limit cycles. As long as there is no perturbation, the system continues to oscillate, but the slightest perturbation makes it jump to another cycle. The system is not an oscillator; it has no stable attractor (neither a point attractor corresponding to equilibrium, nor a cyclic attractor corresponding to a permanent oscillation).

Experiment shows that there is a biological equilibrium and a regular oscillation, i.e., the above model is inadequate. Attempts to improve it rely on a more refined analysis of the underlying elementary phenomena. We have to define at least two varieties of mice. There are, on the one hand, the young, the old and the sick that are defenseless as they face the predators. On the other hand, there are the adult and healthy mice that have learned the dangers of nature and know how to look after themselves. We also have to subdivide the wolves into those that are satiated because they have just eaten several mice and those that are starving because they have not found anything to eat for several days. This means that we have to introduce a delay factor into the law describing the behaviour of the wolves. All this results in a system with a large number of variables of state, which computers can integrate. Comparisons with experiment can then be attempted by following the predator population as a function of time. Similar models with numerous variables of state can be introduced for oscillatory chemical reactions, and even for cyclic economic phenomena, but we then have to be able to evaluate numerically the values of the large number of parameters introduced in the description of the equivalent circuit.

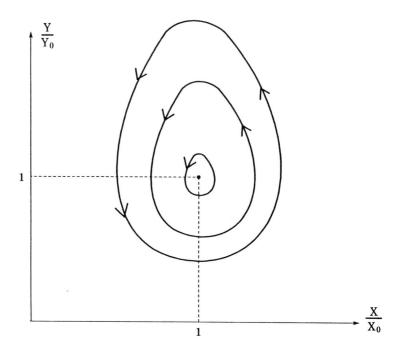

Fig. 9.7 Volterra cycles.

9.3 CAN NATURAL OSCILLATORS BE USED AS STANDARDS?

The great variety of natural oscillators should provide us with a choice of standards for frequency measurement. The motion of the Earth and of the Sun has been used in the measurement of time, and the tuning fork is still used with musical instruments, but all these are weakly damped systems. No other natural oscillators have ever been used as frequency standards because of poor stability, fluctuation mand drift and the sensitivity of frequency to external factors. It has therefore been necessary to design artificial devices, as explained in Chapter 5.

Let first examine a relaxation oscillator, e.g., the intermittent fountain. Let us suppose that the flow rate D_0 from the supply tap is increased by the addition of drops that follow each other periodically at a time intervals θ. It is clear that the priming of the syphon will coincide with the arrival of a drop (which makes the container *overflow*). The period of the oscillator, i.e., the time interval between two switches of the syphon, will thus be equal to an integer multiplied by θ. The frequency of the oscillator locks to a submultiple of the driving frequency, and the volume of the drops plays an essential role in setting the value of the demultiplication factor. We thus encounter the main failing of natural oscillators: they can readily synchronise with (lock on to) external signals, and frequency demultiplication, bifurcation, chaos and so on are all possible. Obviously, this is not very compatible with their eventual use as standards.

This synchronisation (locking) has important consequences for the observation of the system because what we see is not necessarily the effect of the initial oscillator. The biological clock whose period is close to 24 hours (circadian phenomenon) becomes synchronised with the day/night alternation and its apparent period is 24 hours. But if we place the living organism in the dark, the synchronism vanishes, and the fundamental phenomenon and the natural biological rhythm take over again. If we cross different time zones, we experience a resynchronisation (after a transient regime).

The biological clocks whose period is close to 24 hours are nearly always locked to the day/night alternation. The effect of this (sleeping and waking, periodic sensation of hunger and so on) is synchronous with a large-amplitude phenomenon (the night /day alternation) and is readily seen. A 28-day biological cycle could lock itself to the lunar rhythm, and perhaps we should attribute to this the statistical increase in the number of births observed at full Moon. The greatest caution has to be excercised in such a sensitive question. It is no less true, however, that we always observe the same side of the Moon, which means that the rotational period of the Moon around its own axis must be equal to the orbital period of the Moon around the Earth. These two celestial oscillations are synchronised, which represent a coupling. The question then is: can this have a biological effect?

Complex phenomena are also found to arise when a natural resonator is coupled to an oscillator. The mass of water in a closed (or practically closed) basin can oscillate with well defined frequencies. It is enough to shake a basin that is half full with water to realise this. The periodic gravitational effect of the Sun and of the Moon on water causes oscillations that take the form of variations in the level of water when one of the characteristic frequencies of the basin approximately coincides with a stellar frequencies. This is the basic cause of tides, which are quasi-periodical phenomena with numerous periods (including those associated with the rotation of the Earth, the orbiting of the Earth by the Moon, the orbiting of the Sun by the Earth and so on) that are modified by a variety of couplings. At the same time, the amplitudes depend on the dimension of the basin. The tide can be considerable (level changes by 10 m) in certain places and almost zero elsewhere. The period can be half that of a celestial period (on the French coast, the high tide occurs at an interval of approximately twelve hours; elsewhere it repeats itself at an interval of approximately 24 hours).

The marine creatures that live very close to the surface, have to adjust their biological clocks to the day/night alternation and the periodicity of the tide. The periods are quite different, so that their life is not simple. The effect of a few oscillators with well-defined frequencies on a complicated resonator can lead to a signal whose periodic character is very difficult to detect.

We conclude this discussion by considering the influence of a high-Q resonator on a natural oscillator whose frequency can vary over a wide range. This can be for example, a sound pipe with the mouthpiece of a flute at one of its ends. By blowing into the mouthpiece we vary the frequency of the turbulent phenomena that generate the sound in the mouthpiece. This signal resonates with the pipe which reacts on the mouthpiece, and the natural oscillator locks to the frequency of the pipe that is the nearest to its natural frequency. When the difference becomes too large, we observe a sudden jump to another mode. Figure 9.8 illustrates this phenomenon which occurs between plateaus of synchronisation. Since the oscillations of the mouthpiece contain a large number of harmonics, we observe a synchronisation with a subharmonic. The complexity of these phenomena makes them unsuitable for metrological applications.

The extraordinary variety and complexity of phenomena observed in natural oscillators justify the research into artificial sources designed to produce sinusoidal signals with very regular frequency and minimal fluctuations. We have been considering such sources throughout this book. Now that we know how to measure time and frequency, we are equipped to describe the signal of a natural oscillator and then analyse it as a prerequisite

resonator.

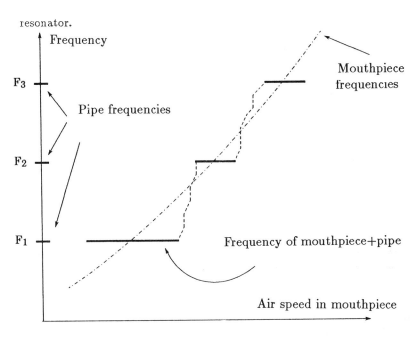

Fig. 9.8 Locking of a natural oscillator to the characteristic frequencies of a resonator.

to the construction of a model. We can, after carrying out the appropriate analysis, perform a computer synthesis of the signal and thus reproduce the sound of, say, a Stradivarius. The extraordinary variety of signals generated by natural oscillators has always been used for the creation of music. The computer now enables us to create *artificial signals* (see Section 8.1), so that we can consider the computer as a synthetic oscillator that can be used to reinvigorate the process of musical creation.

10

Bibliography

10.1 GENERAL BIBLIOGRAPHY

An exhaustive bibliography of the published literature on the topics covered in this book would consist of thousands of references. It would include fundamental publications as well as short conference proceedings, original papers and 'letters to the editor'. Such material would be generally incomprehensible to all except for a few specialists. It would also suffer from a high degree of duplication or even triplication. We have tried to overcome this problem by judicious selection from bibliographic information. We begin with standard monographs on the measurement of time and frequency.

Gerber, E.A. and Ballato, A. (1985) *Precision Frequency Control*, 2 vols., Academic Press, New York.

Kartaschoff, P. (1983) *Frequency and Time*, Academic Press, New York.

Kroupa, V.F. (1983) *Frequency stability: fundamentals and measurements*, IEEE Press, New York.

Blair, B.E. (1974) *Time and frequency: Theory and Fundamentals*, NBS monograph.

Chaleat, R. and Haag, J (1960) *Problèmes de chrométries et de théorie générale des oscillations*, Gauthier Villars, Paris.

Vanier, J. and Audoin, C (1989) *The Quantum Physics of Atomic Frequency Standards*, 2 vols., Adam Hilger, Bristol.

From time to time, the major periodicals publish special issues dedicated to time and frequency determinations. The following may be usefully consulted:

IEEE Proceedings
    vol. 54, February 1966
    vol. 55, June 1967
    vol. 60, May 1972
    vol. 74, January 1986

IEEE Transactions on Ultrasonics, Ferroelectrics and Frequency Control
    vol. 34 - November 1977

Bulletin de Bureau National de Metrologie

L'Onde Electrique

The review periodical *Metrologia* published by Springer-Verlag is dedicated to fundamental problems in measurement techniques. By consulting the tables of contents, one is quickly led to original *papers* on time and frequency topics. Useful sources include the proceedings of periodic meetings at which specialists in the measurement of frequency present their work. Some of these are listed below.

Proceedings of the Annual Frequency Control Symposium, organised by US Army Electronics Command, Ft. Monmouth, N.J.(USA)

Proceedings are available from

National Technical Information Service (NTIS), Sills Buildings, 5285 Post Royal Road, Springfield, Virginia 22161 (since 1973)

Electronic Industries Association (EIAC), c/o Frequency Control Symposium, 2001 Eye Street, Washington DC20006 (1978 and 1983)

Institute of Electrical and Electronic Engineers (IEEE), 445 Hoes Lane, Piscataway, N.J. 08854 (since 1983)

Proceedings of Conference on Precision Electromagnetic Measurements (CPEM), organised annually after 1960 by NBS (NIST), available from IEEE

Proceedings of the Annual Precise Time and Time Interval (PTTI) Applications and Planning Meeting Proceedings, available from US Naval Observatory, Washington, DC

Proceedings of the European Frequency and Time Forum, organised annually in France and Switzerland since 1987

Proceedings are available from

in France: LPMO-CNRS, 32 avenue de l'Observatoire - 25000 Besançon

in Switzerland: F.S.R.M. 8 rue de l'Orangerie, CH-2000 Neuchâtel

Proceedings of l'Union Radioscientifique Internationale, available from URSI, Brussels

Finally, there are the Technical Notes published by the National Institute of Standards and Technology (NIST) which used to be the National Bureau of Standard (NBS). They are as follows.

NBS Technical Note 394 - 1970, Barnes, J.A., Chi, A.R. and Cutler, L.S., *Characterization of Frequency Stability*

NBS Technical Note 616-R, March 1974, *Frequency Standards and Clocks: A Tutorial Introduction*

NBS Technical Note 632, January 1973, Shoaf, J.H. Halford, D. and Rusley, A.S., *Frequency Stability specifications and measurement: high frequency and microwave signals*

NBS Technical Note 662, February 1975, Hellwig H., *A review of precision oscillators*

NBS Technical Note 669, 1985, Allan. D., *The Measurements of Frequency and Frequency Stability of Precision Oscillators*

NBS Technical Note 679, March 1976. Howe, D.A., *Frequency Domain Stability Measurements: A Tutorial Introduction*
NBS Technical Note 683, August 1976, Barnes, J.A., *Models for the Interpretation of Frequency Stability Measurements*
NIST Technical Note 1337, *Characterisation of Clocks and Oscillators*, edited by D.B. Sullivan, D.W. Allan, D.A. Howe, and F.I. Walls, Time and Frequency Division, NIST, Boulder, Colerado 80303-3328

10.2 BIBLIOGRAPHIES FOR INDIVIDUAL CHAPTERS

10.2.1 Chapter 1

Most of the ideas dealt with in this introductory chapter are developed later; the reader should therefore consult the bibliography of the corresponding chapters (more particularly Chapter 5 for Secs. 1.5 and 1.6). For topics concerned with signal processing, introduced in Sections 1.2, 1.3 and 1.4, consult the following:

Blanc-Lapierre A. and Picinbono, B (1981) *Fonctions aléatoires*, Masson, Paris. Dupraz, J (1983) *Probabilités, signaux, bruits*, Eyrolles, Paris.
Goldman, S (1948) *Frequency Analysis, Modulation and Noise*, McGraw Hill, New York.
Oran Brigham, E (1974) *The Fast Fourier Transform*, Prentice-Hall, Englewood Cliffs.
One application of the measurement of frequency is the setting up of the scale of time. Chapter 7 develops the ideas that appear in Section 1.7. On the difficult problem of time, consult

Attali, J (1982) *Histoires du temps*, Fayard.
Daumas, M (1969) *Histoire générale des techniques*, Presses Universitaires de France.
Decaux, B. and Guinot, B (1969) *La mesure du temps*, Presses Universitaires de France.
Landes, D.S (1983) *L'heure qu'il est*, Gallimard.

10.2.2 Chapter 2

The following publications should be consulted for technical details concerning measuring circuits and their components.

Blanchard, A (1976) *Phased-locked loops: Application to coherent receiver design, John Wiley.*
Egan, W.F (1981) *Frequency synthesis by phaselock*, John Wiley.
Gardner, F (1966) *Phaselock techniques*, John Wiley.

Ghauss, M.S. and Laker, M.R. (1981) *Modern filter design*, Pretice Hall.

Girard, M. (1988) *Boucles a verrouillage de phase*, McGraw-Hill.

Klapper, J. and Frankle, J.T. (1972) *Phased-locked and frequency-feedback systems*, Academic Press.

Kroupa, V. (1973) *Frequency synthesis theory design and applications*, John Wiley.

Lindsey, W.C. and Simon, M.K. (1978) *Phased locked loops and their applications*, IEEE Press.

Littauer, R. (1965) *Pulse electronics*, McGraw Hill. Maas, S.A. (1986) *Nonlinear microwave circuits*, Aretch House.

Manasserwitch, V. (1976) *Frequency synthesizers, theory and design*, John Wiley.

Motchenbacher, C.D. and Fitchen, F.C. (1973) *Low noise electronic design*, John Wiley.

Penfield, P. and Rafuse, R.P. (1962) *Varactor applications*, Cambridge, M.A. The MIT Press.

Rohde, V.L. *Digital P.L.L. frequency synthesizers,*, Prentice Hall.

Temes, G. and Mitra. S.K. (1973) *Frequency syntheses theory: design and applications*, John Wiley.

The papers listed below should assist with a better appreciation of the operation of the circuits given as examples in the main text; most of them can be found in the reviews indicated in the general bibliography.

Fest,D. Groslambert.J. and Cagnepain, J.J. (1983) *Individual characterization of an oscillator by means of cross correlation or cross variance method*, IEE Trans. on Instr. and Meas., **32**, 3. 447-450.

Groslambert, J. (1983) *Méthodes expérimentales de caractérisation des oscillateurs dans less domains spectral et temporel*, Bulletin BNM, n. 63-64.

Groslambert, J. Olivier, M. and Uebersfeld, J (1974) *Spectral and short term stability measurements*, CPEM Dig., p. 60-61.

Groslambert, J. Fest, D., Olivier, M. and Gagnepain, J.J. (1981) *Characterization of frequency fluctuations by crosscorrelation and by using three or more oscillators*, Proc. 35th Ann. Freq. Cont. Symp. p. 458-463.

Groslambert, J. Fest, D., Olivier, M. and Hardin, J (1983) *Frequency stability measurements methods. State of the art and recent progress*, Proc. 6th European Conference on Circuit Theory and Design, Stuttgart, p.513-530.

Groslambert, J. Gagnepain, J.J. Vernotte, F. and Walls, F.I. (1989) *A new filtered Allan variance and its application to the identification of phase and frequency noise sources*, Proc. 43rd Ann. Freq. Cont. Symp. p. 326-330.

Halford, D. Wainwright, A.E. and Barnes, J.A. (1968) *Flicker noise of phase in R.F. amplifiers and frequency multipliers: characterisation, cause and cure*, Proc. 22nd Ann. Freq. Cont. Symp. p. 340-342.

Halford, D. Shoaf, J.H. and Risley, A.S. (1973) *Spectral density analysis: frequency domain measurements of frequency stability*, Proc. 27th Ann. Freq. Cont. Symp. p. 421-431.

Kurtz, S.R. (1978) *Specifying mixers on phase detectors*, Microwaves, p.80-87.

Lance, A.L. Seal, W.D. Mendoza, G.G. and Hudson, N.W. (June 1977) *Automated phase noise measurements*, Microwave Journal, **20**, 87-90, 92,94,96,103.

Quinn, T.J. (1989) Metrologia, **26**, 69.

Stein, S. Glaze, J, Levine, J. Gray, J. Hilliard, D. and Howe, D. (1982) *Performance of an automated high accuracy phase measurement system*, Proc. 30th Ann. Freq. Cont. Symp. p. 314-320.

Walls, F.L. Stein, S.R. Gray, J.A. and Glaze, D.J. (1976) *Design consideration in state-of-the-art signal processing and phase noise measurement systems*, Proc. 30th Ann. Freq. Cont. Symp. p. 269-274.

10.2.3 Chapter 3

Basic publications on the measurements of random quantities:

Benda, T. and Piersol, A. (1971) *Random data: analysis and measurements procedures*, J. Wiley.

Blackman, R.B. and Tukey, J.M. (1959) *The measurements of power spectra*, Dover Publications.

Papoulis, A. (1965) *Probability, random variables, and stochastic processes*, McGraw Hill.

Robins, W.P. (1982) *Phase noise in signal sources*, Peter Peregrinus.

Van der Ziel, A. (1970) *Noise: sources, characterisation, measurement* Prentice Hall.

Vanossy, L. (1965) *Theory and practice of the evaluation of measurements*, Clarendon Press, Oxford.

On problems posed by analyses of variance in frequency and time measurements:

Allan, D.W. (1966) *Statistics of atomic frequency standards*, Proc. IEEE, **54**, 221-230.

Allan, D.W. (1974) *The measurements of frequency and frequency stability of precision oscillators*, Proc. 6th.Ann. P.T.T.I. p. 109-142.

Allan, D.W. and Barnes, J.A. (1971) *A modified 'Allan variance' with increased oscillator characterization ability*, Proc. 35th Ann. Freq. Cont. Symp., p. 470-473.

Barnes, J.A. (1971) *Characterisation of frequency stability*, IEEE Trans. on Instru. and Meas. **IM-20**, p.105-120.

Baugh, R.A. (1971) *Frequency modulation analysis with the Hadamard variance*, Proc. 25th. Ann. Freq. Cont. Symp. p. 222-225.

Kartaschoff, P. (1977) *Terms and methods to describe frequency stability*, Tech. Mit. P.T.T.T. **51**, 520-529.

Lesage, P. and Audoin, C. (1973) *Characterisation of frequency stability: uncertainty due to the finite number of measurements*, IEEE Trans. on Instr. and Meas. **IM-22**, 157-161.

Lesage, P. and Audoin, C. (1979) *Effect of dead time on the estimation of the two sample variance*, IEEE Trans on Instr. and Meas. **IM-28**, 6-10.

Rutman, J. (1978) *Characterisation of phase and frequency instabilities on precision frequency sources: fifteen years of progress*, Proc. IEEE, **66**, 1048-1075.

Details of a few special methods of measurement will be found in the following articles:

Barnes, J.A. and Allan, D.W. (1966) *A statistical model of flicker noise*, Proc. IEEE **43**, 176-178.

Gagnepain, J.J. (1983) *Phase and frequency noises in oscillators*, proc. 7th. Int. Conf. on noise in physical systems, Montpellier, 309-317.

Gagnepain, J.J. (1988) *1/f noise in oscillators. Theoretical and experimental progress*, 4th Symp. on Frequency standards and Metrology, Ancona, Italy.

Halford, D.D. (1968) *A general mechanical model for f^a spectral density noise with special reference to flicker noise $1/f$*, Proc. IEEE, **56**, 3, 251-258.

Leeson, D.B. (1966) *A simple model of feedback oscillator noise spectrum*, Proc. IEEE, **54**, 329-330. Rutman, J. and Uebersfeld. J. (1972) *A model for flicker frequency noise of oscillators*, Proc. IEEE, **60**, 2, 233-235.

Details of a few special methods of measurement will be found in the following articles:

Groslambert, J. Gagnepain, J.J. Vernotte, F. and Walls, F.I. (1989) *A new filtered Allan variance and its application to the identification of phase and frequency noise sources*, Proc. 43rd Ann. Freq. Cont. Symp. p. 326-330.

Pierrejean, D. (1987) *Caractérisation de la pureté spectrale et de la stabilité des oscillateurs par une méthode de corrélation*, L'Onde Electrique, **67**, 3.

Vernotte, F. Groslambert, J. and Gagnepain, J.J. (1990) *Mesure du bruit des oscillateurs par une méthode de variances multiples*, Congrés de Chronométrie, Stuttgart, 135-142.

10.2.4 Chapter 4

For microwave techniques, the following may be consulted:

Bailey, A.E. (1985) *Microwave measurements*, Peter Peregrinus.
Chantry, C.W. (1979) *Modern aspects of microwave spectroscopy*, Academic Press. Harvey, R.F. (1963) *Microwave engineering*, Academic Press.
Skolnik, M.I. (1962) *Introduction to radar system*, McGraw Hill Book Company. Slater, J.C. (1950) *Microwave electronics*, Van Nostrand.

Some of the details of circuits mentioned in the text will be found in the following articles:

Alley, G. and Han-Chiu-Wang (1979) *An ultralow noise microwave synthesiser*, IEEE Trans. on Microwave Theory and Techiques , **27**, No. 12, 969-974.

Ashley, J.R. Searles, C.B. and Palka, F.M. (1968) *The measurement of oscillator noise at microwave frequencies*, IEEE Trans. on Microwave Theory and Techniques , **MTT 16**, no. 9, 753-760.

Grauling, C.H. and Healey, D.J. (1960) *Instrumentation for measurements of the short term frequency stability of microwave sources*, Proc. of the IEEE, **54**, 249-257.

Groslambert, J. and Olivier, M. (1984) *Measurements of phase and frequence fluctuations of microwave oscillators*, Proc. 2nd European Frequency and Time Forum, Neuchâtel, 151-163.

Hewlett Packard (1983) *Phase noise characterisat9on of microwave oscillators*, Product Note 11729B-1.

Olivier, M. (1986) *Métrologie des fréquences microndes*, BNM, Proc. Xe Journées 'Mesure du temps et des frequences', 7-15.

Olivier, M. Groslambert, J and Chauvin, J. (1987) *Oscillateur a quartz à 1 GHz utilisé en synthése microonde*, Proc. 1er Forum Eurpéen Temps Fréquence, Besançon, 125-128.

Olivier, M. Groslambert, J. and Valentin, M. (1987) *Mesure du bruit de phase, par des méthodes actives et passives, de sources microondes de haute pureté spectrale*, Proc. 5e Journées nationales microondes, Nice, 56-58.

Ondria, J.G. (1968) *A microwave system for measurements of AM and FM noise spectra*, IEEE Trans. on Microwave Theory and Techniques, **MTT 16**, no. 9, 767-781.

Richardson, J. (1984) *Synthesisers benefit from SAW technology*, Microwave RF, **23**, no. 6, 145-147.

Walls, F.I. Stein, S.R. Gray, J.E. and Glaze, D.J. (1976) *Design considerations in state of the art signal processing and phase noise measurement systems*, Proc. 30th Ann. Freq. Cont. Symp., 269-274.

Basic publications on the optical spectrum:

Borde, CH.J. (1977) *Laser spectroscopy*, Springer.

Button, K.J. (1986) *Infrared electromagnetic waves*, Academic Press.

Button, K.J. Inguscio, M. Strump, F. (1984) *Reviews of infrared and millimeter waves*, Academic Press.

Harper, P.G. and Wherret, B.S. (1977) *Nonlinear optics*, Academic Press.

Levine, A.K. (1968) *Lasers*, Marcel Dekker.

Zernicke, F. and Midwinter, J.E. (1973) *Applied nonlinear optics*, John Wiley.

Details of the circuits for frequency measurements in the optical spectrum:

Barger, R.I. Hall, J.L. (1969) Phys. Rev. Lett, **22**, 4-8.

Bennett, S.J. and Cerez, P. (1978) Optics. Commun, **25**, 343-347.

Borde, CH.J. Camy, G, Decomps, B, Descoubes, J.P. and
Vigue, J. (1981) J. de Phys, **42**, 1393-1411.

Clairon, A, Dahmani, B, Filimon, A, Rutman, J. (1985) IEEE Trans. on Instr. and Meas, **34**, 265-268.

Daneu, V. Sokoloff, D. Sanchez, A. and Javan, A. (1969) Appl. Phys. Lett, **15**, 398.

Daniel, V. Steiner, M. and Walther, H. (1981) Appl. Phys, **B 26**, 19-21.

Domnin, Y.S. Koshieljaevsky, N.B. Tatarenmov, V.M. and Shumjaetsky, P.S. (1980) IEEE Trans. on Instr. and Meas, **29**, 264-267.

Evenson, K.M. Wells, J.S. Petersen, F.R. Danielson, B.L.
Day, G.W. Bargier, R.L. and Hall, J.L. (1972) Phys. Rev. Lett., **29**, 1346-1349.

Freed, C. (1970) Appl. Phys. Lett. **17**, 53-56.

Hall, J.L. and Borde, CH.J. (1976) Appl. Phys. Lett. **29**, 788-790.

Hanes, G.R. and Dahlestrom, C.E. (1969) Appl. Phys. Lett. **14**, 362-364.

Jennings, D.A. Pollock, C.R. Peterson, F.R. Drullingham, R.E. Evenson, K.M. Wells, J.S. Hall, J.L. and Layer, P. (1983) Opt. Lett, **8**, 136-138.

Jolliffe, B.W. Kramer, C. and Chartier, J.M. IEEE Trans. on Instr. and Meas. **25**, 447-450.

Klingenberg, H.H. (1985) Appl. Phys. Lett. **B 37**, 145-149.

Knight, D.J. E. Edwards, G.J. Pearce, P.R. Cross, N.R. IEEE Trans. on Instr. and Meas. **29**, 257-264.

Lee, P.M. and Skolnick, M.I. (1967) Appl. Phys. Lett. **10**, 303-305.

Le Floch, A, Le Naour, R. Lenormand, J.M. and Tachie, J.P. (1980) Phys. Rev. Lett. **45**, 544-547.

Pollack, C.R. Jennings, D.A. Petersen, F.R. Wells, J.R. Drullinger, R.E. Beaty, E.C. Evenson, K.M. (1983) Opt. Lett. **8**, 133-135.

Weiss, C.O. and Godone, A. (1984) Appl. Phys. Lett. **B 35**, 199-200.

10.2.5 Chapter 5

Classical theory of oscillators is part of the theory of nonlinear systems. For this vast subject, we suggest:

Andronov, Vitt, A.A. and Khaikin, S.E. (1966) *Theory of oscillators*, Pergamon Press.

Blaquiere, *Nonlinear system analysis*, Academic Press.

Bogolioubov and Mitropolski (1962) *Les méthodes asymptotiques en théorie des oscillations non linéarires*, Gauthiers Villars.

Chaleat, R. and Hagg, J. (1960) *Problémes de Chronométrie et de théorie générale des oscillations*, Gauthier Villars.

Chua, L.O. (1969) Introduction to nonlinear network theory, McGraw Hill.

Groszkiwsky, J. (1964) *Frequency self oscillators*, Pergamon.

Haag, J. (1955) *Les mouvements vibratoires*, Presses Universitaires de France.

Hassler, H. and Neyrinck, J. (1985) *Circuits non linéaires*, Presses Polytechniques Ro.

Minorski, N. (1962) *Nonlinear oscillations*, Van Nostrand.

Rocard, Y. (1952) *Dynamique générale des vibrations*, Masson.

Stern. T.E. (1975) *Theory of nonlinear networks and systems*, Addison Wesley.

Wilson, A.N. (1975) *Nonlinear networks: theory and applications*, IEEE Press.

Some references on quartz oscillators:

Besson, R. (1984) *Recent evolution and new developmentgs of piezoelectric resonators*, Proc. IEEE Ultrasonics Symposium, 367-377.

Besson, R.J. Groslambert, J.M. and Walls, F.I. (1982) *Quartz crystal resonators and oscillators recent developments and future trends*, Ferroelectrics, **43**, 57-65.

Cady, W.G. (1964) *Piezoelectricity*, volumes 1 and, Dover Publications.

Frerking, M.E. (1978) *Crystal oscillator design and temperature compensation*, Van Nostrand.

Gagnepain, J.J. (1990) *Resonators, detectors, and piezoelectrics*, Advances in electronics and electron physics, **77**, 83-137.

Gagnepain, J.J. and Besson, R. (1975) *Nonlinear effects in piezoelectric quartz crystals*, Physical Acoustics, **XI**, 245-288.

Gerbier, E.A. and Ballato, A. (1985) *Precision frequency control*, **2**, 'Precision Oscillators', Academic Press.

Hafner, E. (1974) *Crystal resonators*, IEEE Transactions on Sonics and Ultrasonics, Vol. 21 (4), 220-237.

Holland, R. and Eernisse, E.P. (1969) *Design of resonant piezoelectric devices*, MIT Press.

Parzen, B. and Ballato A. (1983) *Design of crystal and other harmonic oscillators*, John Wiley.

Microwave oscillators likely to be used as secondary standard:

Guillon, P. (1988) *Dielectric resonators*, Artech House.
Leblond, J. (1972) *Les tubes hyperfréquences*, Masson.
Warnecke, R. and Guneard, P. (1953) *Tubes à modulation de vitesse*, Gauthier- Villard.

Two difficult problems have given rise to an abundant bibliography. On the noise in oscillators, we suggest (see also the references for Chapters 3 and 9):

Brendel, R. Olivier, M. and Marianneau, G. (1975) *Analysis of the internal noise of quartz crystal oscillators*, IEEE Trans. on Instr. and Meas. **IM 24**, 160-170.
Edson, W.A. (1960) *Noise in oscillators*, Proc. I.R.E. **48**, 1454-1466.
Garstens. *Noise in nonlinear oscillators*, J. of Appl. Phys., **28**, 352-356.
Hafner, E. (1966) *The effects of noise in oscillators*, Proc. of the IEEE, **54**, 179-198.
Parker, T.E. (1979) *1/f phase noise in quartz delay line and resonators*. Proc. 24th Ann. Freq. Cont. Symp., 292-301.

On the phenomenon of chaos, we suggest:

Berge, P. Pomeau, Y. and Vidal, C. (1984) *L'ordre dans le chaos*, Hermann.
Mira, C. (1987) *Chaotic dynamics*, World-Scientific.

10.2.6 Chapter 6

Basic knowledge of quantum mechanics and spectroscopy can be acquired from:

Cagnac, B. and Pebay-Peyroula, J.C. (1971) *Physique Atomique*, 2 volumes, Dunod.
Cohen-Tannoudji, C. and Die, B. and Aloe, F.L. *Mécanique Quantique*, 2 volumes, Hermann, (1973).
Kastler, A. (1950) *J. de Phys. et le Radium*, **11**, 255.
Ramsey, N.F. (1956) *Molecular Beams*, Oxford University Press.

On atomic frequency standards, the following general publications can be consulted:

Journal de Physique, **42**, supplement C-8 (Proc. 3rd Symp. on Frequency Standards and Metrology, Aussois, France, 1981).
de Marchi, A (1988) (ed.) Proc. 4th Symp. on Frequency Standards and Metrology, Springer-Verlag.
Audoin, C. and Vanier, J. (1976) Journal of Physics E: Scientific Instruments, **9**, 697.

Hellwig, H. (1985) in *Precision Frequency Control*, vol. 2, edited by E.A. Gerbert and A. Ballato, Academic Press, 113-176.

Vanier, J. and Audoin, C. (1989) *The Quantum Physics of Atomic Frequency Standards*, 2 vols., Adam Hilger.

On the caesium clock:

Ashby, N. and Allan, D.W. (1979) Radio-Science, **14**, 649.

Candelier, V, Giordano, V. Hamel, A. Theobald, G. Cerez, P. and Audoin, C. (1988) Actes du deuxiéme Forum Européen Temps-Fréquence, Neuchâtel, 483-498.

de Marchi, A. Rovera, G.D. and Premoli, A. (1984) *Metrologia*, **20**, 37.

On the hydrogen clock:

Hardy, W.N. Hurliman, M.D. and Cline, R.W. (1987) Jpn. J. Appl. Phys., **26**, supplement 26-3, 2065.

Kleppner, D,. Goldenberg, H.M. and Ramsey, N.F. (1962) Phys. Rev., **126**, 603.

Kleppner, D. Berg, H.C. Crampton, S.B. Ramsey, N.F. Vessot, R.F.C. Peters, H.E. and Vanier, J. (1965) Phys. Rev., **138**, A972.

Petit, P. Desaintfuschien,M. and Audoin, A. (1980) Metrologia, **16**, 7.

On the rubidium clock:

Arditi, M. in *Metrology and Fundamental Constants*, edited by A. Ferro-Milone and P. Giacomo, North-Holland.

Ramsey, N.F. (1983) J. Res. National Bureau of Standards, **88**, 301.

On clocks using confined ions and isolated atoms:

Cutler, L.S. Giffard, R.P. and McGuire, M.D. (1981) *Proc. 13th Ann. Precise Time and Time Interval (PTIT) Meeting*, p. 563.

Dehmelt, H.G., IEEE Trans. Instr. and Meas. **IM-31**, 83.

Jardino, M. Desaintfuscien, M. Barillet, R. Viennet, J. Petit, P. and Audoin, C. (1981) Appl. Phys. **24**, 167.

10.2.7 Chapter 7

The essential references for this chapter were listed in connection with previous chapters (in particular Chapters 3 and 6) as well as in the general bibliography. We now list only a few items referring specifically to to time scales.

Benavente, J. Besson, J. and Parcelier, P. (1979) *Clock comparison by laser in the nanosecond range*, Radio Science, **14**, 701.

Granveaud, M. (1986) *Echelles de temps atomique*, Collection des Monographies du Bureau National de Métrologie, 11, Editions Chiron.

Guinot, B. and Lewandowski, W. (1987) *Use of the GPS time transfer at the Bureau INternational des Poids et Mesures*, Proc. 19th PTTI.

Markowitz, W. Hall, R.G. Essen., L. and Parry, J. (1958) *Frequency of cesium in terms of ephermeris time*, Phys. Rev. Lett, **1**, 105.

Reder, F.H. and Winkler, G.M.R. (1960) *Preliminary flight tests of an atomic clock in preparation of long-range clock synchronization experiments*, Nature, **186**, 592.
Rovera, G. and Angeli, M.T. (1972) *Time scale formation using the Kalman filter technique*, Conf. on Time Scale Algorithms, Boulder.
Stein, S.R. and Filler, R.A. (1988) *Kalman filter analysis for real time. Applications of clocks and oscillators.* Proc. 42nd Ann. Freq. Contl. Symp. 447-452.
Time and Frequency: Theory and Fundamentals, NBS Monograph 140 (1974).
Characterisation of clocks and oscillators, NIST Technical Note 1337 (1990).
Union Internationale des Télécommunications, Avis et Rapports du CCIR, vol. VII, *Fréquences étalons et signaux horaires* (1987).

10.2.8 Chapter 8

Nowadays, $1/f$ noise is studied by many specialists who meet periodically (like the specialists of the measurement of frequency, mentioned in the general bibliography). The proceedings of these conferences are published under the title Noise in Physical Systems in $1/f$ fluctuations.
From among publications that tackle the whole of the subject and examine some of the finer points, we suggest the following.
Clarke, J. and Hawkins, G. (1976) *Flicker noise in Josephson tunnel junctions*, Phys. Rev. **14**, 2826-2831.
Hoodge, F.N. *1/f noise*, Physica, **83B**, 14-23. Keshner, M.S. (1982) *1/f noise*, Proc. IEEE, **70**, 212-218.
Neumke, B. (1978) *1/f noise in membrane*, Biophys. Structure and Mechanism, **4**, 179-199.
Tandon, J.L. and Bilger, H.R. (1976) *1/f noise as a nonstationary process*, J. of Appl. Phys. **47**, 1697-1701.
On 1/f noise in oscillators, we refer the reader to the following, in addition to the references indicated for Chapters 3 and 5.
Gagnepain, J.J. Uebersfeld, J. Goujon, G. and Handel, P. (1981) *Relation between 1/f and Q factor in quartz resonators at room and low temperature*, Proc. of the 35th Ann. Freq. Conf. Symp. 476-483.
Gagnepain, J.J. (1981) *Nonlinear properties of quartz crystal resonators: a review*, Proc. of the 35th Ann. Freq. Cont. Symp. 14-30.

On the concept of fractals, some useful references are the following:
Gagnepain, J.J. Groslambert, J. and Brendel, R. (1985) *The fractal dimension of phase and frequency noises in oscillators*, Proc. of the 39th Ann. Freq. Cont. Symps., 113-118.
Mandelbrot, B. (1983) *The fractal geometry of nature*, Freeman.
Mandelbrot, B. (1984) *Les objets fractals*, Flammarion.
Readers interested in the history of physics may consult a fewbasic articles
Bernamont, J. (1937). Ann. de Physique, **7**, 71.
Johnson, L.B. (1928) Phys. Rev., **29**, 367 and **32**, 97.
Nyquist, J. (1928) Phys. Rev. **32**, 110.
Schottky.W. (1918) Ann. de Physik, **57**, 541. In the last of these, the statistical approach to shot noise was described for the first time. This article was written in Berlin and sent to the Journal on the March 20, 1918... On this day the German offensive smashed through the French lines. The article was refereed, corrections were requested and the article was accepted in July 1918, i.e., the month of the French offensive that pushed the German army back. Unperturbed, the printer continued with his work and the text was published in December 1918 – the month of the revolution in Berlin. A notable example of decoupling between science and society.
Schottky, W. (1926) Ann. der Physik, **28**, 74 and 103.
Finally, on the extensive manifestations of 1/f noise in a variety of signals, we suggest a few unexpected references.
Campbell, M.J. and Jones, B.W. (1972) Science, **177**, 889.
Verveen, A.A. and Derksen, H.E. (1968) Proc. IEEE, **56**, 900.
Voss, R.F. and Clarke, J. (1979) J. Acoust. Soc. Am, **63**, 259.

10.2.9 Chapter 9

Only a few references on the operation of oscillators mentioned in the main text are listed here. Of course, the bibliography of Chapter 5 should also be consulted in the context of general models and oscillators. Excellent popular articles can be found in the review journal La Recherche
Bernard, H. Revue Générale des Sciences, **11**, 1261.
Freeman, C. (1984) *Long waves in the world economy*, Frances Pinter.
Gazis, D.C. Montroll, E.W. and Ryniker, J.E. (1973) *Model of predator-prey populations*, IBM Jurnel of Res. et. Doc.,**17**, 47-53.
Guillon, P. Guillon, D,. Lansal, J. and Soutoul, J.MP. (1986) *Naissance, fertilité, rythmes, cycle lunaire,*, J. Gynecol. Obstetric Biol. Rep. **15**, 265-271.
Reinberg, A. and Ghata, J. (1964) *Rythmes biologiques*, Presses Universitaires de France.
Rossing, T. (1989) *The science of sound*, Addison Wesley.
Van der Pol, B. (1930) *Le coeur considéré comme un oscillateur à relaxation*, L'Onde Electrique, **8**, 365-392.

Vidal, C. and Pacault, A. (1981) *Nonlinear phenomena in chemical dynamics*, Springer.
Volterra, V. (1931) *Théorie mathématique de la lutte pur la vie*, Gauthier-Villars.

Index